ELECTRIC CIRCUITS AND NETWORKS

AUTHORS

Dr. Jamuna K.
Professor,
Vellore Institute of Technology,
Chennai, Tamil Nadu

Dr. Nilanjan Tewari
Assistant Professor,
Vellore Institute of Technology,
Chennai, Tamil Nadu

REVIEWER

Dr. Tapan Pradhan
Assistant Professor,
National Institute of Technology,
Silchar, Assam

KHANNA BOOK PUBLISHING CO. (P) Ltd.
4C/4344, Ansari Road, Darya Ganj,
New Delhi, India

All India Council for Technical Education
Nelson Mandela Marg, Vasant Kunj,
New Delhi, India

Dear Readers,

To prevent the piracy, this book is secured with HIGH SECURITY HOLOGRAM on the front title cover. In case you don't find the hologram on the front cover title, please write us to at contact@khannabooks.com or whatsapp us at +91-99109 09320 and avail special gift voucher for yourself.

Specimen of Hologram on front Cover title:

Moreover, there is a SPECIAL DISCOUNT COUPON for you with EVERY HOLOGRAM.

How to avail this SPECIAL DISCOUNT:

Step 1: Scratch the hologram
Step 2: Under the scratch area, your "coupon code" is available
Step 3: Logon to www.khannabooks.com
Step 4: Use your "coupon code" in the shopping cart and get your copy at a special discount
Step 5: Enjoy your reading!

ISBN: 978-93-5538-943-5
Book Code: DIP450EN

Electric Circuits and Networks
by Dr. Jamuna K. and Dr. Nilanjan Tewari

First Edition: 2025

Originally Published by:
All India Council for Technical Education, Ministry of Education, GOI

Exclusively Published & Distributed by:
Khanna Book Publishing Co. (P) Ltd.
4C/4344, Ansari Road, Daryaganj,
New Delhi (INDIA)

Visit us at: **www.khannabooks.com**
Write us at: **contact@khannabooks.com**

To view complete list of books, Please scan the QR Code: KPH

Printed in Bharat.

प्रो. टी. जी. सीताराम
अध्यक्ष
Prof. T. G. Sitharam
Chairman

सत्यमेव जयते

आज़ादी का अमृत महोत्सव

अखिल भारतीय तकनीकी शिक्षा परिषद्
(भारत सरकार का एक सांविधिक निकाय)
(शिक्षा मंत्रालय, भारत सरकार)
नेल्सन मंडेला मार्ग, वसंत कुंज, नई दिल्ली–110070
दूरभाष : 011–26131498
ई–मेल : chairman@aicte-india.org

ALL INDIA COUNCIL FOR TECHNICAL EDUCATION
(A STATUTORY BODY OF THE GOVT. OF INDIA)
(Ministry of Education, Govt. of India)
Nelson Mandela Marg, Vasant Kunj, New Delhi-110070
Phone : 011-26131498
E-mail : chairman@aicte-india.org

FOREWORD

Engineers are the backbone of any modern society. They are the ones responsible for the marvels as well as the improved quality of life across the world. Engineers have driven humanity towards greater heights in a more evolved and unprecedented manner.

The All India Council for Technical Education (AICTE), have spared no efforts towards the strengthening of the technical education in the country. AICTE is always committed towards promoting quality Technical Education to make India a modern developed nation emphasizing on the overall welfare of mankind.

An array of initiatives has been taken by AICTE in last decade which have been accelerated now by the National Education Policy (NEP) 2020. The implementation of NEP under the visionary leadership of Hon'ble Prime Minister of India envisages the provision for education in regional languages to all, thereby ensuring that every graduate becomes competent enough and is in a position to contribute towards the national growth and development through innovation & entrepreneurship.

One of the spheres where AICTE had been relentlessly working since past couple of years is providing high quality original technical contents at Under Graduate & Diploma level prepared and translated by eminent educators in various Indian languages to its aspirants. For students pursuing 2nd year of their Engineering education, AICTE has identified 87 books, which shall be translated into 12 Indian languages - Hindi, Tamil, Gujarati, Odia, Bengali, Kannada, Urdu, Punjabi, Telugu, Marathi, Assamese & Malayalam. In addition to the English medium, books in different Indian Languages are going to support the students to understand the concepts in their respective mother tongue.

On behalf of AICTE, I express sincere gratitude to all distinguished authors, reviewers and translators from the renowned institutions of high repute for their admirable contribution in a record span of time.

AICTE is confident that these outcomes based original contents shall help aspirants to master the subject with comprehension and greater ease.

T. G. Sitharam

(Prof. T. G. Sitharam)

ACKNOWLEDGEMENT

We are grateful to the authorities of AICTE, particularly Prof. (Dr.) T G Sitharam, Chairman; Dr. Abhay Jere, Vice-Chairman, Prof. Rajive Kumar, Member-Secretary,

Dr. Ramesh Unnikrishnan, Advisor-II and Dr. Sunil Luthra, Director, Training and Learning Bureaufor their planning to publish this book on "Electric Circuits and Networks". I am deeply indebted to Dr.Tapan Pradhan of NIT Silchar who promptly and thoroughly reviewed this book. His review significantly improved the content of the book.

We thank the management of Vellore Institute of Technology and our family members for their constant encouragement and support extended during the book writing process.

Dr. Jamuna K.
Dr. Nilanjan Tewari

ACKNOWLEDGEMENT

PREFACE

This book is titled Electric circuits and Networks. The response of any electric circuit is achieved by many methods. This book includes the explanation of those methods applied to various electric circuits. This book is suitable for those who are searching for the solution methods of an electric circuit in a simpler way by different methods. Most of the units are concentrated to solve the ac circuits and a smaller number of dc circuits.

Unit - I discusses the steady-state response of single-phase ac series and parallel circuits. The circuit consists of different passive components like resistances, inductances and capacitances are considered for the discussion. The concept of representations of sinusoid, phasor representation of sinusoid, concept of impedance, power and power triangle are discussed with suitable examples.

Unit - II describes the source transformations (voltage to current and vice versa), Star-delta, delta – star transformations of three phase connections. Both mesh and node analysis are used for finding the currents in the loops and the node potentials. Many theorems are discussed in detail for ac circuits. The duality of electric circuits is also presented.

Unit - III explains the three-phase systems with star and delta connections under balanced and unbalanced loading conditions. The power calculations for the three-phase systems and power measurement methods are discussed.

Unit - IV discusses the responses of first order and second order systems in the time domain. Source free circuits and step responses of different electric circuits are explained. Application of Laplace transformation for electric circuits with necessary examples is also included.

Unit - V describes the graph theory concepts to the electric circuits. The tie-set analysis and cut-set analysis are discussed. Also, two-port networks and its parameters like impedance parameters, admittance parameters, ABCD parameters and hybrid parameters are deliberated. The relationship between these parameters is also derived and tabulated.

The book consists of the LTspice simulation of an electrical circuit which helps the students to understand the concepts very easily.

Dr. Jamuna K.
Dr. Nilanjan Tewari

OUTCOME BASED EDUCATION

For the implementation of an outcome based education the first requirement is to develop an outcome based curriculum and incorporate an outcome based assessment in the education system. By going through outcome based assessments, evaluators will be able to evaluate whether the students have achieved the outlined standard, specific and measurable outcomes. With the proper incorporation of outcome based education there will be a definite commitment to achieve a minimum standard for all learners without giving up at any level. At the end of the programme running with the aid of outcome based education, a student will be able to arrive at the following outcomes:

Programme Outcomes (POs) are statements that describe what students are expected to know and be able to do upon graduating from the program. These relate to the skills, knowledge, analytical ability attitude and behaviour that students acquire through the program. The POs essentially indicate what the students can do from subject-wise knowledge acquired by them during the program. As such, POs define the professional profile of an engineering diploma graduate.

National Board of Accreditation (NBA) has defined the following seven POs for an Engineering diploma graduate:

PO-1: Basic and Discipline specific knowledge: Apply knowledge of basic mathematics, science and engineering fundamentals and engineering specialization to solve the engineering problems.

PO-2: Problem analysis: Identify and analyses well-defined engineering problems using codified standard methods.

PO-3: Design/development of solutions: Design solutions for well-defined technical problems and assist with the design of systems components or processes to meet specified needs.

PO-4: Engineering Tools, Experimentation and Testing: Apply modern engineering tools and appropriate technique to conduct standard tests and measurements.

PO-5: Engineering practices for society, sustainability and environment: Apply appropriate technology in context of society, sustainability, environment and ethical practices.

PO-6: Project Management: Use engineering management principles individually, as a team member or a leader to manage projects and effectively communicate about well-defined engineering activities.

PO-7: Life-long Learning: Ability to analyse individual needs and engage in updating in the context of technological changes.

COURSE OUTCOMES (CO)

After completion of the course the students will be able to:

CO-1: To understand and solve problems related to single phase A.C series and parallel circuits.

CO-2: To use and apply principles of circuit analysis and network theorems for troubleshooting electric circuits.

CO-3: To interpret and solve problems related to three phase circuits.

CO-4: To understand the transient behaviour of electric circuits using differential equations and Laplace transforms.

CO-5: To analyseelectric circuits using graph theory and concept of two-port networks.

Course Outcomes	EXPECTED MAPPING WITH PROGRAMME OUTCOMES (1- Weak Correlation; 2- Medium correlation; 3- Strong Correlation)											
	PO-1	PO-2	PO-3	PO-4	PO-5	PO-6	PO-7	PO-8	PO-9	PO-10	PO-11	PO-12
CO-1	3	2	2	1	1	1	-	-	-	-	-	3
CO-2	3	3	2	1	1	1	-	-	-	-	-	3
CO-3	3	3	3	3	3	2	-	-	1	2	2	1
CO-4	3	3	3	3	3	3	-	-	1	2	-	3
CO-5	3	3	3	3	3	3	-	-	2	2	2	3

GUIDELINES FOR TEACHERS

To implement Outcome Based Education (OBE) knowledge level and skill set of the students should be enhanced. Teachers should take a major responsibility for the proper implementation of OBE. Some of the responsibilities (not limited to) for the teachers in OBE system may be as follows:

- Within reasonable constraint, they should manoeuvre time to the best advantage of all students.
- They should assess the students only upon certain defined criterion without considering any other potential ineligibility to discriminate them.
- They should try to grow the learning abilities of the students to a certain level before they leave the institute.
- They should try to ensure that all the students are equipped with the quality knowledge as well as competence after they finish their education.
- They should always encourage the students to develop their ultimate performance capabilities.
- They should facilitate and encourage group work and team work to consolidate newer approach.
- They should follow Blooms taxonomy in every part of the assessment.

Bloom's Taxonomy

Level	Teacher should Check	Student should be able to	Possible Mode of Assessment
Creating	Students ability to create	Design or Create	Mini project
Evaluating	Students ability to Justify	Argue or Defend	Assignment
Analysing	Students ability to distinguish	Differentiate or Distinguish	Project/Lab Methodology
Applying	Students ability to use information	Operate or Demonstrate	Technical Presentation/ Demonstration
Understanding	Students ability to explain the ideas	Explain or Classify	Presentation/Seminar
Remembering	Students ability to recall (or remember)	Define or Recall	Quiz

GUIDELINES FOR STUDENTS

Students should take equal responsibility for implementing the Outcome Based Education (OBE). Some of the responsibilities (not limited to) for the students in OBE system are as follows:

- Students should be well aware of each Unit Outcome (UO) before the start of a unit in each and every course.
- Students should be well aware of each Course Outcome (CO) before the start of the course.
- Students should be well aware of each Program Outcome (PO) before the start of the program.
- Students should think critically and reasonably with proper reflection and action.
- Learning of the students should be connected and integrated with practical and real life consequences.
- Students should be well aware of their competency at every level of OBE.

ABBREVIATIONS AND SYMBOLS

List of Abbreviations

General Terms			
Abbreviations	**Full form**	**Abbreviations**	**Full form**
A	Ampere	m	Milli
AC	Alternating Current	Q factor	Quality Factor
BW	Bandwidth	RMS	Root Mean Square
DC	Direct Current	VA	Volt-Ampere
F	Farad	W	Watts
H	Henry	Ω	Ohm
Hz	Hertz	⊠	Mho
KCL	KIrchoff's Current Law	µ	Micro
KVL	KIrchoff's Voltage Law		

LIST OF FIGURES

CONTENTS

1 Semiconductor Diodes

UNIT SPECIFICS

Through this unit the following topics have been discussed:

- Phasors and complex representation of sinusoidal waveforms
- RMS and average value calculation
- Steady state response of pure resistor, inductor and capacitor
- Concept of instantaneous power, average power, reactive power and complex power
- Steady state analysis of series circuits (R-L, R-C, R-L-C)
- Steady state analysis of parallel circuits (R-L, R-C, R-L-C)
- Series and parallel resonance

RATIONALE

This fundamental unit on single phase ac circuit helps the students to get a fundamental knowledge about the phasor representation of sinusoidal, phase delay between different sinusoidal waveforms, calculation of rms and average value of alternating quantity. This knowledge is further applied to analyse different ac circuits under steady-state conditions. The steady-state response also includes the behaviour of resistance, capacitance and inductance for different circuit configurations. The idea about complex power in ac circuit also gives clear idea about the energy dissipating and energy storage elements. The knowledge of resonance is acquired for series as well as parallel circuits, which is extremely useful for filter design, frequency selection for radio communication and other applications. All these fundamental knowledge are helpful to understand the concepts of power engineering. The related problems are solved and included for insight knowledge of this unit. Circuit theory is the foundation to understand and acquire the knowledge in the domain of power engineering.

PRE-REQUISITES

Trigonometry

Vector calculus

UNIT OUTCOMES

List of outcomes of this unit is as follows:

U1-O1: Understand the phasor representation of sinusoidal waveform

U1-O2: Apply mathematical formula to calculate the rms and average value of alternating waveforms

U1-O3: Apply phasor concepts to solve the ac circuits under steady state conditions

U1-O4: Realize the power distribution and their role in circuit components

U1-O5: Apply vector calculus to solve complex problems

Unit-1 Outcome	EXPECTED MAPPING WITH COURSE OUTCOMES (1 - Weak Correlation; 2 - Medium Correlation; 3 - Strong Correlation)				
	CO-1	CO-2	CO-3	CO-4	CO-5
U1-O1	3	3	-	-	-
U1-O2	3	2	-	-	-
U1-O3	3	3	-	-	-
U1-O4	1	1	-	-	-
U1-O5	2	2	-	-	-

1.1 GENERATION OF ALTERNATING VOLTAGE

The common ways of manifestation of energy are light, heat and sound. The term 'energy' means the ability of doing work. The use of energy became popular after the invention of electricity. Fossils fuels, potential energy of water and nuclear energy are the major source of the generation of electricity. Though renewable energy sources are also getting popularity to cater the problems of environmental hazard and rapid depletion of fossil fuels.

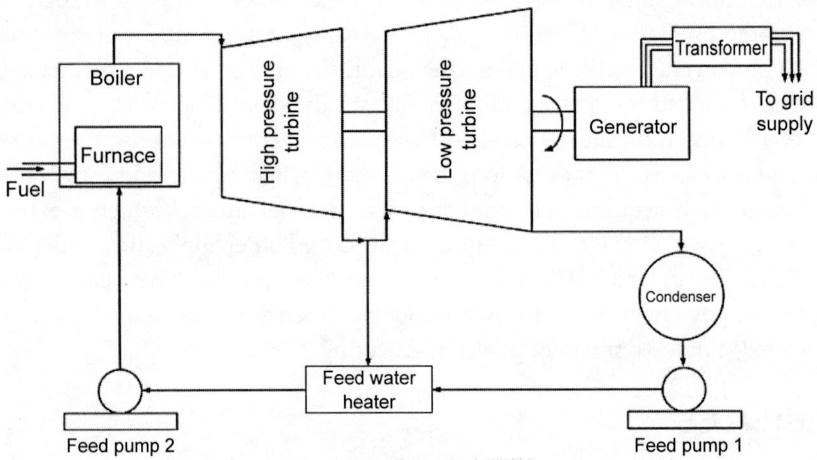

Fig.1.1: Schematic of thermal power plant

Fig.1.1 depicts the schematic diagram of fossil fuel based power plant. In a fossil fuel based power plant or commonly known as thermal power plant the main source of energy is coal, gas or oil. Chemical energy of these fuels are converted in to heat energy in furnace. This heat energy is used to generate super-heated steam in boiler chamber. The steam then enters in turbine which resembles a giant fan with hundreds of blades. High pressure steams are discharged against the turbine blades to rotate it. The potential energy of the steams is converted in to the mechanical energy which is responsible for the rotation of the turbine. The turbine shaft is connected to the alternator or synchronous generator to feed the mechanical energy into it. This mechanical energy is then converted into electrical energy by the principle of electromagnetic induction in the alternator. Usuallystep up transformers which are also known as generating transformer, are connected immediately after the alternator to step up the alternating voltage in order to reduce the transmission loss and synchronise it with utility grid.

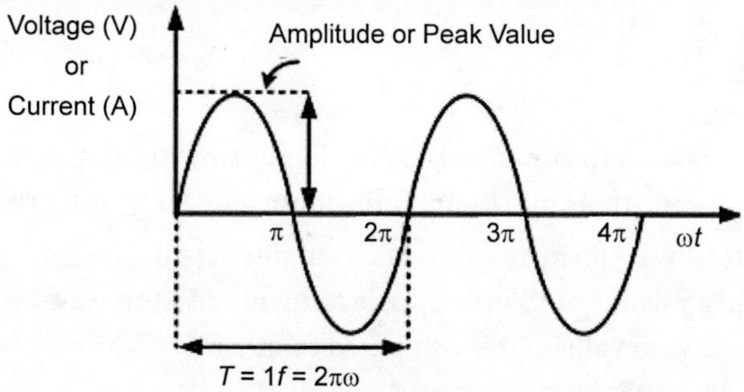

Fig.1.2: An alternating voltage or current waveform

Alternating voltage is generated by a rotating magnetic field linked with a stationary coil of an alternator. It can also be produced by the interaction of rotating coil and static magnetic field. Alternating voltage and current produced by the alternator are represented by sinusoidal wave as shown in Fig.1.2. It is very common type of alternating voltage and alternating current. An alternating waveform changes its magnitude and direction periodically. Fig.1.3 represents few other types of alternating waveforms to representalternating voltage.

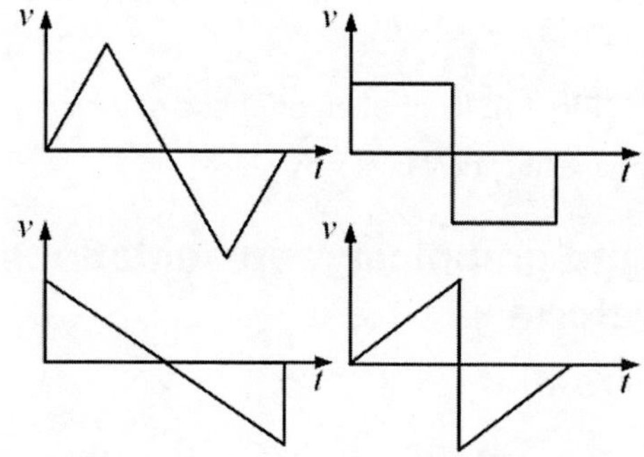

Fig.1.3: Different alternating voltage waveforms

1.2 IMPORTANT TERMS RELATED TO ALTERNATING WAVEFORM

Instantaneous value: It is the value of any alternating quantity or waveform at any instant of time.

Cycle: When the alternatingwaveform completes one set of positive and negative values, it is called one cycle.

Amplitude: The maximum positive and negative value of an alternating waveform is known as amplitude or peak value of it.

Time period: Time taken by any alternating waveform to complete one cycle is called time period. It is measured in seconds.

Frequency: The number of cycles completed per second by any alternating waveform is known as frequency of that waveform. It is measured in hertz (Hz) or cycles per second (c/sec).

The relation between frequency (f) and time period (T) is given by,

$$f = \frac{1}{T} \tag{1.1}$$

Phase: It is the angular displacement of any alternating waveform, which denotes the position of the waveform.

Phase difference: It is used to compare the position of two alternating waveforms. Two alternating waveforms are in phase means their maximum, minimum and zero values reaches at the same time.

A leading alternating waveform is one which attains its maximum, minimum and zero values earlier compared to other waveform. Similarly, a lagging alternating waveform is one which attains its maximum, minimum and zero values later compared to other waveform. A plus sign is used to indicate leading and minus sign indicates lagging phase difference.

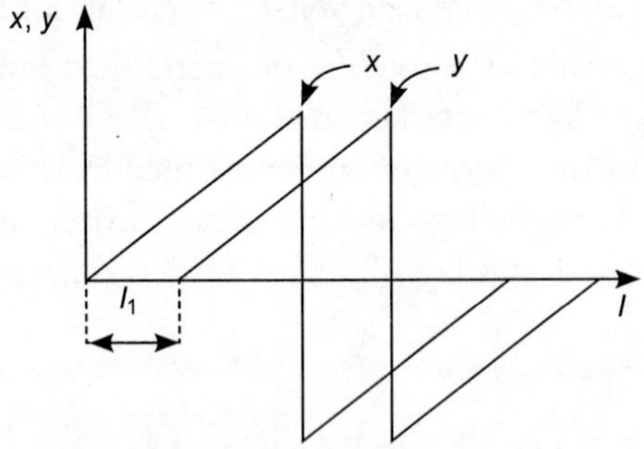

Fig.1.4: Illustration of phase difference

In Fig.1.4, *x* leads *y* by a phase of t_1 or y lags x by t1.

1.2.1 Mathematical and graphical representations of sinusoidalwaveform

A sinusoidal voltage can be expressed as,

$$v(t) = V_m \sin\theta = V_m \sin\omega t = V_m \sin 2\pi ft = V_m \sin\frac{2\pi}{T}t \tag{1.2}$$

Where

v(t)	=	instantaneous value in volts.
Vm	=	amplitude or peak value in volts.
θ	=	angular displacement or phase in degree or radians.
T	=	time period in sec
f	=	frequency of the sinusoidal voltage in Hz.

ω = angular frequency in radians /sec.

ωt = Angular displacement or argument in radians.

T = time period in sec.

From the sinusoidal voltage waveform as indicated in Fig.1.2 it is clear that ωT = 2π. Sinusoidal waveforms repeat itself after every 'T'seconds. The waveforms repeat itself after certain time are known as periodic waveforms. These type of waveforms are represented as,

$$f(t) = f(t + nT) \tag{1.3}$$

Where

$f(t)$ = periodic waveform.

n = integer value.

If $n = 1$, then using equation (1.2), it can be written as

$$v(t+T) = V_m \sin\omega(t+T) = V_m \sin\omega\left(t+\frac{2\pi}{\omega}\right) = V_m \sin(\omega t+2\pi) = V_m \sin\omega t \tag{1.4}$$

Hence in general, equation (1.4) is expressed as,

$$v(t) = v(t + nT) \tag{1.5}$$

$v(t)$ is periodic because $v(t)$ and $v(t + T)$ have the same value as indicated in equation (1.4).

Equation (1.2) represents a sinusoidal voltage waveform with no phase delay. In general, it is expressed as,

$$v_1 (t) = V_m \sin(\omega t + \varphi) \tag{1.6}$$

Or

$$v_2 (t) = V_m \sin(\omega t + \varphi) \tag{1.7}$$

Where

φ = phase.

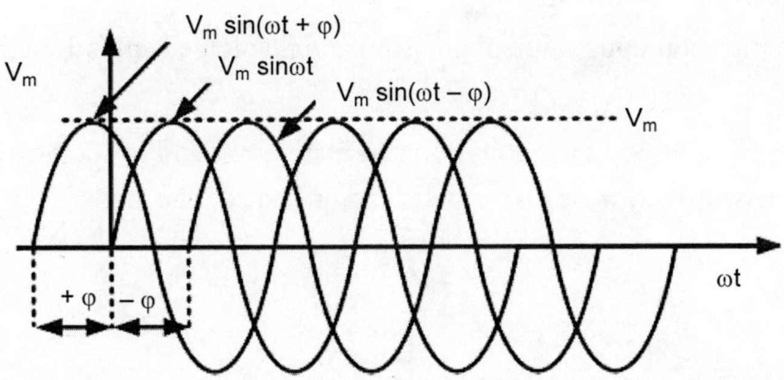

Fig.1.5: Three sinusoidal voltage waveforms with different phases

Fig.1.5 shows three sinusoidal voltage waveforms with different phases and same amplitude and frequency. $V_m\sin(\omega t + \varphi)$ leads Vmsinωt and $V_m\sin(\omega t - \varphi)$ by φ and 2φ respectively. This means the maximum value of $V_m \sin(\omega t + \varphi)$ reaches first compared to other two waveforms. Similarly, $V_m\sin(\omega t - \varphi)$ lags $V_m\sin \omega t$ by φ that's why its phase is represented by '–φ'.

Sinusoidal waveforms are mathematically represented in either sine or cosine form. While comparing two or more sinusoidal waveforms, it is convenient to represent all of them either in sine or in cosine. In this regard the following trigonometric relations are useful to transfer sine to cosine or vice versa.

$$\sin(\omega t \pm 90°) = \pm\cos\omega t \tag{1.8}$$

$$\sin(\omega t \pm 180°) = -\sin\omega t \tag{1.9}$$

$$\cos(\omega t \pm 90°) = \sin\omega t \tag{1.10}$$

$$\cos(\omega t \pm 180°) = -\cos\omega t \tag{1.11}$$

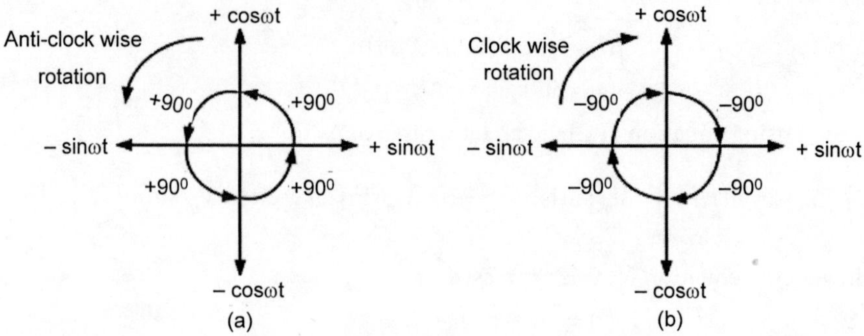

Fig.1.6: Relation between sine and cosine: A graphical representation

Equations (1.8) to (1.11) can be illustrated graphically through Fig.1.6 also. In this figure, angles are taken positive when they are measured anti-clock wise from any axis. Magnitude of sine is considered along horizontal axis and magnitude of cosine is considered along vertical axis. Using this graphical approach, it is possible to co-relate sinusoidals which are represented as a function of sine as well as cosine.For an example, if 900 is added with sinωt i.e. sin(ωt + 900) then it becomes cosωt by having a rotation of 900 in anti-clock wise direction from +sinωt, which is mentioned in Fig.1.6 (a). Similarly, as per Fig.1.6 (b) if 900 is subtracted from sinωt i.e. sin(ωt – 900) then it will be – cosωt having a rotation of 900 from +sinωt in clock wise direction.

For example, the instantaneous value of any sinusoidal voltage is mentioned by

$$v(t) = 10\sin(\omega t - 45°) \tag{1.12}$$

The graphical representation of the equation (1.12) is shown in Fig.1.7.An angle of 450 is measured from +sinωt in clock wise direction as '–ve' sign is present before the angle.

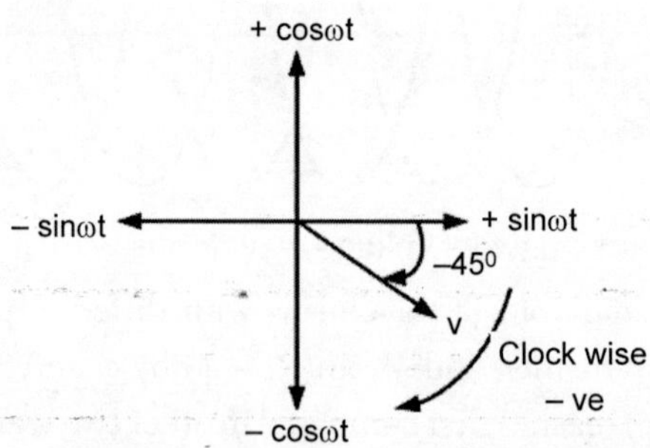

Fig.1.7: Graphical representation of the voltage mentioned in equation (1.12)

Example 1.1:

Find out the phase angle between the currents i_1 (t) = 5 sin(ωt + 30°) in A and i_2 (t) = –6 cos(ωt – 60°) in A. Also comment which current is lagging.

Solution:

Approach 1 (Mathematical)

To find out the phase angle between i_1 and i_2, it is required to represent them in either sine or cosine form and amplitudes with same sign. Here both the quantities are represented in sine form and amplitudes with positive sign as,

$$i_1(t) = 5\sin(\omega t + 30°)$$

$$i_2(t) = -6\cos(\omega t - 60°) = 6\sin(\omega t - 60° - 90°) \text{ [Using equation (1.8)]}$$

$$i_2 = 6\sin(\omega t - 150°)$$

As i_1 is leading by 300 and i_2 is lagging by 1500, hence it is clear that the phase difference between them is (300 + 1500) i.e. 1800. It is also evident from the expression that i_1 leads i_2 by 1800.

Approach 2 (Graphical)

i_1 and i_2 are presented graphically as shown in Fig.1.8. i_1 is represented by rotating 300 in anti-clock wise direction from + sinωt as '+'ve sign is associated with this angle. Similarly, i_2 is represented by rotating 600 in clock wise direction from –cosωt as '-'ve sign is associated with this angle. From this figure, it is clear that the phase difference between i1 and i2 is 1800.

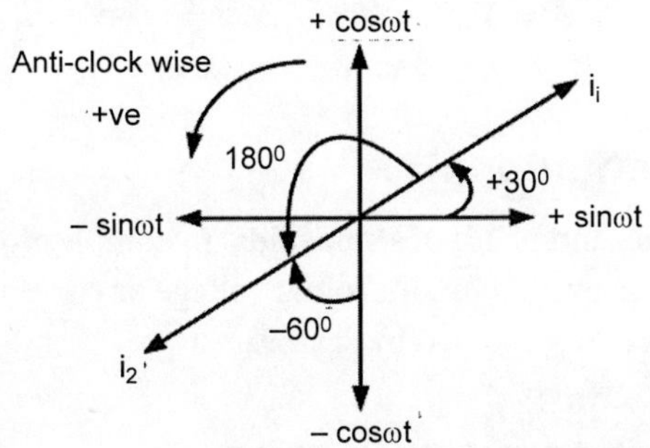

Fig.1.8: Graphical representation of the i_1 and i_2 as mentioned in Example 1.1

Example 1.2:

Find the amplitude, phase, time period and frequency of the voltage $v(t)$ = 15 sin(50t + 30°) in Volts.

Solution:

The instantaneous value of the voltage is given by,

$$v(t) = 15\sin(50t + 30°)$$

The standard form of the sinusoidal voltage is,

$$v(t) = V_m \sin(2\pi ft + \varphi) = V_m \sin(\omega t + \varphi)$$

Comparing the above two equations, we can get,

Amplitude or Maximum value $V_m = 15$ volts

$$\text{Phase} = \varphi = 300$$

Frequency in Hz is given by,

$$f = \frac{50}{2\pi} = \frac{25}{\pi}$$

Angular frequency = $\omega = 2\pi f = 50$ rad / sec.
Time Period in sec. is given by,

$$T = \frac{1}{f} = \frac{\pi}{25}$$

Example 1.3

A sinusoidal alternating voltage waveform takes 0.6 ms to become 10 A for the first time after being instantaneously zero. The frequency of the voltage is 50 Hz. Find the maximum value of the voltage waveform considering zero phase delay.

Solution:

$$v(t) = 10 \text{ A}, t = 0.6 \text{ ms}, f = 50 \text{ Hz}$$

The instantaneous voltage waveform is represented as,

$$v(t) = V_m \sin 2\pi ft$$

$$10 = V_m \sin(2 \times \pi \times 50 \times 0.6 \times 10^{-3})$$

$$10 = V_m \times 0.187$$

$$V_m = 53.39 \text{ Volts.}$$

1.2.2 Phasor representations

Phasor represents the amplitude and phase of a sinusoidal in complex form. The concept of phasor yields an easier way to analyse a circuit excited by sinusoidal voltage or current source. The plot of a sinusoidal voltage of on a complex plane is represented in Fig.1.9.

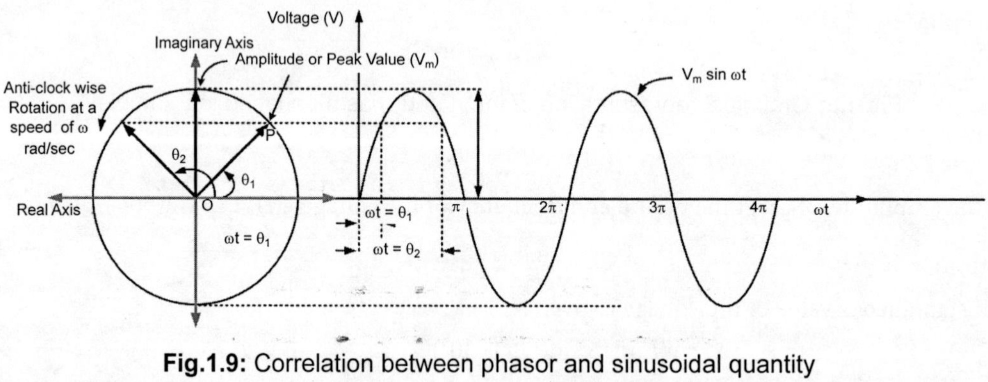

Fig.1.9: Correlation between phasor and sinusoidal quantity

As the phase passes, the tip of the arrow rotates on a circle of radius V_m (maximum value of the sinusoidal) with the same angular velocity or frequency (ω) as that of the sinusoidal voltage in anti-clock wise direction. The arrow is represented as a rotating phasor with an amplitude of the maximum value of the sinusoidal. The direction or position of this phasor is determined by it's angle with the real axis measured in anti-clock wise direction. As this phasor has both magnitude and phase, it can be represented as a phase. Hence, for this sinusoidal voltage in phasor form (polar) it can be represented as,

$$V = V_m \angle 0° \tag{1.13}$$

In general, sinusoidal voltages with an amplitude of V_m and phase lead of θ or phase lag of θ are represented in phasors (polar form) by equations (1.14) and (1.15) respectively.

$$V = V_m \angle \theta \tag{1.14}$$

$$V = V_m \angle -\theta \tag{1.15}$$

Phasor representation of the sinusoidal voltage and current is shown Table 1.1 and Fig.1.10.

Table 1.1: Time domain to phasor domain transformation of sinusoidal

Time domain representation of sinusoidal	**Phasor domain representation of sinusoidal (Polar)**

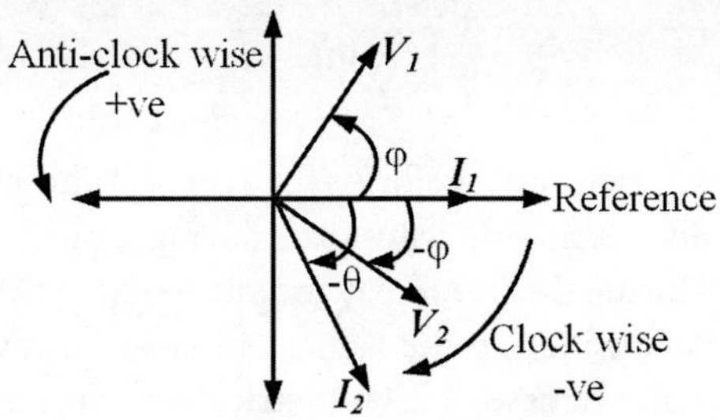

Fig.1.10: Phasor representation of the sinusoidal quantities mentioned in Table 1.1.

In rectangular form or trigonometric form, the same phasors of equations (1.14) and (1.15) can be written as,

$$V = V_m \; [\cos\theta \pm j\sin\theta] \tag{1.16}$$

Where, $j = \sqrt{-1}$ is a unit imaginary in complex plane. j is a rotational operator and $+j$ denotes rotation of 90^0 in anti-clock wise direction. Similarly, any phasor in rectangular form e.g. can be represented in polar form and exponential form by the following relations,

$$x + jy = \sqrt{x^2 + y^2} \angle \tan^{-1} \tan^{-1}\frac{y}{x} = r\angle\theta = \sqrt{x^2 + y^2}\, e^{\mathrm{j}\tan^{-1}\mathrm{j}\tan^{-1}\frac{y}{x}} = re^{j\theta} \tag{1.17}$$

Example 1.4

Two sinusoidal currents have amplitudes 10 A and 15 A respectively. Second current is leading the first one by 60^0. Find the resultant sum of these currents in polar form.

Solution:

If the phase of first current is taken as 0^0, then the currents phasors can be written as,

$$I_1 = 10\angle 0°A$$

$$I_2 = 15\angle 60°\ A$$

Resultant current or sum of these currents are given by,

$$I = I_1 + I_2 = 10°\angle 0 + 15\angle 60°$$

$$= 10[\cos 0° + j \sin 0°] + 15[\cos 60° + j \sin 60°]$$

$$= 10+7.5+j\ 0.866$$

$$= 15+j\ 0.866$$

$$= 15+j\ 0.866$$

$$= 15.025\angle 86.69°\ A$$

[Using the formula mentioned in equation (1.17)]

1.3 AVERAGE AND ROOT MEAN SQUARE (RMS) VALUE OF ALTERNATING WAVEFORMS

1.3.1 Average value of alternating waveforms

Average value of any periodic waveform, *x(t)* with a period of *T* is calculated as,

$$X_{avg} = \frac{1}{T}\int_0^T x(t)\,dt \tag{1.18}$$

In case of any symmetrical alternating waveforms the average value over a time period is zero because the area under the symmetrically alternating waveform during a positive half cycle is cancelled by the area under the same waveform during the immediate negative half cycle. Hence, for these quantities the average values are calculated by considering the values either in positive half of the time period or in negative half of the time period. But in case of asymmetrical waveforms, the average value is calculated by considering the total time period.

Example 1.5

The instantaneous value of any sinusoidal voltage is given by, Volts. Find the average value.

Solution:

As, sinusoidal waveforms have the symmetry in their positive and negative half cycles, then the average value of this voltage is calculated over positive or negative half cycle as follows,

$$V_{avg} = \frac{1}{\pi}\int_0^\pi V_m \sin \omega t\, d\omega t$$

$$= \frac{1}{\pi}\left[-V_m \cos \omega t\right]_0^\pi$$

$$= \frac{2V_m}{\pi}\ Volts$$

Example 1.6

Find the average value of the waveform shown in Fig.1.11 considering sinusoidal variation of the waveform.

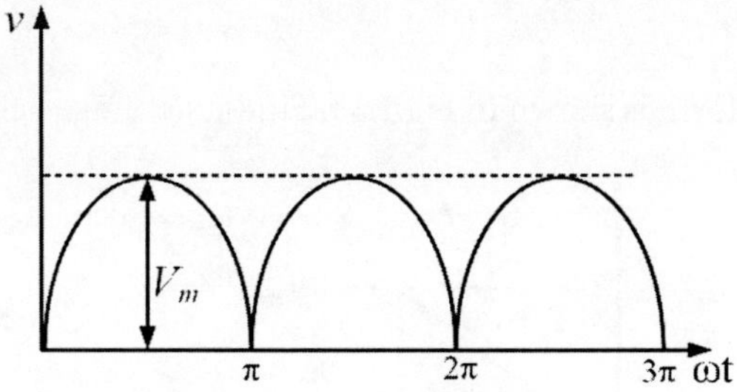

Fig.1.11: Rectified voltage waveform

Solution:

The mathematical equation of the waveform shown in Fig.1.11 is given by,

$$v(t) = Vm\ \sin\omega t \quad 0 \le \omega t \le \pi$$

The waveform is not symmetrical alternating waveform, so the average value of the waveform is obtained as,

$$V_{avg} = \frac{1}{\pi}\int_0^{\pi} V_m \sin\omega t\, d\omega t = \frac{2V_m}{\pi}$$

Example 1.7

Find the average value of the waveform shown in Fig.1.12. Consider sinusoidal variation of the waveform indicated.

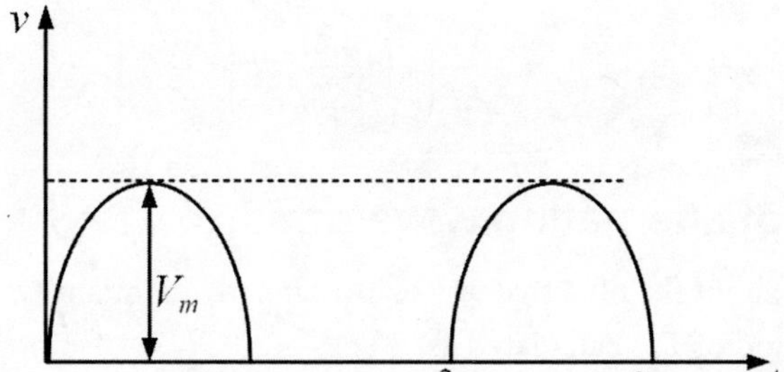

Fig.1.12: Voltage waveform

Solution:

The mathematical representation for the waveform shown in Fig.1.12 is given by,

$$v(t) = V_m \sin\omega t \quad 0 \le \omega t \le \pi$$

$$= 0 \quad \pi < \omega t < 2\pi$$

The average value of the waveform is as mentioned below,

$$V_{avg} = \frac{1}{2\pi}\left(\int_0^{\pi} V_m \sin\omega t\, d\omega t + \int_{\pi}^{2\pi} 0 d\omega t\right) = \frac{V_m}{2\pi}\left[-\cos\omega t\right]_0^{\pi} = \frac{V_m}{\pi}$$

Example 1.8

An alternating current waveform is shown in Fig.1.13. Find it's average value.

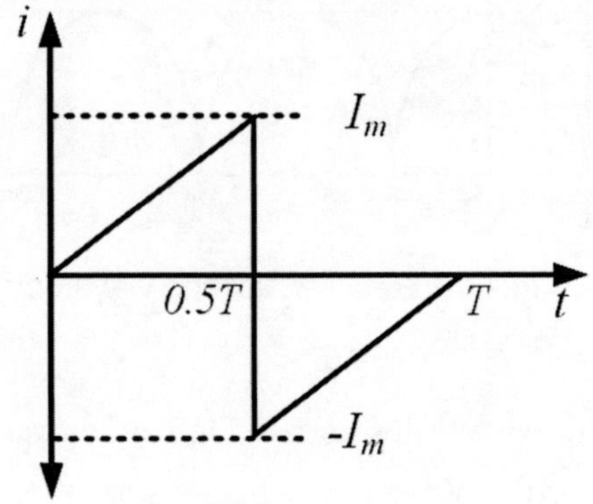

Fig.1.13: Current waveform

Solution:

The mathematical representation for the current waveform shown in Fig.1.13 is given by,

$$i(t) = \frac{I_m - 0}{0.5T - 0}t = \frac{2I_m}{T}t \quad 0 \le t \le 0.5T$$

$$= \frac{0-(-I_m)}{T-0.5T}t = \frac{2I_m}{T}t \quad 0.5T \le t \le T$$

As the waveform is symmetrical, then the average value of the current waveform is written as,

$$I_{avg} = \frac{1}{0.5T}\left(\int_0^{0.5T} \frac{2I_m}{T} t dt\right) = \frac{4I_m}{T^2}\left[\frac{t^2}{2}\right]_0^{0.5T} = \frac{I_m}{2}$$

1.3.2 RMS value of alternating waveforms

RMS value gives the idea about the effectiveness of any alternating quantity. RMS value of any periodic waveform, $x(t)$ with a period of T is calculated as,

$$X_{rms} = \sqrt{\frac{1}{T}\int_0^T x^2(t)dt} \tag{1.19}$$

In case of any symmetrical alternating waveform the rms value over a full cycle is equal to the rms value of over a half cycle due to symmetry. Hence, for these quantities the rms value is calculated by considering the values either in positive half of the time period or in negative half of the time period. But in case of asymmetrical waveforms, the rms value is calculated by considering the total time period.

Example 1.9

The instantaneous value of any sinusoidal voltage is given by, $v(t) = V_m \sin \omega t$. Find the rms value.

Solution:

As, sinusoidal waveforms have the symmetry in their positive and negative half cycles, then the rms value of this voltage over the half cycle is enough to get the actual rms value as follows,

$$V_{rms} = \sqrt{\frac{1}{\pi}\int_0^{\pi} V_m^{\,2} \sin^2 \omega t \, d\omega t}$$

$$= \sqrt{\frac{V_m^{\,2}}{2\pi}\int_0^{\pi}(1-\cos 2\omega t)\, d\omega t}$$

$$= \sqrt{\frac{V_m^{\,2}}{2\pi}\left[\omega t - \frac{\sin 2\omega t}{2}\right]_0^{\pi}}$$

$$= \sqrt{\frac{V_m^{\,2}}{2\pi}\left[(\pi-0)-(\frac{\sin 2\pi}{2}-\frac{\sin 0}{2})\right]}$$

$$= \frac{V_m}{\sqrt{2}}$$

Example 1.10

Find the rms value of the waveform shown in Fig.1.11.

Solution:

The mathematical equation of the waveform shown in Fig.1.11 is given by,

$$v(t) = V_m \ \sin\omega t \quad 0 < \omega t < \pi$$

The waveform is not symmetrically alternating waveform during the entire time period, so the rms value of the waveform is as follows,

$$V_{rms} = \sqrt{\frac{1}{\pi}\int_0^{\pi} V_m^{\,2} \sin^2 \omega t \, d\omega t} = \sqrt{\frac{V_m^{\,2}}{2\pi}\int_0^{\pi}(1-\cos 2\omega t)\, d\omega t} == \frac{V_m}{\sqrt{2}}$$

Example 1.11

Find the rms value of the current waveform shown in Fig.1.14.

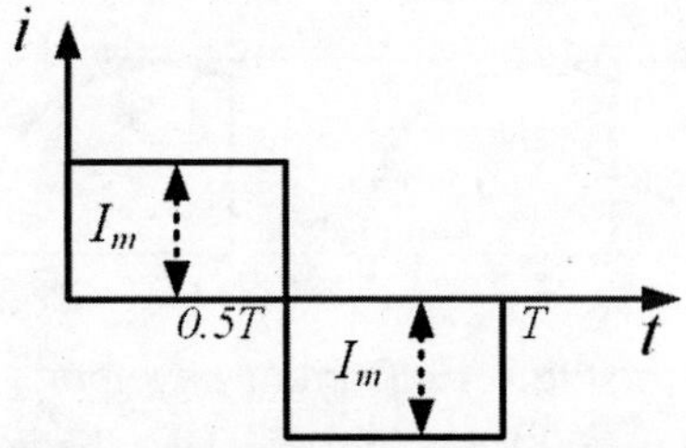

Fig. 1.14: Current waveform

Solution:

The mathematical representation for the current waveform shown in Fig.1.14 is given by,

$$i(t) = I_m \quad 0 < t < 0.5T$$
$$= -I_m \quad 0.5T < t < T$$

As the waveform is symmetrically alternating waveform in the positive half as well as negative half of the entire time period, so the rms value of the waveform is given by,

$$I_{rms} = \sqrt{\frac{1}{0.5T}\left(\int_0^{0.5T} I_m^2 dt\right)}$$

$$= \sqrt{\frac{I_m^2}{0.5T}[t]_0^{0.5T}} = I_m$$

1.4 FORM FACTOR AND PEAK FACTOR OF ALTERNATING WAVEFORMS

Form factor of an alternating waveform is defined as the ratio of its rms value to its average value.

$$Form\ Factor = \frac{rms\ value}{average\ or\ mean\ value}$$

Peak factor of an alternating waveform is defined as the ratio of its maximum value to its rms value.

$$Peak\ Factor = \frac{maximum\ value}{rms\ value}$$

Example 1.12

An alternating voltage waveform as shown in Fig. 1.15 is given by,

$$v(t) = 0 0 \le \omega t < \frac{\pi}{3}$$

$$= V_m \sin \omega t \frac{\pi}{3} \le \omega t \le \pi$$

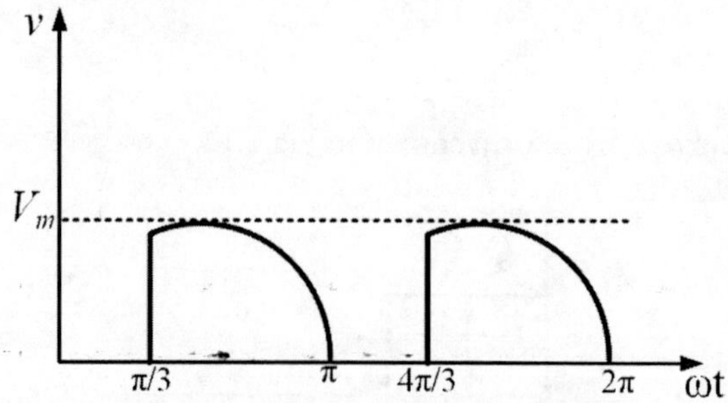

Fig.1.15: Current waveform

Find the average and rms value of this waveform. Also calculate the form factor and peak factor of the waveform shown in Fig. 1.15

Solution:

The average value of the waveform is calculated as,

$$V_{avg} = \frac{1}{\pi}\int_{\frac{\pi}{3}}^{\pi} V_m \sin\omega t d\omega t$$

$$= \frac{V_m}{\pi}[1+\cos\frac{\pi}{3}]$$

$$= \frac{3V_m}{2\pi}$$

The rms value of the waveform is given by,

$$V_{rms} = \sqrt{\frac{1}{\pi}\int_{\frac{\pi}{3}}^{\pi} V_m{}^2 \sin^2\omega t d\omega t}$$

$$= \sqrt{\frac{V_m{}^2}{2\pi}\int_{\frac{\pi}{3}}^{\pi}(1-\cos 2\omega t)d\omega t}$$

$$= \sqrt{\frac{V_m{}^2}{2\pi}\left[\omega t - \frac{\sin 2\omega t}{2}\right]_{\frac{\pi}{3}}^{\pi}}$$

$$= \sqrt{\frac{V_m{}^2}{2\pi}\left[\omega t - \frac{\sin 2\omega t}{2}\right]_{\frac{\pi}{3}}^{\pi}}$$

$$= \sqrt{\frac{V_m{}^2}{2\pi}\left[(\pi-\frac{\pi}{3})-(\frac{\sin 2\pi}{2}-\frac{\sin\frac{2\pi}{3}}{2})\right]}$$

$$= V_m\sqrt{\frac{1}{3}+0.0689}$$

$$=0.63V_m$$

The form factor of the waveform is calculated as,

$$Form\ Factor = \frac{0.63V_m}{0.477V_m} = 1.32$$

The form factor of the waveform is calculated as,

$$Peak\ Factor = \frac{V_m}{0.63V_m} = 1.587$$

1.5 STEADY STATE RESPONSE OF PURE RESISTOR, INDUCTOR AND CAPACITOR FOR A SINUSOIDAL AC VOLTAGE SOURCE

1.5.1 Pure Resistive Circuit

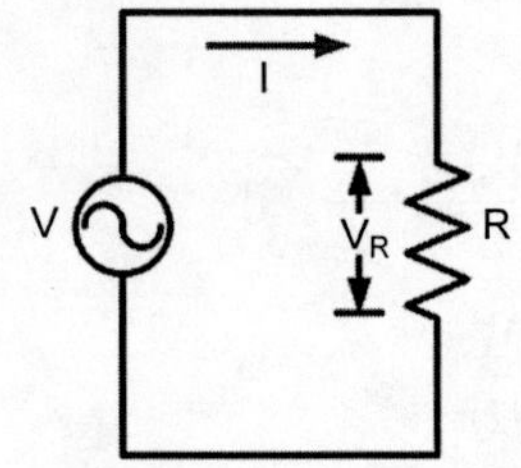

Fig.1.16: Pure resistive circuit

A pure resistive circuit of resistance 'R' is connected to a sinusoidal ac voltage source of $v(t) = V_m$ sinωt as shown in Fig.1.16. The phasor form of this voltage is expressed as,

$$V = Vm \angle° \qquad (1.20)$$

The instantaneous value of the current flowing through the circuit can be expressed as,

$$i(t) = \frac{V_m \sin \omega t}{R} = I_m \sin \omega t \qquad (1.21)$$

Where, $I_m = \frac{V_m}{R}$ is the maximum value of alternating current flowing through the circuit. The current phasor (I) is given by,

$$I = \frac{V_m \angle 0^0}{R} = I_m \angle 0^0 \qquad (1.22)$$

The voltage across the resistor in phasor form can be expressed as,

$$V_R = I_R = V \qquad (1.23)$$

1.5.1.1 Phase, impedance and power factor of pure resistive circuit

Phase:

The current and voltage waveforms and their phasor representations are shown in Fig.1.17 and Fig.1.18 respectively.

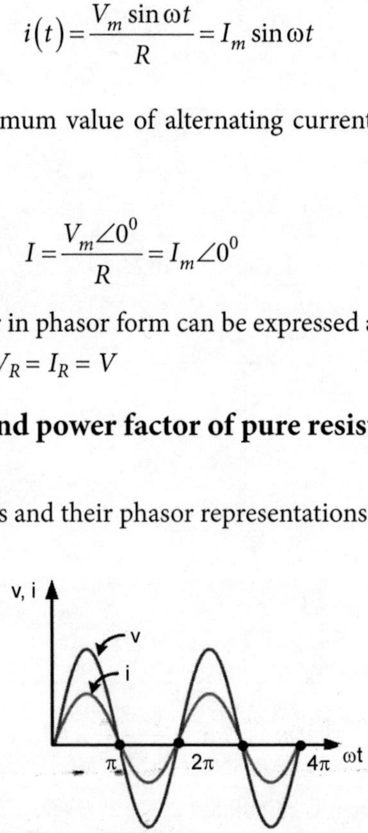

Fig.1.17: Voltage and current waveforms of pure resistive circuit excited by sinusoidal ac voltage source

Fig.1.18: Voltage and current phasors of pure resistive circuit excited by sinusoidal ac voltage source

It is clear from both the diagrams and equations (1.20) and (1.21) that the current and voltage in a pure resistive circuit excited by sinusoidal ac voltage source are in phase. In other word, the phase difference between the current flowing through the circuit and voltage across the resistor is zero.

Impedance:

Impedance of any ac circuit is defined as the resistance offered by the circuit to the flow of current through it. It is measured in ohm (Ω). In a pure resistive circuit excited by ac voltage source, the impedance of the circuit is given by,

$$Z = \frac{V}{I} = \frac{V_m \angle 0^0}{I_m \angle 0^0} = \frac{V_m}{V_m} \times R = R \tag{1.24}$$

Admittance of any ac circuit is just reciprocal of impedance. It is measured in siemens (S). In case of pure resistive ac circuit, it is just the conductance (G) i.e reciprocal of resistance.

$$Y = \frac{1}{R} = G \tag{1.25}$$

Power factor:

Power factor (*PF*) is defined as the cosine of the angle between the current phasor and voltage phasor. Sometimes, this *PF* is called as displacement *PF*. But in general, for power system applications, this is termed as just *PF*. In this case voltage and current phasors are in phase and the *PF* is obtained as,

$$\text{PF} = \cos 0° = 1 \tag{1.26}$$

1.5.1.2 Instantaneous and average power of pure resistive circuit

Instantaneous power:

The instantaneous power absorbed by any electric circuit element is the product of instantaneous current flowing through that element and instantaneous voltage across the same element. It is measured in watts (W). The expression of instantaneous power is given by,

$$p(t) = v(t)\, i(t) \tag{1.27}$$

For this circuit the instantaneous power is calculated as,

$$p(t) = V_m \sin\omega t \times I_m \sin\omega t = V_m I_m \sin^2 \omega t = \frac{1}{2} V_m I_m [1 - \cos 2\omega t]$$

$$p(t) = \frac{1}{2} V_m I_m - \frac{1}{2} V_m I_m \cos 2\omega t \tag{1.27}$$

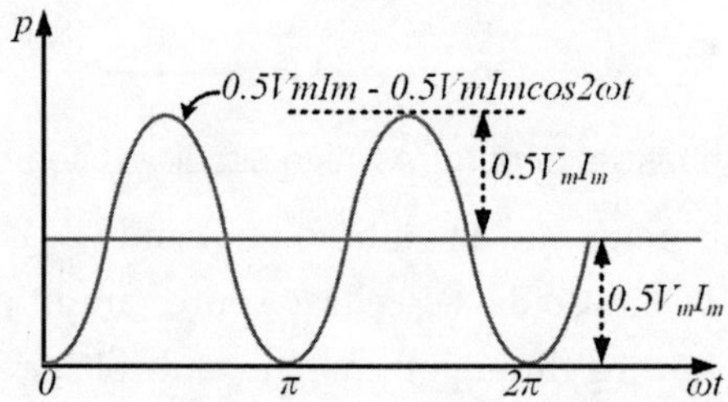

Fig.1.19: Power waveform of pure resistive circuit excited by sinusoidal ac voltage source

From equation (1.28), it is well understood that the first term of the instantaneous power is constant and time independent. Whereas the second term of the instantaneous power is sinusoidal with a frequency double the frequency of voltage and current. Fig.1.19 depicts the instantaneous power of the pure resistive circuit excited by ac voltage source. It is clear from the diagram that it is periodic function with frequency of 2ω or 2*f*, if the ω or *f* is the frequency of voltage and current waveforms. If the time period of the voltage and current waveform is indicated as T then the time period of the instantaneous power is 0.5T. The instantaneous power is positive for every cycle of it except every interval of π from ωt = 0. The positive instantaneous power indicates that the entire power is being absorbed by the circuit element (R for this circuit).

Average power:

Average power is the average or mean of the instantaneous power over a complete time period of it. The unit of average power is also watts (W). The expression of the average power is obtained as,

$$P_{avg} = \frac{1}{0.5T}\int_0^{0.5T} p(t)d\omega t$$

$$= \frac{1}{0.5T}\int_0^{0.5T} [\frac{1}{2}V_m I_m - \frac{1}{2}V_m I_m \cos 2\omega t]d\omega t$$

$$P_{avg} = \frac{V_m I_m}{2} = \frac{VI}{2} = \frac{V_m}{\sqrt{2}} \times \frac{I_m}{\sqrt{2}} = V_{rms} I_{rms} \qquad (1.29)$$

Equation (1.29) can be further re-written by using equation (1.24) as,

$$P_{avg} = \frac{V_m{}^2}{2R} = \frac{I_m{}^2 R}{2} = \frac{V_{rms}{}^2}{R} = I_{rms}{}^2 R \qquad (1.30)$$

1.5.2 Pure Inductive Circuit

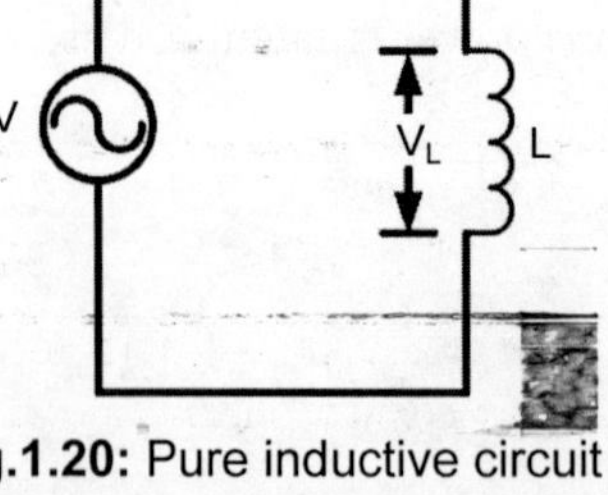

Fig.1.20: Pure inductive circuit

A pure inductive circuit of inductance 'L' is connected to a sinusoidal ac voltage source of $v = V_m \sin\omega t$ as shown in Fig.1.20. The phasor form of this voltage is expressed as,

$$V = V_m \angle 0° \tag{1.31}$$

The instantaneous value of the current flowing through the circuit can be obtained as,

$$i(t) = \frac{1}{L}\int v\,dt = \frac{V_m}{L}\int \sin\omega t dt$$

$$= \frac{V_m}{\omega L}[-\cos\omega t] = \frac{V_m}{\omega L}\sin\left(\omega t - \frac{\pi}{2}\right) = I_m \sin\left(\omega t - \frac{\pi}{2}\right) \tag{1.32}$$

Where, $I_m = \frac{V_m}{\omega L}$ is the maximum value of alternating current flowing through the circuit. Using the concept of Table 1.1 in equation (1.31), the current phasor (I) is given by,

$$I = \frac{V_m}{\omega L}\angle -90^0 = I_m \angle -90^0 = -jI_m = I_m e^{-j90^0} \tag{1.33}$$

The phasor form of voltage across the inductor can be calculated as,

$$V_L = V_m \angle 0° = I_m\,\omega L \angle 0° \tag{1.34}$$

In phasor form, the voltage across the inductor denotes anti-clock wise 90^0 rotation of it w.r.t current phasor.

1.5.2.1 Phase, impedance and power factor of pure inductive circuit

Phase:

The current and voltage waveforms and their phasor representations are shown in Fig.1.21 and Fig.1.22 respectively based on the equations (1.30) and (1.32). It is clear from both the diagrams and equations (1.30), (1.32) and (1.33) that the current and voltage in a pure inductive circuit excited by sinusoidal ac voltage source have a phase difference of 90^0. In other word, it can be stated that the current flowing the pure inductive circuit lags the supply voltage by 90^0.

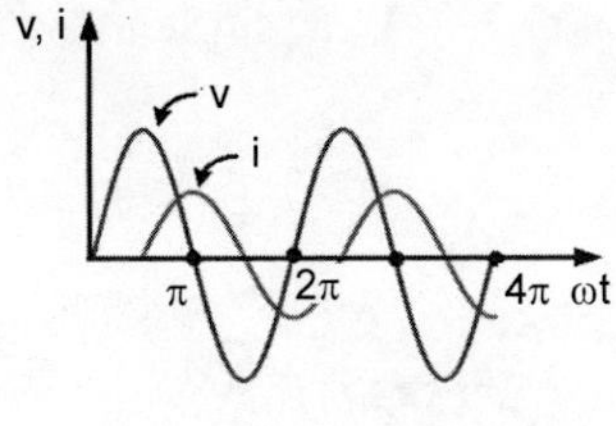

Fig.1.21: Voltage and current waveforms of pure inductive circuit excited by sinusoidal ac voltage source

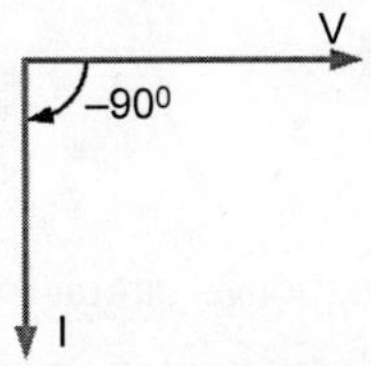

Fig.1.22: Voltage and current phasors of pure inductive circuit excited by sinusoidal ac voltage source

Impedance:

The impedance of a pure inductive circuit excited by sinusoidal ac voltage source is given by,

$$Z=\frac{V}{I}=\frac{V_m\angle 0^0}{I_m\angle -90^0}=\frac{V_m}{V_m}\times\omega L\angle 90^0=j\omega L=jX_L \tag{1.35}$$

The term ωL is known as inductive reactance and it is denoted as X_L. Impedance in this type of circuit depends on the frequency of the supply. If the frequency of supply is zero then the inductive reactance term becomes zero. In case of DC supply, the frequency is zero that means the inductive reactance value will be zero. Hence, inductor behaves as a short circuit in DC supply due to the absence of X_L. Admittance in purely inductive ac circuit is calculated as,

$$Y=\frac{1}{j\omega L} \tag{1.36}$$

Power factor:

Power factor (*PF*) for a pure inductive circuit, which is the cosine of the angle between the current phasor and voltage phasor is obtained as,

$$\text{PF} = \cos(90°) = 0 \tag{1.37}$$

1.5.2.2 Instantaneous and average power of pure inductive circuit

Instantaneous power:

The expression of instantaneous power for a pure inductive circuit is given by using the equations (1.30) and (1.31),

$$p(t) = v(t)\, i(t) \tag{1.38}$$

$$p(t)=V_m\sin\omega t\times I_m\sin(\omega t-\frac{\pi}{2})=-V_mI_m\sin\omega t\cos\omega t=-\frac{1}{2}V_mI_m\sin 2\omega t$$

$$p(t)=-\frac{1}{2}V_mI_m\sin 2\omega t \tag{1.38}$$

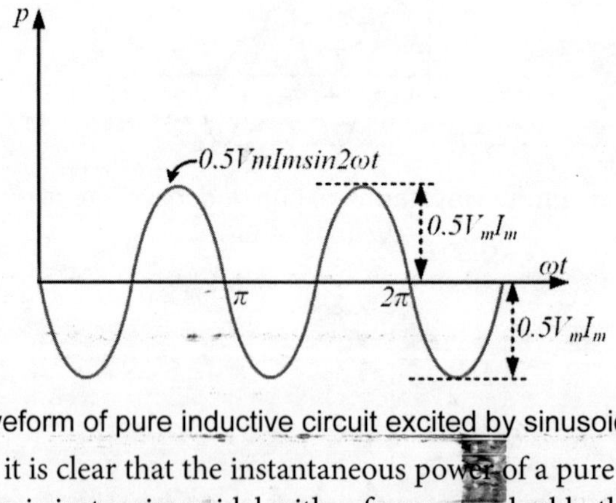

Fig.1.23: Power waveform of pure inductive circuit excited by sinusoidal ac voltage source

From equation (1.38), it is clear that the instantaneous power of a pure inductive circuit excited by sinusoidal ac voltage source is just a sinusoidal with a frequency double the frequency of voltage and current. Fig.1.23 depicts the instantaneous power of the pure inductive circuit excited by ac voltage source. It is clear from the diagram that it is periodic function with frequency of *2ω* or *2f*, if the ω or *f* is the frequency of voltage and current waveforms. If the time period of the voltage and current waveform is indicated as *T* then the time period of the instantaneous power is *0.5T*. The positive half cycle in instantaneous power indicates that the absorption of power by the inductor from the source and the negative half cycle in instantaneous power depicts absorption of power by the source from the charged inductor. That means in one half cycle inductor charges by absorbing energy from the source and in the next half cycle it throws back the stored energy to the source by discharging itself.

Average power:

The expression of the average power is obtained as,

$$P_{avg} = \frac{1}{0.5T}\int_{0}^{0.5T} p(t)d\omega t$$

$$= \frac{1}{0.5T}\int_{0}^{0.5T} [\frac{1}{2}V_m I_m \sin 2\omega t]d\omega t$$

$$P_{avg} = 0 \qquad (1.40)$$

The average power absorption by an inductive element over a time period is zero. This means inductor does not consume any average power. Fig.1.23 indicates the power only oscillates between source and inductor (load) and it reacts upon itself. It does not do any useful work. Hence it is known as watt-less power or re-active power. Reactive power is measured in VAr. From equations (1.35) and (1.39) the peak value or maximum value of this reactive power is derived as,

$$|Q_m| = \frac{V_m}{\sqrt{2}}\frac{I_m}{\sqrt{2}} = V_{rms}I_{rms} = \frac{V_m^2}{2X_L} = \frac{I_m^2 X_L}{2} \qquad (1.40)$$

1.5.3 Pure Capacitive Circuit

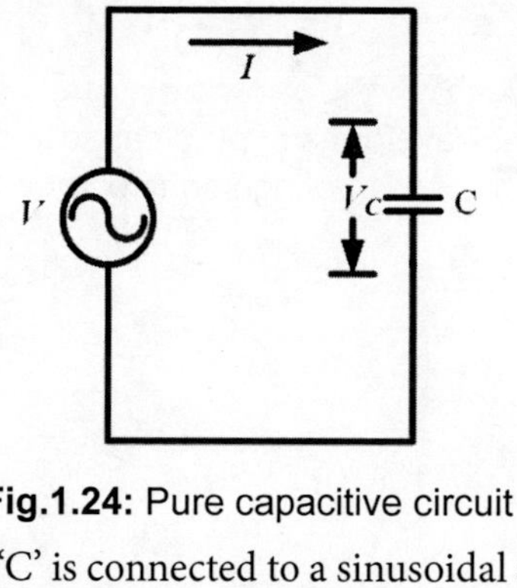

Fig.1.24: Pure capacitive circuit

A pure capacitor of capacitance 'C' is connected to a sinusoidal ac voltage source of $v(t) = V_m \sin\omega t$ as shown in Fig.1.24. The phasor form of this voltage is expressed as,

$$V = V_m\angle 0° \qquad (1.42)$$

The instantaneous value of the current flowing through the circuit can be obtained as,

$$i(t) = C\frac{dv}{dt} = C\frac{d(V_m \sin\omega t)}{dt}$$

$$= \omega C V_m \cos\omega t = \omega C V_m \sin\left(\omega t + \frac{\pi}{2}\right) = I_m \sin\left(\omega t + \frac{\pi}{2}\right) \quad (1.43)$$

Where, $I_m = \omega C V_m$ is the maximum value of alternating current flowing through the circuit. Using the concept of Table 1.1 in equation (1.43), the current in phasor form (I) is given by,

$$I = \omega C V_m \angle + 90° = I_m \angle + 90° = jI_m = \text{Im } e^{j90°} \quad (1.44)$$

The phasor form of voltage across the capacitorcan be calculated as,

$$V_C = V_m \angle 0^0 = \frac{I_m}{\omega C} \angle 0^0 \quad (1.45)$$

The phasor form of voltage across capacitor denotes clock wise 90^0 rotation of it w.r.t current phasor.

1.5.3.1 Phase, impedance and power factor of pure capacitive circuit

Phase:

The current and voltage waveforms and their phasor representations are shown in Fig.1.25 and Fig.1.26 respectively based on the equations (1.42), (1.43), (1.44) and (1.45). It is clear from both the figures (Fig.1.25 and Fig.1.26) and equations (1.42), (1.43), (1.44) and (1.45) that the current and voltage in a pure capacitive circuit excited by sinusoidal ac voltage source have a phase difference of 90^0. In other words, it can be stated that the current flowing the pure capacitive circuit leads the supply voltage by 90^0.

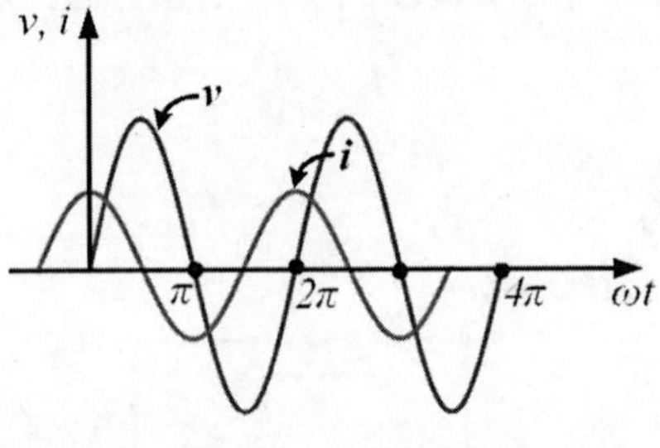

Fig.1.25: Voltage and current waveforms of pure capacitive circuit excited by sinusoidal ac voltage source

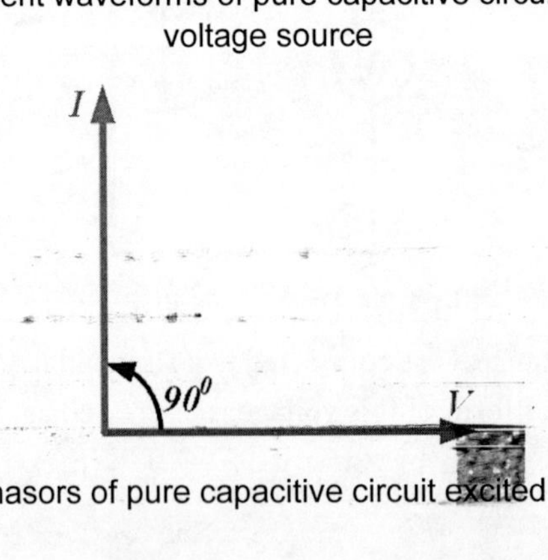

Fig.1.26: Voltage and current phasors of pure capacitive circuit excited by sinusoidal ac voltage source

Impedance:

The impedance of a pure capacitive circuit excited by sinusoidal ac voltage source is given by,

$$Z = \frac{V}{I} = \frac{V_m \angle 0^0}{I_m \angle 90^0} = \frac{V_m}{V_m} \times \frac{1}{\omega C} \angle -90^0 = -j\frac{1}{\omega C} = -jX_C \quad (1.46)$$

The term $\frac{1}{\omega C}$ is known as capacitive reactance and it is denoted as X_C. Impedance in this type of circuit also depends on the frequency of the supply. If the frequency of supply is zero then the capacitive reactance term becomes infinity. In case of DC supply, the frequency is zero that means the capacitive reactance value will be infinity. Hence, capacitor behaves as an open circuit in DC supply due to the very high value of X_C. Admittance in purely capacitive ac circuit is calculated as,

$$Y = j\omega C \quad (1.47)$$

Power factor:

Power factor (*PF*) for a pure capacitive circuit, which is the cosine of the angle between the current phasor and voltage phasor is obtained as,

$$\text{PF} = \cos 90° = 0 \quad (1.48)$$

1.5.3.2 Instantaneous and average power of pure capacitive circuit

Instantaneous power:

The expression of instantaneous power for a pure capacitive circuit is given by

$$p(t) = v(t)\ i(t) \quad (1.49)$$

$$p(t) = V_m \sin \omega t \times I_m \sin(\omega t + \frac{\pi}{2}) = V_m I_m \sin \omega t \cos \omega t = \frac{1}{2} V_m I_m \sin 2\omega t$$

$$p(t) = \frac{1}{2} V_m I_m \sin 2\omega t \quad (1.50)$$

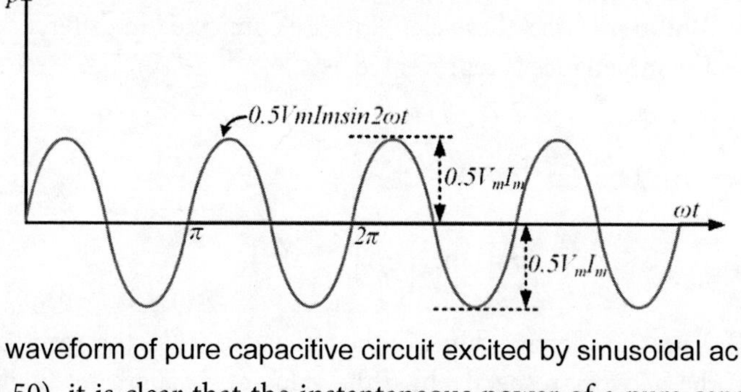

Fig.1.27: Power waveform of pure capacitive circuit excited by sinusoidal ac voltage source

From equation (1.50), it is clear that the instantaneous power of a pure capacitive circuit excited by sinusoidal ac voltage source is just a sinusoidal with a frequency double the frequency of voltage and current. Fig.1.27 shows the instantaneous power of the pure capacitive circuit excited by ac voltage source. It is clear from the diagram that it is periodic function with frequency of *2ω* or *2f*, if the ω or *f* is the frequency of voltage and current waveforms. If the time period of the voltage and current waveform is indicated as T then the time period of the instantaneous power is $0.5T$. The positive instantaneous power indicates that the absorption of power by the capacitor from the source and negative instantaneous power depicts the absorption of power by the source from the charged capacitor.

Average power:

The expression of the average power is obtained as,

$$P_{avg} = \frac{1}{0.5T} \int_0^{0.5T} p(t) d\omega t$$

$$= \frac{1}{0.5T} \int_0^{0.5T} [\frac{1}{2} V_m I_m \sin 2\omega t] d\omega t$$

$$P_{avg} = 0 \tag{1.51}$$

The average power absorption by a pure capacitive element over a time period is zero. As shown in Fig.1.27, the instantaneous power oscillates between source and capacitor (load) and it reacts upon itself. It does not do any useful work. From equations (1.46) and (1.50) the different formula of peak value of this reactive power is derived as,

$$|Q_m| = \frac{V_m}{\sqrt{2}} \frac{I_m}{\sqrt{2}} = V_{rms} I_{rms} = \frac{V_m^2}{2X_C} = \frac{I_m^2 X_C}{2} \tag{1.52}$$

In conclusion, it can be stated that the pure resistive load absorbs power from the source but the pure reactive (pure inductive or pure capacitive) load consumes zero average power.

1.5.4 General Concept of Impedance, Impedance Triangle and Admittance

1.5.4.1 Impedance and Impedance Triangle

The concept of impedance for pure resistive, pure inductive and pure capacitive circuits are discussed in the previous sections. But in practice, these elements are connected in different combinations to form a network. In that case, the impedance is represented as,

$$Z = R + j(X_L - X_C) = R + jX = |Z| \angle\theta \tag{1.53}$$

$$|Z| = \sqrt{R^2 + X^2} \tag{1.54}$$

$$\theta = \tan^{-1}\left(\frac{X}{R}\right) \tag{1.55}$$

Where R, X_L, X_C, X are the effective values of the resistance, inductive reactance, capacitive reactance and net reactance of the network respectively. If the value of X_L is greater than X_C then the imaginary term of the impedance becomes positive indicating net reactance as inductive reactance and if the value of X_L is less than X_C then the imaginary term of the impedance becomes negative indicating net reactance as capacitive reactance. Impedance is measured in ohms (Ω).

Impedance angle:

θ is known as impedance angle. It is positive if the net reactance is inductive and it is negative if the net reactance is capacitive. It is to be noted that impedance angle is phase angle between voltage and current or power factor angle.

As the impedance is expressed in complex form, so it can be represented in complex plane as shown in Fig. 1.28.

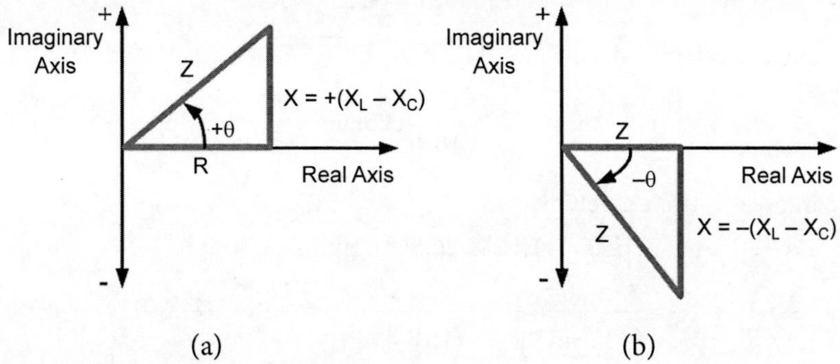

Fig. 1.28: Impedance triangle for (a) net reactance as inductive, (b) for net reactance as capacitive

Impedance triangle:

Impedance in complex plane form a triangle which is known as impedance triangle. Impedance triangle illustrates the relationship between resistance, reactance and impedance. But it is to be remembered that these quantities are not in phasor form.

1.5.4.2 Admittance

Admittance (*Y*) is reciprocal of impedance in AC circuits. It is measured in Siemens (S). The expression for admittance is given by,

$$Y = \frac{1}{Z} = \frac{1}{R + jX} = \frac{R - jX}{(R + jX)(R - jX)}$$

$$Y = \frac{R}{R^2 + X^2} - j\frac{X}{R^2 + X^2} = G + jB \tag{1.56}$$

The real part of the admittance is known as conductance (G) and it is measured in mho (℧) or Siemens (S). The imaginary term in the admittance is susceptance (B) and it is measured in mho (℧) or Siemens (S).

Example 1.13

The equivalent resistance, equivalent inductive reactance and equivalent capacitive reactance of a circuit is given by 10 ohm, 15 ohm and 5 ohm respectively. Find out the equivalent impedance and admittance of the circuit. Also calculate the impedance angle.

Solution:

We know,

$$Z = R + j(X_L - X_C) = R + jX = |Z|\angle\theta$$
$$R = 10 \text{ ohm.}$$
$$X = X_L - X_C = (15 - 10) = 5 \text{ ohm}$$

Impedance is obtained as,

$$Z = 10 + j5 \text{ ohm}$$
$$|Z| = \sqrt{(10)^2 + (5)^2} = \sqrt{125} = 11.18$$

Impedance angle is calculated as,

$$\theta = \tan^{-1}\left(\frac{5}{10}\right) = 26.56^0$$

Impedance in polar form is given by,

$$|Z|\angle\theta = 11.18 \angle 26.56° \text{ ohm}$$

$$Y = \frac{R}{R^2 + X^2} - j\frac{X}{R^2 + X^2} = \frac{10}{(10)^2 + (5)^2} - j\frac{5}{(10)^2 + (5)^2} = \frac{10}{125} - j\frac{5}{125}$$

$$Y = 0.08 - j\,0.04 \text{ S.}$$

1.5.5 Concept of power and power triangle

The steady-state voltage-current relationships and concept of power for pure resistive, pure inductive and pure capacitive are discussed in the previous sections. But in practice, all these passive elements are connected in different series-parallel combinations. Hence, it is essential to know the concept of generalized power where a sinusoidal ac voltage source is connected to a load combined with *R*, *L* and *C*.

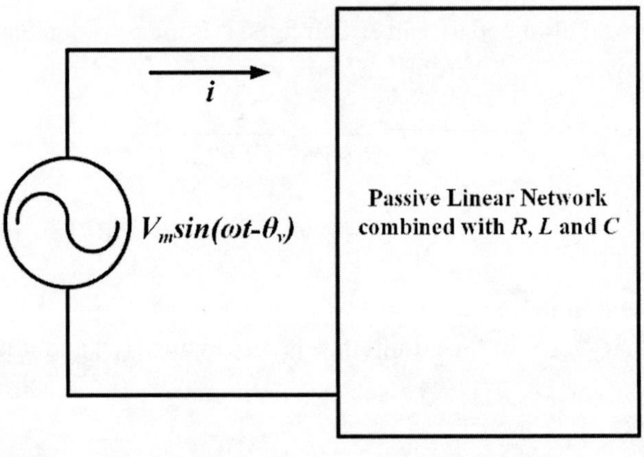

Fig. 1.29: Passive linear network excited by sinusoidal ac voltage source

Let us assume a passive linear network combined with *R*, *L* and *C* is excited by a sinusoidal ac voltage source with a magnitude of V_m, angular frequency of ω and a phase delay of θ_v as shown in Fig. 1.29. Due to this voltage source, let us consider a sinusoidal current of magnitude of I_m and phase delay of θ_i is flowing through the circuit. The instantaneous voltage and current waveforms are indicated in Fig.1.30 and their expressions are given as,

$$v(t) = V_m \sin(\omega t - \theta_v) \quad (1.57)$$

$$i(t) = I_m \sin(\omega t - \theta i) \quad (1.58)$$

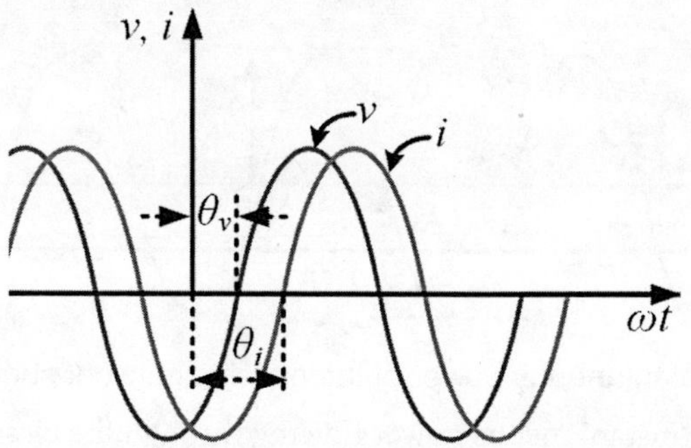

Fig.1.30: Instantaneous voltage and current waveforms of the circuit mentioned in Fig.1.29

1.5.5.1 Instantaneous power, average power and complex power

Instantaneous power:

The instantaneous power flowing through this circuit is given by

$$p(t) = v(t)\, i(t) \quad (1.59)$$

Equation (1.59) can be modified by using equations (1.57) and (1.58) as,

$$p(t) = V_m \sin(\omega_t - \theta_v) \times I_m \sin(\omega_t - \theta_i)$$

$$p(t) = \frac{1}{2} V_m I_m \{\cos(\theta_i - \theta_v) - \cos(2\omega t - \theta_v - \theta_i)\}$$

$$p(t) = \frac{1}{2} V_m I_m \{\cos(\theta_v - \theta_i) - \cos(2\omega t - \theta_v - \theta_i)\}$$

$$p(t) = \frac{1}{2} V_m I_m \cos(\theta_v - \theta_i) - \frac{1}{2} V_m I_m \cos(2\omega t - \theta_v - \theta_i) \quad (1.60)$$

Let us consider the phase difference between the voltage and current phasor is $\theta = (\theta_v - \theta_i)$. Then the expression of the instantaneous power can be modified as,

$$p(t) = \frac{1}{2} V_m I_m \cos\theta - \frac{1}{2} V_m I_m \cos(2\omega t - \theta_v - \theta_i) \quad (1.61)$$

First term of equation (1.61) is time independent and it depends on the cosine of the phase angle between the voltage and current. As the phase angle between the voltage and current is constant, the first term is also a constant. The second term is a time varying sinusoidal function. The frequency of this term is double the frequency of supply voltage and current through the circuit. Hence the time period of this second term is $\frac{\pi}{\omega}$ second if the time period of the voltage and current signal is $\frac{2\pi}{\omega}$ second. If the time period of the voltage and current is denoted as T then the time period of the second term in the instantaneous power will be $0.5T$.

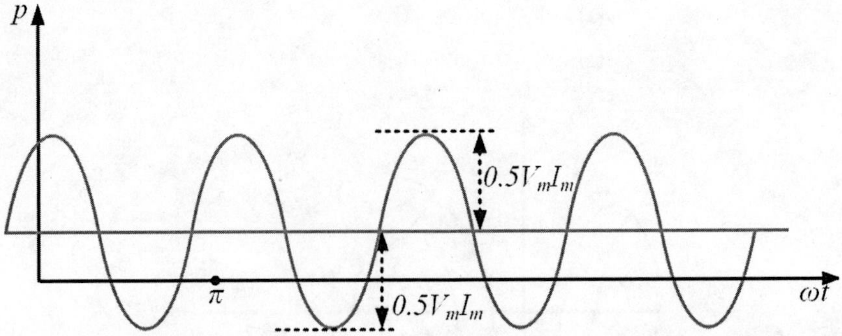

Fig. 1.31: Instantaneous power of the passive network shown in Fig. 1.29

Fig. 1.31 represents the instantaneous power flowing through the circuit considered. It is clear from the figure that the average value of this quantity is positive as it has positive values in all the cycles. The positive value of the power waveform indicates the absorption of power by the load from the source and the negative power indicates the delivery of power to the source by the load. Power delivery to the source is possible by discharging inductor or capacitor or both capacitor and inductor.

Average power:

Instantaneous power is changing w.r.t time. So, it is very difficult to measure it. Rather it is convenient to measure the average power. Usually, the instrument 'wattmeter' is used to measure the average power in Watts. The expression of the average power for this circuit is obtained as,

$$P_{avg} = \frac{1}{0.5T}\int_0^{0.5T} p(t)d\omega t$$

$$P_{avg} = \frac{1}{0.5T}\int_0^{0.5T} [\frac{1}{2}V_mI_m\cos\theta - \frac{1}{2}V_mI_m\cos(2\omega t - \theta_v - \theta_i)d\omega t$$

$$P_{avg} = \frac{1}{2}V_mI_m\cos\theta \tag{1.62}$$

Equation (1.62) represents the average power flowing through a passive network consists of *R*, *L* and *C* excited by sinusoidal ac voltage source. In the expression of average power, the second term of the instantaneous power disappears as it is time varying sinusoidal function. The average power is also constant and it depends on the phase difference between voltage and current. It is interesting to note that the instantaneous power is time varying but the average power is time independent. Hence the calculation of instantaneous power requires the knowledge of instantaneous voltage and instantaneous current in time domain. But average power can be evaluated in time domain as well as in frequency domain (in phasor form). The instantaneous voltage and current mentioned in equations (1.57) and (1.58) are expressed in phasor as,

$$V = V_m\angle\theta v \tag{1.63}$$

$$I = I_m\angle\theta_i \tag{1.64}$$

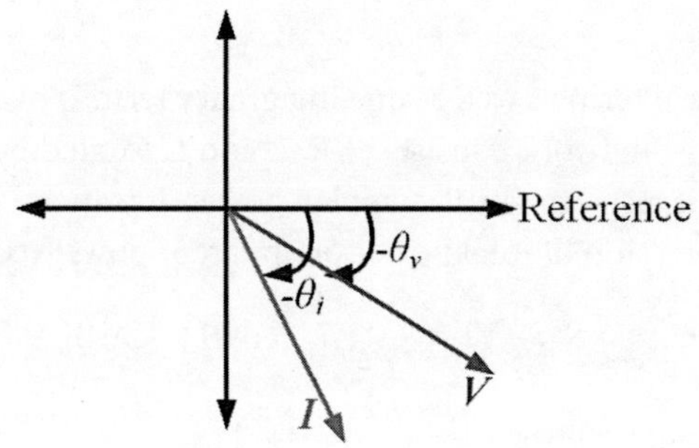

Fig.1.32: Voltage and Current phasors as mentioned in equations (1.63) and (1.64)

One point is to be remembered while calculating the average power in phasor form that the expression of average power includes the phase difference between the voltage and current phasor. Considering this, if we take the conjugate of current phasor and multiply it with 0.5 times of voltage phasor then the expression becomes,

$$\frac{1}{2}VI^* = \frac{1}{2}V_m\angle\theta_v \times I_m\angle -\theta_i = \frac{1}{2}V_m I_m \angle(\theta_v - \theta_i)$$

$$\frac{1}{2}VI^* = \frac{1}{2}V_m I_m[\cos(\theta_v - \theta_i) + j\sin(\theta_v - \theta_i)]$$

$$\frac{1}{2}VI^* = \frac{1}{2}V_m I_m[\cos\theta + j\sin\theta] \tag{1.65}$$

The real term of the above equation is the average power flowing through a passive network consists of *R*, *L* and *C* excited by sinusoidal ac voltage source. So, it can be concluded that the real part of the equation (1.65) is the average power, which can be expressed as,

$$P_{avg} = \text{Re}[\frac{1}{2}VI^*] = \frac{1}{2}V_m I_m \cos\theta \tag{1.66}$$

When voltage and current are in phase, i.e. θ = 0 then the circuit becomes pure resistive circuit. In that case, the average power becomes,

$$P_{avg} = \frac{1}{2}V_m I_m = \frac{V_m^{\ 2}}{2R} = \frac{I_m^{\ 2}R}{2} = \frac{|V|^2}{2R} = \frac{|I|^2 R}{2} \tag{1.67}$$

Where |V| and |I| are the absolute values of voltage and current phasors respectively. When the phase difference between the voltage and current phase is 900, i.e. θ = ±90° then the circuit becomes either pure capacitive or pure inductive. In that case, the average power becomes,

$$P_{avg} = 0 \tag{1.68}$$

Equations (1.67) and (1.68) again conform that the pure resistive circuit consumes power and pure inductive or capacitive circuit does not consume any average power. These satisfy the concept discussed in equations (1.30),(1.40) and (1.51).

Complex power:

Equation (1.64) includes one real term as well as one imaginary term. This is the total power or apparent power consumed by any passive network consists of *R*, *L* and *C* excited by sinusoidal ac source. As it is expressed in complex form, it is also known as complex power. It is measured in Volt-Ampere (VA) and usually denoted as '*S*'. The expression of total power or apparent power or complex power is given by,

$$S = \frac{1}{2}VI^* = \frac{1}{2}V_m I_m[\cos\theta + j\sin\theta]$$

In terms of rms values the complex power is expressed as,

$$S = V_{rms}\, I_{rms}{}^* = |V_{rms}||I_{rms}|[\cos\theta + j\sin\theta] \tag{1.70}$$

Where the rms values of voltage and current in phasor forms are as follows,

$$V_{rms} = \frac{V}{\sqrt{2}} = \frac{V_m}{\sqrt{2}}\angle\theta_v = |V_{rms}|\angle\theta_v \tag{1.71}$$

$$I_{rms} = \frac{I}{\sqrt{2}} = \frac{I_m}{\sqrt{2}}\angle\theta_i = |I_{rms}|\angle\theta_i \tag{1.72}$$

Real power or active power:

It is already discussed that the real part of equation (1.69) or (1.70) is the average power consumed by the circuit. As it is the real part of the equation, it is also known as real power or true power. It involves in active work done so sometimes it is also termed as active power. It is the useful power consumed or dissipated by the load. It is measured in Watts and denoted as *P*. Real power is consumed by resistive element of the circuit. Inductive or capacitive element does not consume active power.

$$P = \mathrm{Re}[S] = \mathrm{Re}\left[\frac{1}{2}VI^*\right] = \frac{1}{2}V_m I_m\cos\theta = |V_{rms}||I_{rms}|\cos\theta \tag{1.73}$$

Reactive power:

The imaginary part of the equation (1.69) or (1.70) is the reactive power of the circuit. It is the measure of power exchange between the source and reactive part of the network. This power transfers back and forth between the source and reactive part of the circuit. The reactive power in a pure resistive circuit is zero. Whatever power is transferred to the reactive elements by the source in one half cycle, the same is thrown back to the source from those reactive elements in the next half cycle. It is known as reactive power as it reacts upon itself by lossless interchanging of power. It is measured in VAr and denoted as *Q*.

$$Q = Im[S] = \mathrm{Re}\left[\frac{1}{2}VI^*\right] = \frac{1}{2}V_m I_m\sin\theta = |V_{rms}||I_{rms}|\sin\theta \tag{1.74}$$

The expression of total power or apparent power is mentioned in terms of active power and reactive power as follows.

$$S = P + jS \tag{1.75}$$

$$|S| = \sqrt{P^2 + Q^2} = |V_{rms}||I_{rms}| \tag{1.76}$$

Power factor can also be calculated by apparent power and active power as,

$$PF = \cos\theta = \frac{P}{S} = \frac{\text{Active Power}}{\text{Apparent Power}}$$

Impedance can be evaluated by using equations (1.63), (1.64), (1.71) and (1.72) as,

$$Z = \frac{V}{I} = \frac{V_m \angle\theta_v}{I_m \angle\theta_i} = \frac{V_m}{I_m}\angle(\theta_v - \theta_i) = \frac{|V_{rms}|\angle\theta_v}{|I_{rms}|\angle\theta_i} = \frac{|V_{rms}|}{|I_{rms}|}\angle(\theta_v - \theta_i) \quad (1.78)$$

As the phase angle $(\theta_v - \theta_i)$ is already denoted as θ, then the equation (1.78) is modified as,

$$Z = \frac{V_m}{I_m}\angle\theta = \frac{|V_{rms}|}{|I_{rms}|}\angle\theta = |Z|\angle\theta = \sqrt{R^2 + X^2}\angle\theta = R + jX \quad (1.79)$$

Where R and X are effective or equivalent resistance and reactance of the network considered in Fig. 1.29. By comparing equations (1.17) and (1.79), R and X can be expressed as,

$$R = |Z| \cos\theta \quad (1.80)$$

$$X = |Z| \sin\theta \quad (1.81)$$

From the equations (1.73), (1.74), (1.80) and (1.81) the active, reactive power and apparent power can also expressed as,

$$P = |V_{rms}||I_{rms}|\cos\theta = \frac{|V_{rms}|}{|Z|}|I_{rms}|R = |I_{rms}|^2 R = \frac{|V_{rms}|^2}{R} \quad (1.82)$$

$$Q = |V_{rms}||I_{rms}|\sin\theta = \frac{|V_{rms}|}{|Z|}|I_{rms}|X = |I_{rms}|^2 X = \frac{|V_{rms}|^2}{X} \quad (1.83)$$

$$|S| = \sqrt{P^2 + Q^2} = |I_{rms}|^2\sqrt{R^2 + X^2} = |I_{rms}|^2|Z| = \frac{|V_{rms}|^2}{|Z|} \quad (1.84)$$

Example 1.14

The voltage across a circuit is given by *120 sin(100 π t + 30°)* V. The current through the circuit is *5 sin(100 π t + 45°)* A. Find out the active power, reactive power, apparent power and power factor.

Solution:

$$V = 120\angle 30°$$

$$I = 5\angle 45°$$

Power factor angle is given by,

$$\theta = -15°$$

Active power is given by,

$$P = \frac{1}{2} \times 120 \times 5 \times \cos\left(-15^0\right) = 289.77W$$

Reactive power is calculated as,

$$Q = \frac{1}{2} \times 120 \times 5 \times \sin\left(-15^0\right) = -77.645 VAr$$

Total power and apparent power is given by,

$$S = 289.77 - j77.645 \text{ VA}$$

Power factor is obtained as,

$$PF = \cos(-15°) = 0.966 \text{ leading}$$

1.5.5.2 Power triangle

The total power is expressed in complex form and it can be represented in complex plane. The relationships between the active power, reactive power and total or apparent power are shown in Fig.1.33. It forms a triangular shape that is why it is known as power triangle.

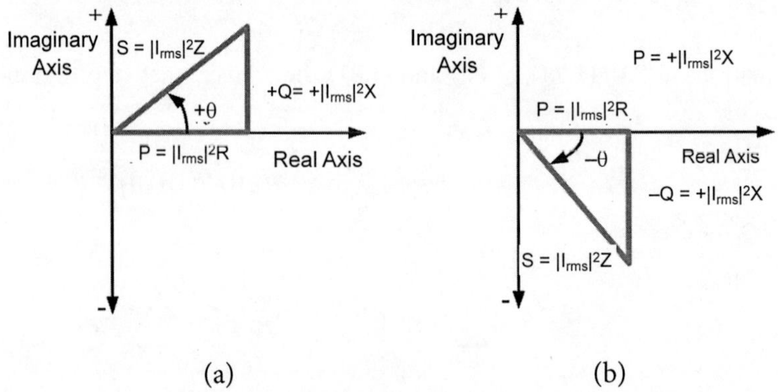

Fig.1.33: Power triangle for (a) positive reactive power, (b) for negative reactive power

When the phase angle ($\theta = \theta_v - \theta_i$) between voltage and current is positive, the reactive power is positive. Positive phase angle is possible when current lags the voltage i.e in a circuit with a dominance of inductive reactance. During this time the PF is lagging. On the other hand, reactive power is negative in a circuit where capacitive reactance is dominating. During this time, current lead the supply voltage and PF is leading. Equation (1.79) and comparison of Fig.1.28 and Fig.1.33 also confirm that the impedance angle is the phase angle between the voltage and current as discussed in section 1.5.4.Another interesting fact is to be noted that if all the sides of impedance triangle is extended by $|I_{rms}|^2$ times without altering their direction then it will become power triangle.

1.6 STEADY STATE RESPONSE OF SERIES CIRCUITS FOR A SINUSOIDAL AC VOLTAGE SOURCE

In this section, the steady state response of networks consist of different series combinations of R, L and C excited by sinusoidal ac voltage source are discussed.

1.6.1 R-L Series Circuit

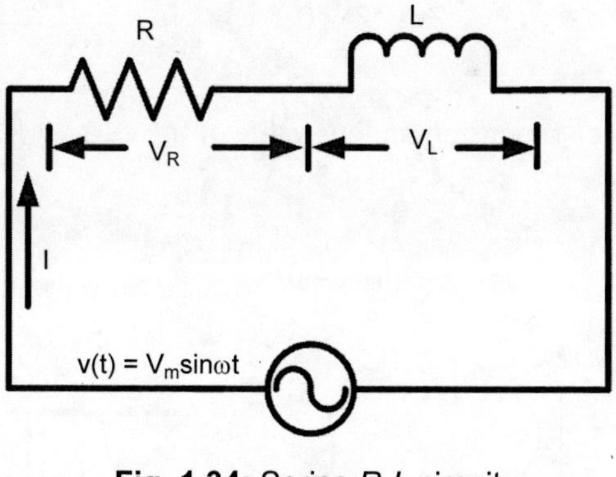

Fig. 1.34: Series *R-L* circuit

A circuit consists of series combination of pure resistor (R) and pure inductor (L) is connected to a sinusoidal ac voltage source of $v(t) = V_m$ sinωt as shown in Fig.1.34. The phasor form of the supply voltage is expressed as,

$$V = V_m \angle 0° \tag{1.85}$$

The phasor form of voltage across R and L are mentioned as V_R and V_L in Fig.1.34. Applying the concept of phasor sum the relation between V, V_R and V_L is given by,

$$V = V_R + V_L \tag{1.86}$$

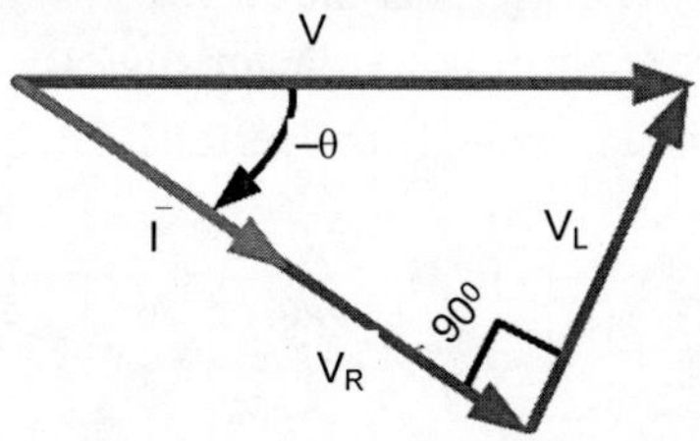

Fig. 1.35: Voltage and current phasors of series *R-L* circuit

The phasor sum of all the voltages are also represented in Fig.1.35. Using the concept mentioned in equations (1.24) and (1.34), the above equation can be modified as,

$$V = I_R + jIX_L \tag{1.87}$$

$$\mathrm{V} = \mathrm{I}\,(\mathrm{R} + jX_L) = IZ \tag{1.88}$$

Where the impedance Z is given by,

$$\mathrm{Z} = |\mathrm{Z}|\angle\theta \tag{1.89}$$

$$|Z| = \sqrt{R^2 + {X_L}^2} \tag{1.90}$$

$$\theta = \tan^{-1}\left(\frac{X_L}{R}\right) = \tan^{-1}\left(\frac{\omega L}{R}\right) \tag{1.91}$$

This impedance angle is also power factor angle for this circuit. The value of current flowing through the circuit can be found out by using equations (1.85), (1.88) and (1.89) as,

$$I = \frac{V}{Z} = \frac{V_m\angle 0^0}{|Z|\angle\theta} = \left(\frac{V_m}{|Z|}\right)\angle -\theta = I_m\angle -\theta \tag{1.92}$$

Where, I_m is $\left(\frac{V_m}{|Z|}\right)$. Hence the instantaneous value of the current flowing through the circuit is as follows,

$$i = I_m \sin(\omega t - \theta) \tag{1.93}$$

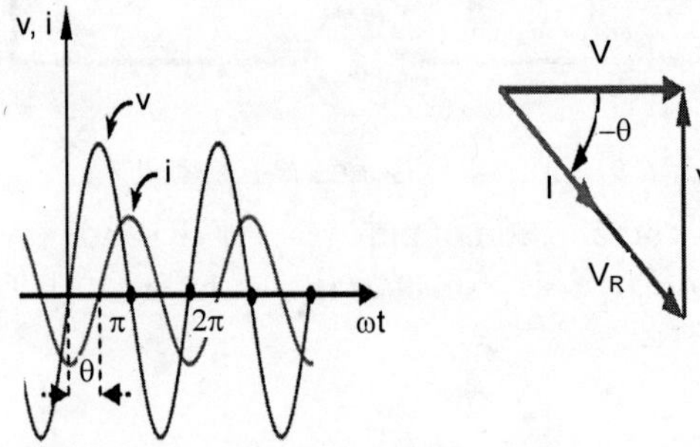

Fig. 1.36: Waveforms of supply voltage and current of series *R-L* circuit

The relation between instantaneous value of voltage and current is represented in Fig.1.36. Fig.1.36 and equations (1.85) and (1.92) clearly depict that the current through a series R-L is lagging the supply voltage by an angle of θ (impedance angle or power factor angle).

The instantaneous power flowing through the series R-L circuit is obtained by using (1.61) as follows,

$$p(t) = \frac{1}{2}V_m I_m \cos\theta - \frac{1}{2}V_m I_m \cos(2\omega t - \theta) \tag{1.94}$$

The first term of equation (1.94) is the active power or average power and it is given by,

$$P = \frac{1}{2}V_m I_m \cos\theta = |V_{rms}||I_{rms}|\cos\theta = |I_{rms}|^2 R = \frac{|V_{R,rms}|^2}{R} \tag{1.95}$$

The second term of the instantaneous power is a time varying sinusoidal whose average value over a complete cycle is zero. It represents only charging of inductor in one half cycle and discharging of it in the next half cycle by pumping the power back to the source. This circulating power reacts upon itself and it is known as reactive power. It is the measure of oscillating power between the source and inductive reactance. The reactive power of the circuit in different forms are expressed as,

$$Q = \frac{1}{2}V_m I_m \sin\theta = |V_{rms}||I_{rms}|\sin\theta = |I_{rms}|^2 X_L = \frac{|V_{X,rms}|^2}{X_L} \tag{1.96}$$

The reactive power for a series R-L circuit is positive and it represents the lagging reactive power as indicated in Fig.1.33 (a). The expression of total power or apparent power for the series *R-L* circuit is mentioned as follows,

$$S = \frac{1}{2}VI^* = \frac{1}{2}V_m I_m[\cos\theta + j\sin\theta] \tag{1.97}$$

$$S = P + jQ \tag{1.98}$$

$$|S| = \sqrt{P^2 + Q^2} = \frac{1}{2}V_m I_m = |V_{rms}||I_{rms}| = |I_{rms}|^2 Z = \frac{|V_{rms}|^2}{Z} \tag{1.99}$$

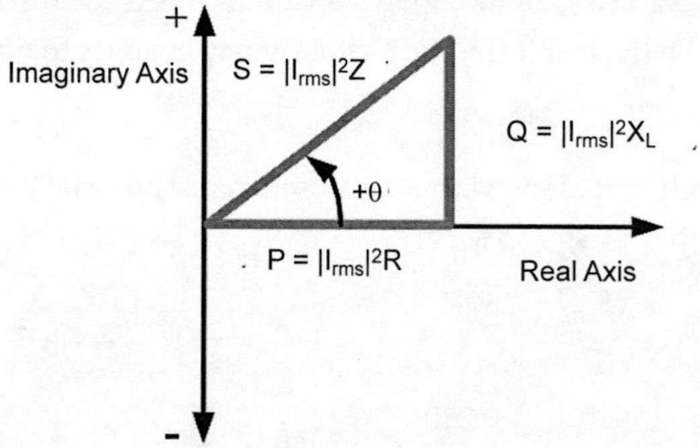

Fig. 1.37: Power triangle for a series *R-L* circuit

The power triangle for this circuit is presented in Fig.1.37. The power triangle and equation (1.96) confirm the positive and lagging reactive power and positive power factor angle.

Power factor can also be calculated by apparent power and active power or resistance and impedance as,

$$PF = \cos\theta = \frac{P}{S} = \frac{R}{Z} \tag{1.100}$$

The power factor for a series *R-L* circuit is lagging in nature. It can also be evaluated from voltage phasors as indicated in Fig.1.35 as,

$$PF = \cos\theta = \frac{V_R}{V} \tag{1.101}$$

Example 1.15

A series R-L circuit is connected to 230V, 50Hz AC supply. It draws a total current *4 -j3 A*. Find out the active power, reactive power and apparent power.

Solution:

$$V = 230\angle 0°$$

$$I = 4 - j3A = \sqrt{16+9}\angle\tan^{-1}\left(\frac{-3}{4}\right) = 5\angle -36.86^0$$

AC quantities are specified in terms of rms value. Hence the voltage and current given are in rms values. Active power is given by,

$$P = 230 \times 5 \times \cos(0° - (-36.86°)) = 920.12 \text{ W}$$

Reactive power is calculated as,

$$Q = 230 \times 5 \times \sin(0° - (-36.86°)) = 689.84 \text{ VAr}$$

Total power and apparent power is given by,

$$S = 920.12 + j\,689.84 \text{ VA}$$

Example 1.16

A practical coil connected to a 230V, 50Hz supply consumes 6 A while dissipating 500 W across it. Find the value of current drawn by the coil if the supply frequency changes to 60 Hz from 50 Hz.

Solution:

Practical coil includes an inductor along with a small series resistor. Hence, the coil can be modelled as,

$$Z = R + jX_L$$

$$V_{rms} = 230\text{V}$$

$$I_{rms} = 6\text{A}$$

$$|Z| = \frac{230}{6} = 38.33 ohm$$

$$P = |I_{rms}|^2 R$$

$$500 = (6)^2 \times R$$

$$R = \frac{500}{36} = 13.89 ohm$$

$$|Z| = 38.33 = \sqrt{(13.89)^2 + X_L{}^2}$$

$$1469.189 = 192.93 + X_L{}^2$$

$$X_L = 35.724 \text{ ohm}$$

$$X_L = 2\pi fL$$

$$L = \frac{35.724}{2 \times 3.14 \times 50} = 0.1137H$$

The inductive reactance changes with the change in frequency. Now the new reactance at 60 Hz is calculated as,

$$X_{L,\,new} = 2\pi fL = 2 \times 3.14 \times 60 \times 0.1137 = 42.86 \text{ ohm}$$

$$|Z_{,new}| = \sqrt{(13.89)^2 + (42.86)^2} = 45.05 ohm$$

$$\theta = \tan^{-1}\left(\frac{42.86}{13.89}\right) = 72.04^0$$

$$I = I_m \angle\theta = \frac{|\sqrt{2}V_{rms}| \angle 0^0}{|Z| \angle 72.04^0} = 7.22\angle -72.04^0\, A$$

1.6.2 R-C Series Circuit

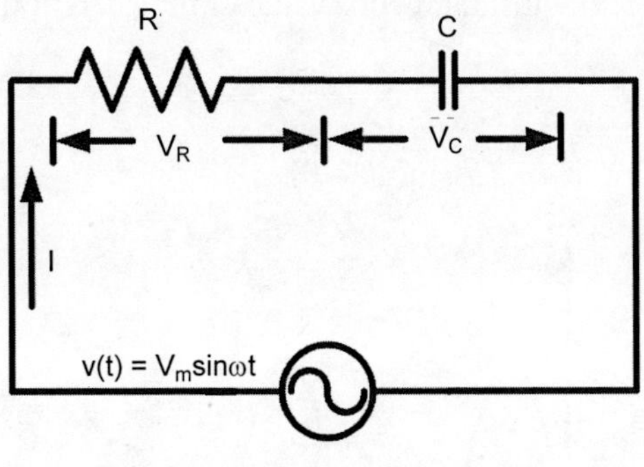

Fig. 1.38: Series *R-C* circuit

A circuit consists of series combination of pure resistor (R) and pure capacitor (C) is connected to a sinusoidal ac voltage source of $v(t) = V_m \sin\omega t$ as shown in Fig.1.38. The phasor form of the supply voltage is expressed as,

$$V = V_m \angle 0° \tag{1.102}$$

The phasor form of voltage across R and C are mentioned as VR and VC in Fig.1.38. Applying the concept of phasor sum the relation between V, VR and VL is given by,

$$V = V_R + V_C \tag{1.103}$$

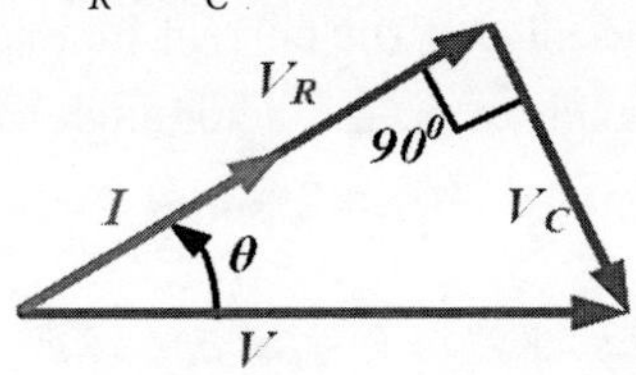

Fig. 1.39: Voltage and current phasors of series *R-C* circuit

The phasor sum of all the voltages are also represented in Fig.1.39. Using the concept mentioned in equations (1.24) and (1.45), the above equation can be modified as,

$$V = IR - jIX_C \tag{1.104}$$

$$V = I(R - jX_C) = IZ \tag{1.105}$$

Where the impedance Z is given by,

$$Z = |Z|\angle\theta \tag{1.106}$$

$$|Z| = \sqrt{R^2 + X_C{}^2}$$

$$\theta = \tan^{-1}\left(\frac{-X_C}{R}\right) = \tan^{-1}\left(\frac{-1}{R\omega C}\right) \tag{1.108}$$

From equations (1.106) and (1.108), it is clear the impedance angle for a series *R-C* circuit is negative. The value of current flowing through the circuit can be found out by using equations (1.102), (1.106) as,

$$I = \frac{V}{Z} = \frac{V_m\angle 0^0}{|Z|\angle -\theta} = \left(\frac{V_m}{|Z|}\right)\angle\theta = I_m\angle\theta \tag{1.109}$$

Where, I_m is $\left(\frac{V_m}{|Z|}\right)$. Hence the instantaneous value of the current flowing through the circuit is as follows,

$$i = I_m \sin(\omega t + \theta) \tag{1.110}$$

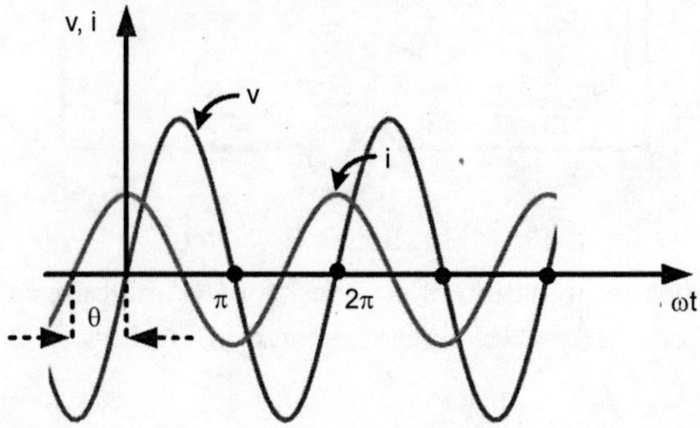

Fig. 1.40: Waveforms of supply voltage and current of series *R-C* circuit

The relation between instantaneous value of voltage and current is represented in Fig.1.40. Equations (1.102) and (1.109) and Fig. 1.40 present that the current in a series *R-C* circuit is leading the supply voltage by an angle of θ (impedance angle or power factor angle).

The instantaneous power flowing through the series *R-C* circuit is obtained by using (1.61) as follows,

$$p(t) = \frac{1}{2}V_m I_m \cos(-\theta) - \frac{1}{2}V_m I_m \cos(2\omega t + \theta) \tag{1.111}$$

The first term of equation (1.111) is the active power or average power and it is given by,

$$P = \frac{1}{2}V_m I_m \cos\theta = |V_{rms}||I_{rms}|\cos\theta = |I_{rms}|^2 R = \frac{|V_{R,rms}|^2}{R} \tag{1.112}$$

The second term of the instantaneous power is a time varying sinusoidal whose average value over a complete cycle is zero. It represents only charging of the capacitor in one half cycle and discharging of it in the next half cycle by pumping the power back to the source. This reactive is the measure of oscillating power between the source and capacitive reactance. The reactive power of the series *R-C* circuit in different forms are expressed as,

$$Q = \frac{1}{2}V_m I_m \sin(-\theta) = -\frac{1}{2}V_m I_m \sin\theta \tag{1.113}$$

$$Q = -|V_{rms}||I_{rms}|\sin\theta = -|I_{rms}|^2 X_C = -\frac{|V_{X,rms}|^2}{X_C} \tag{1.114}$$

Equations (1.113) and (1.114) show that the reactive power is negative. It is because of negative impedance angle or negative and leading power factor angle. This reactive power is the leading reactive power. The expression of total power or apparent power for the series *R-C* circuit is mentioned as follows,

$$S = P - jQ \tag{1.115}$$

$$|S| = \sqrt{P^2 + Q^2} = \frac{1}{2}V_m I_m = |V_{rms}||I_{rms}| = |I_{rms}|^2 Z = \frac{|V_{rms}|^2}{Z} \tag{1.116}$$

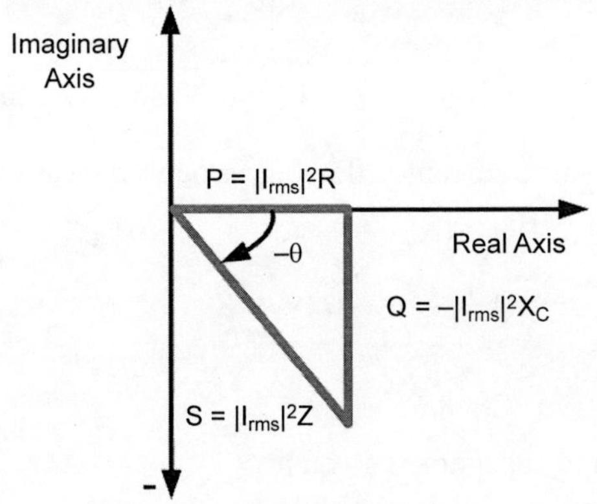

Fig. 1.41: Power triangle for a series *R-C* circuit

The power triangle for this circuit is presented in Fig.1.41. The power triangle also represents the negative and leading reactive power and negative power factor angle.

Power factor can also be calculated by apparent power and active power or resistance and impedance as,

$$PF = \cos(-\theta) = \frac{P}{S} = \frac{R}{Z} \tag{1.117}$$

The power factor for a series *R-C* circuit is leading in nature. It can also be evaluated from voltage phasors as indicated in Fig.1.39 as,

$$PF = \cos(-\theta) = \frac{V_R}{V} \tag{1.118}$$

Example 1.17

A series R-C circuit is connected to a 220V, 50Hz supply. The resistor dissipates 250W at 110V. Find the value of C and calculate the maximum energy stored by the capacitor.

Solution:

$$V = 220 \times \sqrt{2}\angle 0^0$$

The formula for active power in terms of voltage and resistance is given by,

$$P = \frac{|V_{R,rms}|^2}{R}$$

$$R = \frac{(110)^2}{250} = 48.4ohm$$

The formula for active power in terms of current and resistance is given by,

$$P = |I_{rms}|^2 R$$

$$I_{rms} = \sqrt{\frac{P}{R}} = \sqrt{\frac{250}{48.4}} = 5.16A$$

$$|V_c| = \sqrt{V^2 - V_R{}^2} = \sqrt{96800 - 24200} = 269.44V$$

As it is series circuit, the same current is flowing through the capacitor also. So,

$$|V_c| = IX_C$$

$$269.44 = (5.16 \times \sqrt{2}) \times \frac{1}{2\pi fC}$$

$$C = 8.62 \times 10^{-5} \text{ F}$$

Maximum energy stored in capacitor is given by,

$$E = \frac{C|V_C|^2}{2} = 3.129Joule$$

Example 1.18

A series R-C circuit is connected to a $110 \sin(100 \pi t + 30^o)$ V supply and current flowing through the circuit is $20 \sin(100 \pi t + 60^o)$ A. Find the power factor, active power and reactive power.

Solution:

$$V = 110\angle 30°$$

$$I = 20\angle 60°$$

Power factor is given by,

$$PF = \cos(\theta_v - \theta_i) = \cos(-30°) = 0.866 \text{ (leading)}$$

Active Power is given by,

$$P = \frac{1}{2} V_m I_m \cos\left(-30^0\right) = 110 \times 20 \times 0.866 = 1905.25W$$

Reactive Power is given by,

$$Q = \frac{1}{2} V_m I_m \sin\left(-30^0\right) = -110 \times 20 \times 0.5 = -1100VAr$$

1.6.3 R-L-C Series Circuit

A circuit consists of series combination of pure resistor (R), pure inductance (L) and pure capacitor (C) is connected to a sinusoidal ac voltage source of $v(t) = V_m \sin\omega t$ as shown in Fig.1.42.

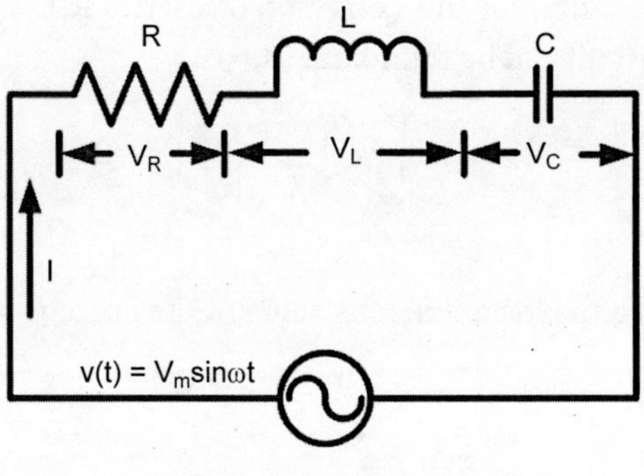

Fig. 1.42: Series *R-L-C* circuit

The phasor form of the supply voltage is expressed as,

$$V = V_m \angle 0° \tag{1.119}$$

The phasor form of voltage across *R*, *L* and *C* are mentioned as V_R, V_L and V_C in Fig.1.42. Applying the concept of phasor sum the relation between *V*, V_R, V_L and V_C is given by,

$$V = V_R + V_L + V_C \tag{1.120}$$

The series *R-L-C* circuit is analysed here by considering two conditions. First one is $X_L > X_C$ and second one is $X_L < X_C$.

Condition 1: $X_L > X_C$

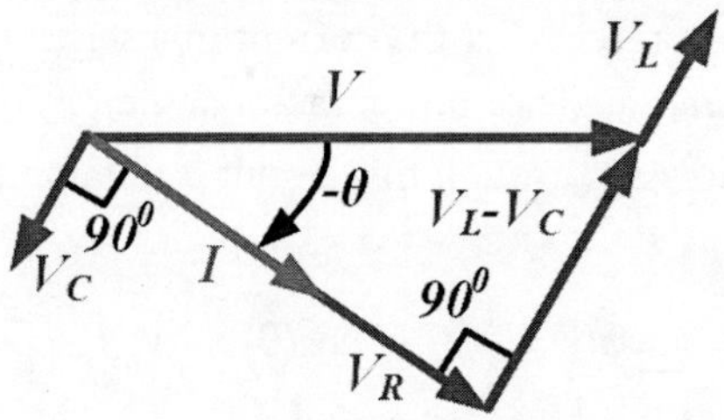

Fig. 1.43: Voltage and current phasors of series *R-L-C* circuit for $X_L > X_C$

The phasor sum of all the voltages for this condition are represented in Fig.1.43. Using the concept mentioned in equations (1.24), (1.34) and (1.45) the equation (1.120) can be modified as,

$$V = IR + jI(X_L - X_C) \tag{1.121}$$

$$V = I\,(R + j(X_L - X_C) = I(R + jX) = IZ \tag{1.122}$$

Where $X_L - X_C = X$. Value of net reactance i.e *X* is positive as $X_L > X_C$. Considering net reactance as positive the impedance *Z* is given by,

$$Z = |Z| \angle \theta \tag{1.123}$$

$$|Z| = \sqrt{R^2 + (X_L - X_C)^2} \tag{1.124}$$

$$\theta = \tan^{-1}\left(\frac{X_L - X_C}{R}\right) \tag{1.125}$$

As, $X_L > X_C$, the impedance angle for this condition of a series R-L-C circuit is positive. The value of current flowing through the circuit can be found out as,

$$I = \frac{V}{Z} = \frac{V_m \angle 0^0}{|Z| \angle \theta} = \left(\frac{V_m}{|Z|}\right) \angle -\theta = I_m \angle -\theta \quad (1.126)$$

Where, I_m is $\left(\frac{V_m}{|Z|}\right)$. Hence the instantaneous value of the current flowing through the circuit is as follows,

$$i = I_m \sin(\omega t - \theta) \quad (1.127)$$

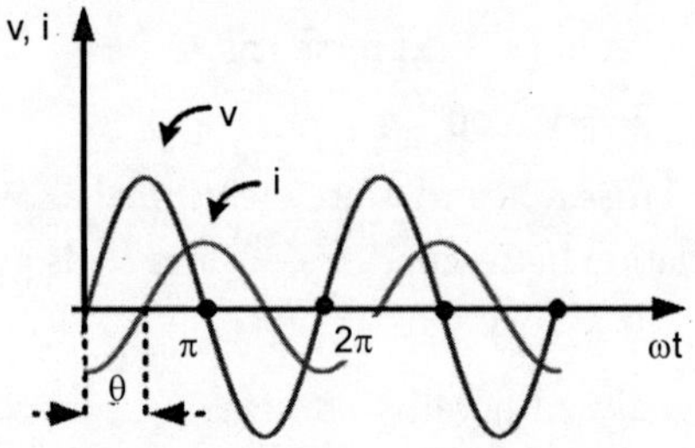

Fig. 1.44: Waveforms of supply voltage and current of series *R-L-C* circuit for $X_L > X_C$

The relation between instantaneous value of voltage and current is represented in Fig.1.44. Equations (1.119) and (1.126) and Fig. 1.44 present that the current in a series *R-L-C* circuit is lagging the supply voltage by an angle of θ (impedance angle or power factor angle), if $X_L > X_C$.

The instantaneous power flowing through this circuit is obtained by using the concept of equation (1.61) as follows,

$$p(t) = \frac{1}{2} V_m I_m \cos(\theta) - \frac{1}{2} V_m I_m \cos(2\omega t - \theta) \quad (1.128)$$

The first term of equation (1.128) is the active power or average power and it is given by,

$$P = \frac{1}{2} V_m I_m \cos\theta = |V_{rms}||I_{rms}|\cos\theta = |I_{rms}|^2 R = \frac{|V_{rms}|^2}{R} \quad (1.129)$$

The second term of the instantaneous power is a time varying sinusoidal whose average value over a complete cycle is zero. It represents only charging of the inductor and capacitor in one half cycle and discharging of them in the next half cycle by pumping the power back to the source. This power is the reactive power which is the measure of oscillating power between the source and capacitive reactance. The reactive power of the series *R-L-C* circuit for $X_L > X_C$ in different forms are expressed as,

$$Q = \frac{1}{2} V_m I_m \sin\theta \frac{1}{2} V_m I_m \sin\theta = |V_{rms}||I_{rms}|\sin\theta \quad (1.130)$$

$$Q = |I_{rms}|^2 (X_L - X_C) = \frac{|V_{rms}|^2}{(X_L - X_C)} \quad (1.131)$$

Equations (1.130) and (1.131) show that the net reactive power is positive as $X_L>X_C$. This net reactive power is lagging reactive power. The expression of total power or apparent power for the series *R-C* circuit for $X_L>X_C$ is mentioned as follows,

$$S = P + jQ \tag{1.132}$$

$$|S| = \sqrt{P^2 + Q^2} = \frac{1}{2}V_m I_m = |V_{rms}||I_{rms}| = |I_{rms}|^2 Z = \frac{|V_{rms}|^2}{Z} \tag{1.133}$$

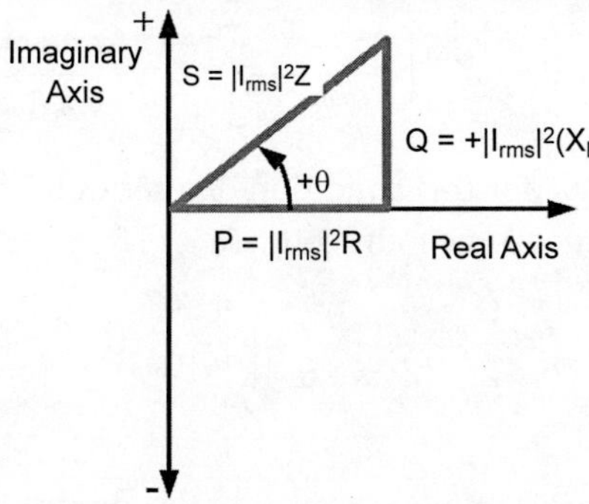

Fig. 1.45: Power triangle for a series *R-L-C* circuit for $X_L>X_C$

The power triangle for this circuit is presented in Fig.1.45. The power triangle also represents the positive and lagging net reactive power.

Power factor can also be calculated by apparent power and active power or resistance and impedance as,

$$PF = \cos\theta = \frac{P}{S} = \frac{R}{Z} \tag{1.134}$$

The power factor for a series *R-L-C* circuit for $X_L>X_C$ is lagging in nature. It can also be evaluated from voltage phasors as indicated in Fig.1.43 as,

$$PF = \cos\theta = \frac{V_R}{V} \tag{1.135}$$

Condition 2: $X_L<X_C$

The phasor sum of all the voltages for this condition are represented in Fig.1.46.

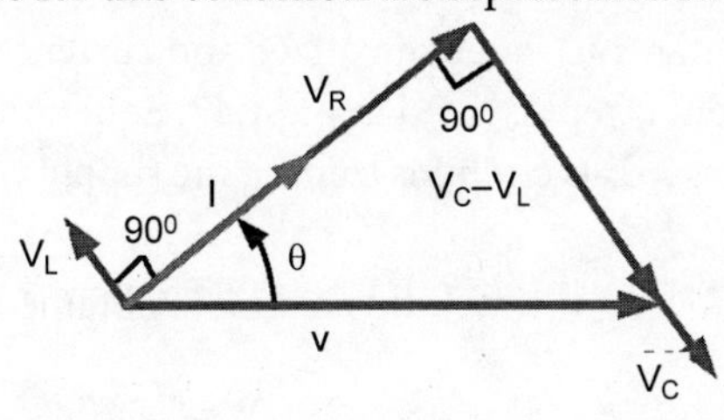

Fig. 1.46: Voltage and current phasors of series *R-L-C* circuit for $X_L<X_C$

Using the concept mentioned in equations (1.24), (1.34) and (1.45) the equation (1.120) can be modified as,

$$V = IR + jI(X_L - X_C) \quad (1.136)$$

$$V = I(R + j(X_L - X_C) = I(R - jX) = IZ \quad (1.137)$$

Where $X_L - X_C = X$. The value of net reactance i.e X is negative as $X_L < X_C$ which is also indicated in equation (1.37). Considering net reactance as negative, the impedance Z is given by,

$$Z = |Z|\angle\theta \quad (1.138)$$

$$|Z| = \sqrt{R^2 + (X_L - X_C)^2} \quad (1.139)$$

$$\theta = \tan^{-1}\left(\frac{X_L - X_C}{R}\right) = \tan^{-1}\left(\frac{L}{RC}\right) \quad (1.140)$$

As, XL<XC, the impedance angle for this condition of a series R-L-C circuit is negative. The value of current flowing through the circuit can be found out as,

$$I = \frac{V}{Z} = \frac{V_m\angle 0^0}{|Z|\angle -\theta} = \left(\frac{V_m}{|Z|}\right)\angle\theta = I_m\angle\theta \quad (1.141)$$

Where, I_m is $\left(\frac{V_m}{|Z|}\right)$. Hence the instantaneous value of the current flowing through the circuit is as follows,

$$i = I_m \sin(\omega t + \theta) \quad (1.142)$$

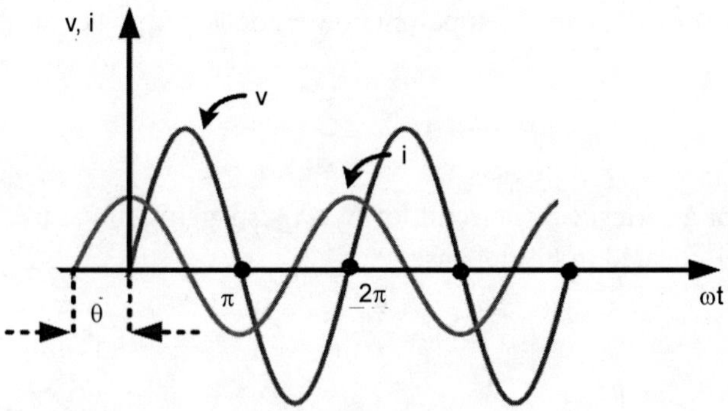

Fig. 1.47: Waveforms of supply voltage and current of series *R-L-C* circuit for $X_L<X_C$

The relation between instantaneous value of voltage and current as well as the relation between the phasor form of voltage and current is represented in Fig.1.47. Equations (1.119) and (1.142) and Fig. 1.47 present that the current in a series *R-L-C* circuit is leading the supply voltage by an angle of θ (impedance angle or power factor angle), if $X_L<X_C$.

The instantaneous power flowing through this circuit is obtained by using the concept of equation (1.61) as follows,

$$p(t) = \frac{1}{2}V_m I_m \cos(-\theta) - \frac{1}{2}V_m I_m \cos(2\omega t + \theta) \quad (1.143)$$

The first term of equation (1.143) is the active power or average power and it is given by,

$$P = \frac{1}{2} V_m I_m \cos\theta = |V_{rms}||I_{rms}|\cos\theta = |I_{rms}|^2 R = \frac{|V_{rms}|^2}{R} \quad (1.144)$$

The second term of the instantaneous power is a time varying sinusoidal whose average value over a complete cycle is zero. It represents only charging of the inductor and capacitor in one half cycle and discharging of them in the next half cycle by pumping the power back to the source. This power is the reactive power which is the measure of oscillating power between the source and capacitive reactance. The reactive power of the series *R-L-C* circuit for $X_L<X_C$ in different forms are expressed as,

$$Q = -\frac{1}{2} V_m I_m \sin\theta = -|V_{rms}||I_{rms}|\sin\theta \quad (1.145)$$

$$Q = |I_{rms}|^2 (X_L - X_C) = -|I_{rms}|^2 X = \frac{|V_{rms}|^2}{(X_L - X_C)} = -\frac{|V_{rms}|^2}{X} \quad (1.146)$$

Equations (1.145) and (1.146) show that the net reactive power is negative as $X_L<X_C$. This net reactive power is leading reactive power. The expression of total power or apparent power for the series *R-C* circuit for $X_L<X_C$ is mentioned as follows,

$$S = P - jQ \quad (1.147)$$

$$|S| = \sqrt{P^2 + Q^2} = \frac{1}{2} V_m I_m = |V_{rms}||I_{rms}||I_{rms}|^2 Z = \frac{|V_{rms}|^2}{Z} \quad (1.148)$$

The power triangle for this circuit is presented in Fig.1.48. The power triangle also represents the negative and leading net reactive power.

Power factor can also be calculated by apparent power and active power or resistance and impedance as,

$$PF = \cos\theta = \frac{P}{S} = \frac{R}{Z}$$

The power factor for a series *R-L-C* circuit for $X_L<X_C$ is leading in nature. It can also be evaluated from voltage phasors as indicated in Fig.1.43 as,

$$PF = \cos\theta = \frac{V_R}{V} \quad (1.150)$$

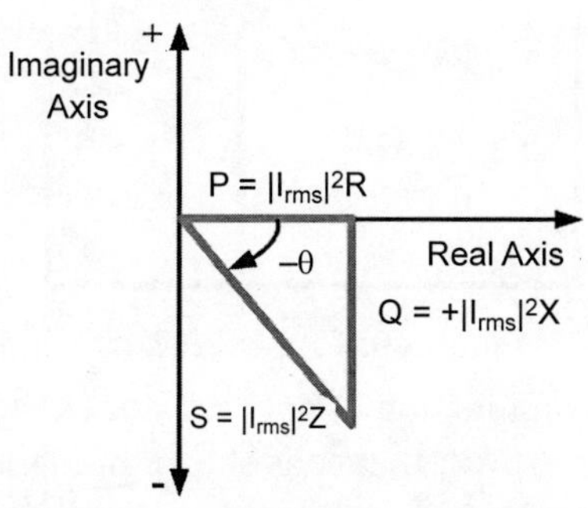

Fig. 1.48: Power triangle for a series *R-L-C* circuit for $X_L<X_C$

Example 1.19

A coil is connected in series with a practical capacitor. The coil inductance and resistance are 0.1H and 5 ohm respectively. The equivalent impedance of the circuit is given by 6+j5 ohm. If the supply voltage for this circuit is single phase, 230V, 50Hz then find the impedance of the practical capacitor.

Solution:

The practical capacitor includes one capacitor in series with a small resistance. Hence the impedance of the practical capacitor becomes,

$$Z_C = R_C - jX_C \text{ ohm}$$

Reactance of the coil is given by,

$$X_L = 2 \times 3.14 \times 50 \times 0.1 = 31.4 \text{ ohm}$$

The equivalent impedance of the circuit is given by,

$$Z_{eq} = 6 + j5 = (5 + R_C) + j(31.4 - X_C) \text{ ohm}$$

Comparing the real and imaginary term of the above equation the value of the impedance of the capacitor is obtained as,

$$ZC = 1 - j26.4 \text{ ohm}$$

1.7 STEADY STATE RESPONSE OF PARALLEL CIRCUITS FOR A SINUSOIDAL AC VOLTAGE SOURCE

In this section, the steady state response of networks consist of different parallel combinations of *R*, *L* and *C* excited by sinusoidal ac voltage source are discussed.

1.7.1 R-L Parallel Circuit

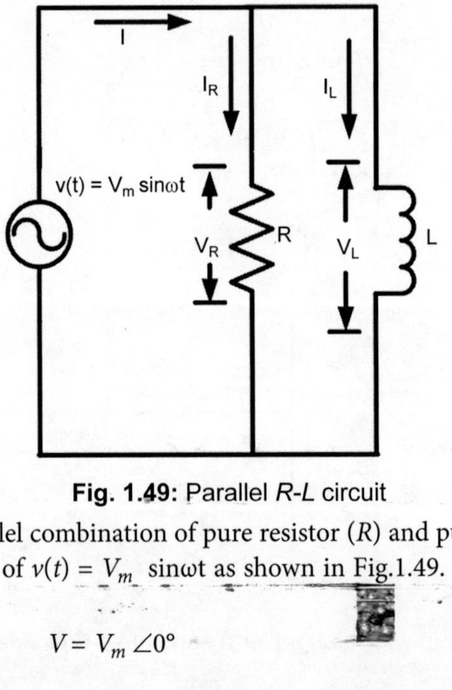

Fig. 1.49: Parallel *R-L* circuit

A circuit consists of parallel combination of pure resistor (*R*) and pure inductor (*L*) and excited by a sinusoidal ac voltage source of $v(t) = V_m \sin\omega t$ as shown in Fig.1.49. The phasor form of the supply voltage is expressed as,

$$V = V_m \angle 0° \tag{1.151}$$

The phasor form of voltage across R and L are same as supply voltage as they are in parallel. The voltage across R and L are mentioned as V_R and V_L in Fig.1.49. The relation between V, V_R and V_L is given by,

$$V = V_R = I_R R = V_L = jI_L \omega L \tag{1.152}$$

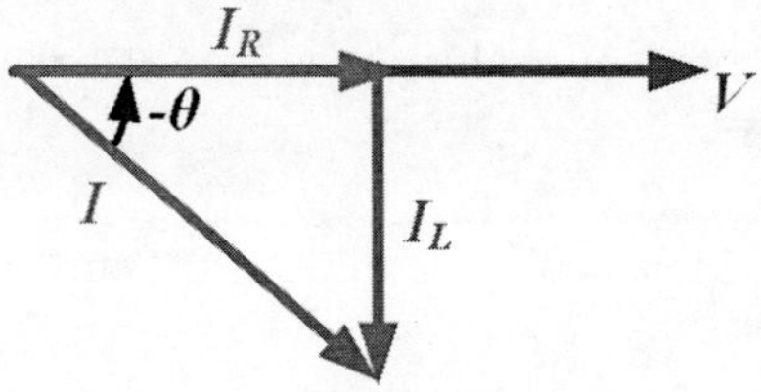

Fig. 1.50: Voltage and current phasors of parallel *R-L* circuit

As R and L are in parallel, the input current is divided in to two branches as shown in Fig.1.49. The phasor sum of the current flowing through R and L are shown in Fig.1.50. Their relation is mentioned in the equation below as,

$$I = I_R + I_L \tag{1.153}$$

If the equivalent impedance of this circuit is taken as Z, the equation (1.153) is modified by putting the values of currents from equation (1.152) as,

$$\frac{V}{Z} = \frac{V_R}{R} - j\frac{V_L}{\omega L} \tag{1.154}$$

As the values of voltage across all the elements are same as supply voltage then the impedance of the circuit is given by,

$$\frac{1}{Z} = \frac{1}{R} - j\frac{1}{\omega L} \tag{1.155}$$

$$\frac{1}{Z} = \frac{1}{R} - j\frac{1}{X_L} \tag{1.156}$$

$$Z = \frac{RjX_L}{R + jX_L} \tag{1.157}$$

Separating the real and imaginary term of equation (1.157), it becomes,

$$Z = \frac{RX_L{}^2}{R^2 + X_L{}^2} + j\frac{R^2X_L}{R^2 + X_L{}^2} \tag{1.158}$$

$$\mathrm{Z} = |\mathrm{Z}|\angle\theta \tag{1.159}$$

$$|Z| = \sqrt{\left(\frac{RX_L{}^2}{R^2 + X_L{}^2}\right)^2 + \left(\frac{R^2X_L}{R^2 + X_L{}^2}\right)^2} \tag{1.160}$$

$$\theta = \tan^{-1}\left(\frac{R}{X_L}\right) = \tan^{-1}\left(\frac{R}{\omega L}\right) \tag{1.161}$$

The value of total input current flowing through the circuit can be found out by using equations (1.151) and (1.159) as,

$$I=\frac{V}{Z}=\frac{V_m\angle 0^0}{|Z|\angle\theta}=\left(\frac{V_m}{|Z|}\right)\angle-\theta=I_m\angle-\theta \tag{1.162}$$

Where, I_m is . Hence the instantaneous value of the input current flowing through the circuit is as follows,

Where, I_m is $\left(\frac{V_m}{|Z|}\right)$. Hence the instantaneous value of the current flowing through the circuit is as follows,

$$i=I_m\sin(\omega t+\theta) \tag{1.163}$$

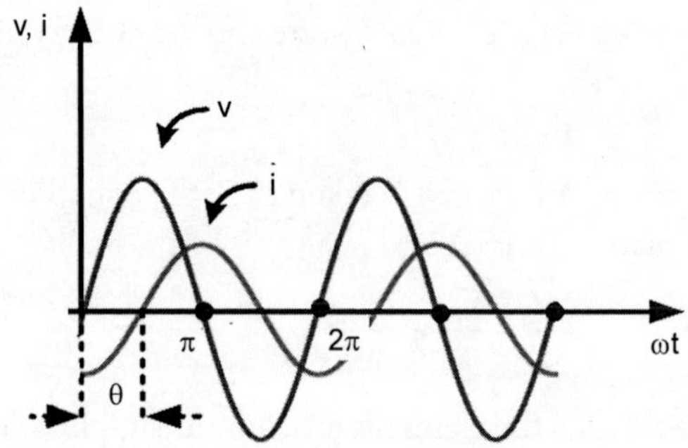

Fig. 1.51: Waveforms of supply voltage and current of parallel *R-L* circuit

The relation between instantaneous value of supply voltage and current is presented in Fig.1.51. Fig.1.51 and equations (1.151) and (1.162) clearly depict that the input current through a parallel *R-L* is lagging the supply voltage by an angle of θ (impedance angle or power factor angle).

The active power or average power of this circuit is consumed by only *R* and it is given by,

$$P=\frac{1}{2}V_mI_m\cos\theta=|V_{rms}||I_{rms}|\cos\theta=|I_{R,rms}|^2R=\frac{|V_{R,rms}|^2}{R} \tag{1.164}$$

$I_{R,rms}$ and $V_{R,rms}$ are the rms value of current through *R* and rms value of voltage across *R*. Similarly, the reactive power is only associated with reactive elements. Here, the reactive power oscillates between the inductor and source and its value is given by,

$$Q=\frac{1}{2}V_mI_m\sin\theta=|V_{rms}||I_{rms}|\sin\theta=|I_{L,rms}|^2X_L=\frac{|V_{L,rms}|^2}{X_L} \tag{1.165}$$

$I_{L,rms}$ and $V_{L,rms}$ are the rms value of current through *L* and rms value of voltage across *L*. The expression of total power or apparent power for the parallel *R-L* circuit is mentioned as follows,

$$S=P+jQ \tag{1.166}$$

$$|S|=\sqrt{P^2+Q^2}=\frac{1}{2}V_mI_m=|V_{rms}||I_{rms}|=|I_{rms}|^2Z=\frac{|V_{rms}|^2}{Z} \tag{1.167}$$

The power triangle for this circuit is presented in Fig.1.52. The power triangle confirms the positive and lagging reactive power and positive power factor angle. Power factor can also be calculated by apparent power and active power or resistance and impedance as,

$$PF = \cos\theta = \frac{P}{S} \tag{1.168}$$

The power factor for a series *R-L* circuit is lagging in nature. It can also be evaluated from current phasors as indicated in Fig.1.50 as,

$$PF = \cos\theta = \frac{I_R}{I} \tag{1.169}$$

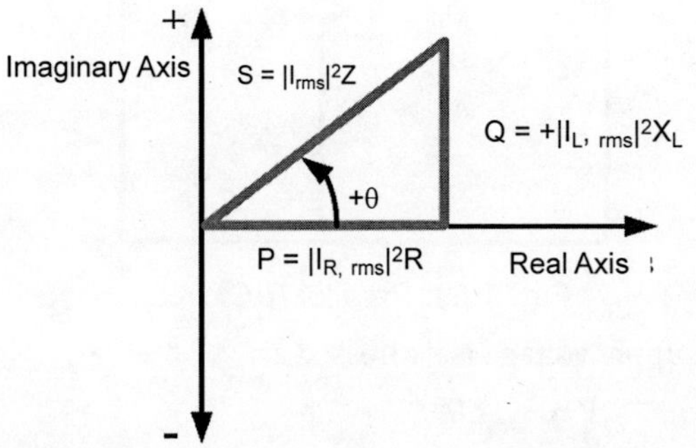

Fig. 1.52: Power triangle for a parallel *R-L* circuit

Example 1.20

A resistor of 50 ohm is connected in parallel to a 2H inductor. This parallel circuit is connected to a 220V, 50 Hz single phase sinusoidal voltage supply. Find the active power dissipated in the resistor and reactive power in the inductor branch.

Solution:

Current through the resistor is given by,

$$I_R = \frac{220}{50} = 4.4A$$

Current through the inductor is given by,

$$I_L = -j\frac{220}{X_L} = -j\frac{220}{628} = -j0.350A[X_L = 2\pi fL]$$

The active power dissipated in the resistor is given by,

$$P = I_{R,rms}{}^2\, R = (4.4)^2 \times 50 = 968 \text{ W}$$

The reactive power in the inductor is given by,

$$Q = I_{L,\, rms}{}^2\, X_L = (0.35)^2 \times 628 = 76.93\ VAr$$

1.7.2 R-C Parallel Circuit

A circuit consists of parallel combination of pure resistor (R) and pure capacitor (C) and excited by a sinusoidal ac voltage source of $v(t) = V_m \sin \omega t$ as shown in Fig.1.53.

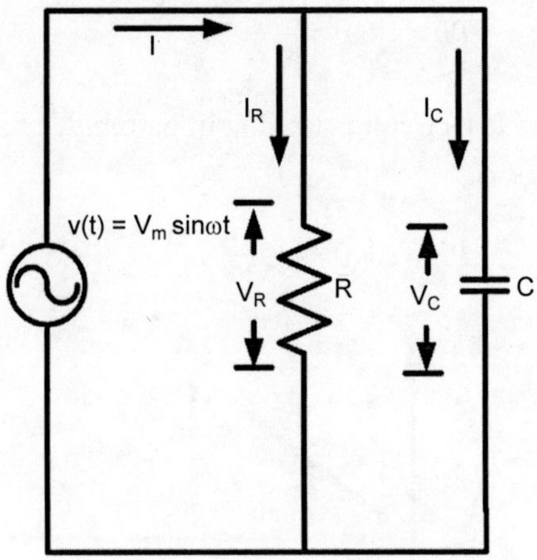

Fig. 1.53: Parallel *R-C* circuit

The phasor form of the supply voltage is expressed as,

$$V = V_m\angle 0° \tag{1.170}$$

The phasor form of voltages across R and C are same as supply voltage as they are in parallel, the voltage across R and C are mentioned as V_R and V_C in Fig.1.53. The relation between V, V_R and V_C is given by,

$$V = V_R = I_R R = V_C = \frac{-jI_C}{\omega C} \tag{1.171}$$

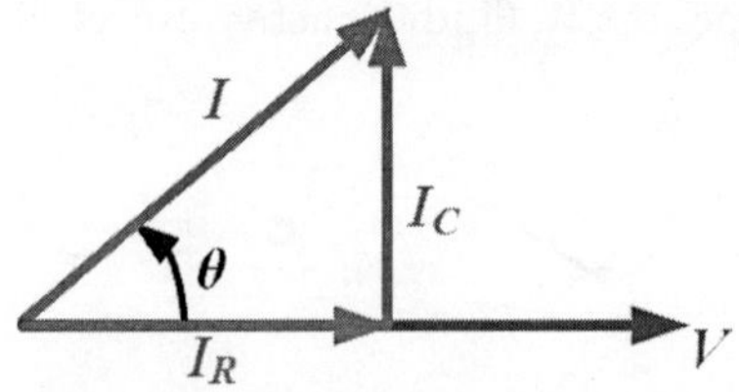

Fig. 1.54: Voltage and current phasors of parallel *R-C* circuit

As R and C are in parallel, the input current is divided in to two branches as shown in Fig.1.53. The phasor sum of the current flowing through R and C are shown in Fig.1.54. Their relation is mentioned in the equation below as,

$$I = I_R + I_C \tag{1.172}$$

If the equivalent impedance of this circuit is taken as Z, then the equation (1.172) is modified by the putting the values of currents from equation (1.171) as,

$$\frac{V}{Z} = \frac{V_R}{R} + jV_C\omega C \tag{1.173}$$

As the values of voltage across all the elements are same as supply voltage then the impedance of the circuit is given by,

$$\frac{1}{Z}=\frac{1}{R}+j\omega C \tag{1.174}$$

$$\frac{1}{Z}=\frac{1}{R}+j\frac{1}{X_C} \tag{1.175}$$

The value of Z in complex form is obtained as ,

$$Z=\frac{R}{1+\omega^2C^2R^2}-j\frac{\omega CR^2}{1+\omega^2C^2R^2} \tag{1.176}$$

$$Z = |Z|\angle\theta$$

$$|Z|=\sqrt{\left(\frac{R}{1+\omega^2C^2R^2}\right)^2+\left(-\frac{\omega CR^2}{1+\omega^2C^2R^2}\right)^2} \tag{1.177}$$

$$\theta=\tan^{-1}(-\omega C)=\tan^{-1}\left(-\frac{1}{X_C}\right) \tag{1.178}$$

It is clear from equation (1.178) that the impedance angle is negative. The value of total input current flowing through the circuit can be found out by using equations (1.170) and (1.176) as,

$$I=\frac{V}{Z}=\frac{V_m\angle 0^0}{|Z|\angle-\omega c}=\left(\frac{V_m}{|Z|}\right)\angle\theta=I_m\angle\theta \tag{1.179}$$

Where, I_m is $\left(\frac{V_m}{|Z|}\right)$. Hence the instantaneous value of the current flowing through the circuit is as follows,

$$i = I_m\ \sin(\omega t + \theta) \tag{1.180}$$

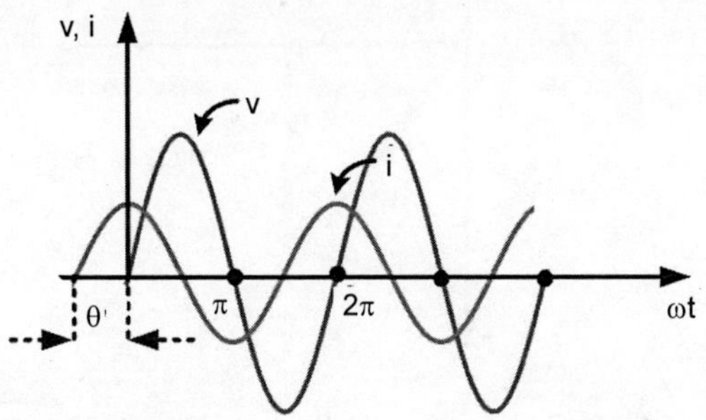

Fig. 1.55: Waveforms of supply voltage and input current of parallel *R-C* circuit

The relation between instantaneous value of supply voltage and input current is presented in Fig.1.55. Fig.1.55 and equations (1.170) and (1.179) clearly depict that the input current through a parallel *R-C* is leading the supply voltage by an angle of θ (impedance angle or power factor angle).

The active power or average power of this circuit is consumed by only R and it is given by,

$$P = \frac{1}{2}V_m I_m \cos\theta = |V_{rms}||I_{rms}|\cos\theta = |I_{R,rms}|^2 R = \frac{|V_{R,rms}|^2}{R} \tag{1.181}$$

$I_{R,rms}$ and $V_{R,rms}$ are the rms value of current through R and rms value of voltage across R. Similarly, the reactive power is only associated with reactive elements. Here the reactive power oscillates between the capacitor and source and its value is given by,

$$Q = -\frac{1}{2}V_m I_m \sin\theta = -|V_{rms}||I_{rms}|\sin\theta = -|I_{C,rms}|^2 X_C = -\frac{|V_{rms}|^2}{X_C} \tag{1.182}$$

The expression of total power or apparent power for the parallel R-C circuit is mentioned as follows,

$$S = P - jQ \tag{1.183}$$

$$|S| = \sqrt{P^2 + Q^2} = \frac{1}{2}V_m I_m = |V_{rms}||I_{rms}| = |I_{rms}|^2 Z = \frac{|V_{rms}|^2}{Z} \tag{1.184}$$

The power triangle for this circuit is presented in Fig.1.56. The power triangle confirms the negative and leading reactive power and negative power factor angle.

Power factor can also be calculated by apparent power and active power as,

$$PF = \cos\theta = \frac{P}{S} \tag{1.185}$$

The power factor for a parallel R-C circuit is leading in nature. It can also be evaluated from current phasors as indicated in Fig.1.54 as,

$$PF = \cos\theta = \frac{I_R}{I} \tag{1.186}$$

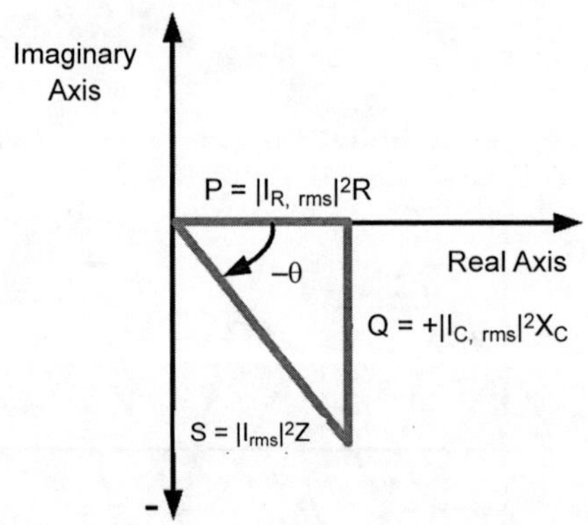

Fig. 1.56: Power triangle for a parallel *R-C* circuit

Example 1.21

A resistor of 10 ohm is connected in parallel to a 20μF capacitor. This parallel circuit is connected to a 200V, 50 Hz single phase sinusoidal voltage supply. Find the total current flowing through the circuit and equivalent impedance of the circuit.

Solution:

Current through the resistor is given by,

$$I_R = \frac{200\sqrt{2}}{10} = 20\sqrt{2}A$$

Current through the capacitor is given by,

$$I_C = j\frac{200\sqrt{2}}{X_C} = j\frac{200\sqrt{2}}{159.23} = j1.77A[X_C = \frac{1}{\omega C}]$$

$$I = I_R + I_C = 28.28 + j1.77 = 28.33 \angle 3.58° \text{ A}$$

Impedance of the circuit is given by,

$$Z = \frac{V}{I} = \frac{200\sqrt{2}}{28.33\angle 3.58^0} = 9.98\angle -3.58^0 ohm$$

1.7.3 R-L-C Parallel Circuit

A circuit consists of parallel combination of pure resistor (R), pure inductor (L) and pure capacitor (C) and excited by a sinusoidal ac voltage source of $v(t) = V_m \sin \omega t$ as shown in Fig.1.57.

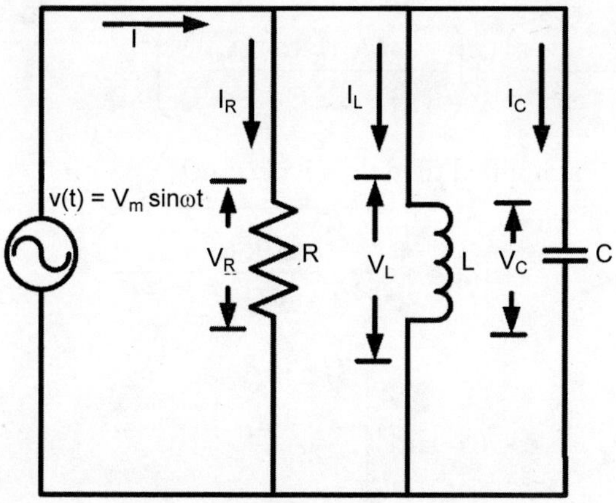

Fig. 1.57: Parallel *R-L-C* circuit

The phasor form of the supply voltage is expressed as,

$$V = V_m\angle 0° \qquad (1.187)$$

The phasor form of voltage across *R*, *L* and *C* are same as supply voltage as they are in parallel. The voltage across *R*, *L* and *C* are mentioned as V_R, V_L and V_C in Fig.1.57. The relation between *V*, V_R V_L and V_C is given by,

$$V = V_R = I_R R = V_L = jI_L \omega L = V_C = \frac{jI}{\,} \qquad (1.188)$$

As *R*, *L* and *C* are in parallel, the input current is divided in to three branches as shown in Fig.1.57. Their relation is mentioned in the equation below as,

$$I = I_R + I_L + I_C \qquad (1.189)$$

If the equivalent impedance of this circuit is taken as Z, then the equation (1.189) is modified by the putting the values of currents from equation (1.188) as,

$$\frac{V}{Z}=\frac{V_R}{R}-j\frac{V_L}{\omega L}+jV_C\omega C \tag{1.190}$$

As the values of voltage across all the elements are same as supply voltage then the impedance of the circuit is given by,

$$\frac{1}{Z}=\frac{1}{R}-j\frac{1}{\omega L}+j\omega C \tag{1.191}$$

$$\frac{1}{Z}=\frac{1}{R}+j\left(\frac{1}{X_C}-\frac{1}{X_L}\right) \tag{1.192}$$

The value of Z in complex form is obtained as,

$$Z=\frac{\omega^2L^2R}{R^2(1-\omega^2LC)^2+\omega^2L^2}+j\frac{\omega LR^2(1-\omega^2LC)}{R^2(1-\omega^2LC)^2+\omega^2L^2} \tag{1.193}$$

$$Z=|Z|\angle\theta \tag{1.194}$$

$$|Z|=\sqrt{\left(\frac{\omega^2L^2R}{R^2(1-\omega^2LC)^2+\omega^2L^2}\right)^2+\left(\frac{\omega LR^2(1-\omega^2LC)}{R^2(1-\omega^2LC)^2+\omega^2L^2}\right)^2} \tag{1.195}$$

If the equation (1.195) is simplified then then it becomes,

$$|Z|=\sqrt{\left(\frac{\omega^2L^2R^2}{R^2(1-\omega^2LC)^2+\omega^2L^2}\right)^2} \tag{1.196}$$

$$\theta=\tan^{-1}\left(\frac{\frac{1}{X_L}-\frac{1}{X_C}}{\frac{1}{R}}\right) \tag{1.197}$$

The parallel R-LC circuit is also analysed considering two conditions. First one is $X_L>X_C$ and second one is $X_L<X_C$.

Condition 1: $X_L>X_C$

If the inductive reactance is larger than the capacitive reactance then,

$$X_L>X_C$$

$$\omega L>\frac{1}{\omega C}$$

$$\omega^2LC>1$$

$$1-\omega^2LC<0 \tag{1.198}$$

From the above inequality, it is clear that the imaginary term of the equation (1.193) is negative for $X_L > X_C$. Hence the net reactance (X) is capacitive. Similarly,

$$X_L > X_C$$

$$\frac{1}{X_L} < \frac{1}{X_C}$$

$$\frac{1}{X_L} - \frac{1}{X_C} < 0 \tag{1.199}$$

From the inequality mentioned in (1.199), it is clear that the impedance angle in equation (1.197) is negative for $X_L > X_C$.

The value of total input current flowing through the circuit can be found out by considering negative impedance angle as,

$$I = \frac{V}{Z} = \frac{V_m \angle 0^0}{|Z| \angle -\theta} = \left(\frac{V_m}{|Z|}\right) \angle \theta = I_m \angle \theta \tag{1.200}$$

Where, I_m is $\left(\frac{V_m}{|Z|}\right)$. Hence the instantaneous value of the input current flowing through the circuit is as follows,

$$i = I_m \sin(\omega t + \theta) \tag{1.201}$$

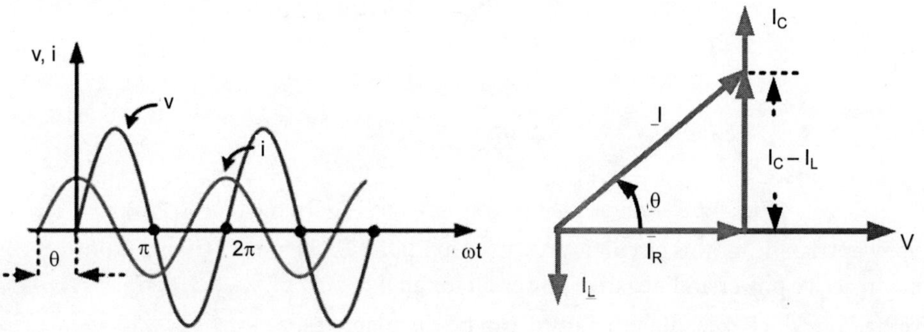

Fig. 1.58: Parallel *R-L-C* circuit for $X_L > X_C$ (a) waveforms; (b) phasor diagram

The relation between instantaneous value of supply voltage and input current as well as phasor diagram of of supply voltage and input current are represented in Fig.1.58. Fig.1.58 and equations (1.187) and (1.200) clearly depict that the input current through a parallel *R-L-C* for $X_L > X_C$ is leading the supply voltage by an angle of θ (impedance angle or power factor angle).

The active power or average power of this circuit is consumed by only R and it is given by,

$$P = \frac{1}{2} V_m I_m \cos\theta = |V_{rms}||I_{rms}|\cos\theta = |I_{R,rms}|^2 R = \frac{|V_{R,rms}|^2}{R} \tag{1.202}$$

$I_{R,rms}$ and $V_{R,rms}$ are the rms value of current through R and rms value of voltage across R. Similarly, the reactive power is only associated with reactive elements. Here the reactive power oscillates between the inductor, capacitor and source and its value is given by,

$$Q = -\frac{1}{2}V_m I_m \sin\theta = -|V_{rms}||I_{rms}|\sin\theta = -|I_{X,rms}|^2 X = -\frac{|V_{rms}|^2}{X} \tag{1.203}$$

Where X is the net reactance of the circuit which is nothing but the imaginary term of equation (1.193). The value of I_X are given by,

$$I_X = \frac{V}{X} \tag{1.204}$$

The expression of total power or apparent power for the parallel R-C circuit is mentioned as follows,

$$S = P - jQ \tag{1.205}$$

$$|S| = \sqrt{P^2 + Q^2} = \frac{1}{2}V_m I_m = |V_{rms}||I_{rms}| = |I_{rms}|^2 Z = \frac{|V_{rms}|^2}{Z} \tag{1.206}$$

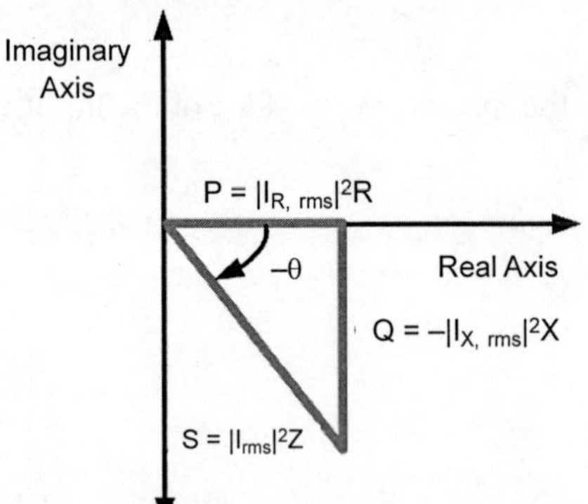

Fig. 1.59: Power triangle for parallel *R-L-C* circuit for $X_L > X_C$

The power triangle for this circuit is presented in Fig.1.59. The power triangle confirms the negative and leading reactive power and negative power factor angle.

Power factor, which is leading in nature, can be calculated as,

$$PF = \cos\theta = \frac{P}{S} = \frac{I_R}{I} \tag{1.207}$$

Condition 2: $X_L < X_C$

If the inductive reactance is smaller than the capacitive reactance then,

$$X_L > X_C$$

$$\omega L > \frac{1}{\omega C}$$

$$\omega^2 LC > 1$$

$$1 - \omega^2 LC < 0 \tag{1.208}$$

From the above inequality, it is clear that the imaginary term of the equation (1.193) is positive for $X_L<X_C$. Hence the net reactance is inductive. Similarly,

$$X_L > X_C$$

$$\frac{1}{X_L} < \frac{1}{X_C}$$

$$\frac{1}{X_L} - \frac{1}{X_C} < 0 \tag{1.208}$$

From the inequality mentioned in (1.208), it is clear that the impedance angle in equation (1.197) is positive for $X_L>X_C$.

The value of total input current flowing through the circuit can be found out by considering negative impedance angle as,

$$I = \frac{V}{Z} = \frac{V_m\angle 0^0}{|Z|\angle\theta} = \left(\frac{V_m}{|Z|}\right)\angle -\theta = I_m\angle -\theta \tag{1.209}$$

Where, I_m is $\left(\frac{V_m}{|Z|}\right)$. Hence the instantaneous value of the input current flowing through the circuit is as follows,

$$i = I_m \sin(\omega t - \theta) \tag{1.210}$$

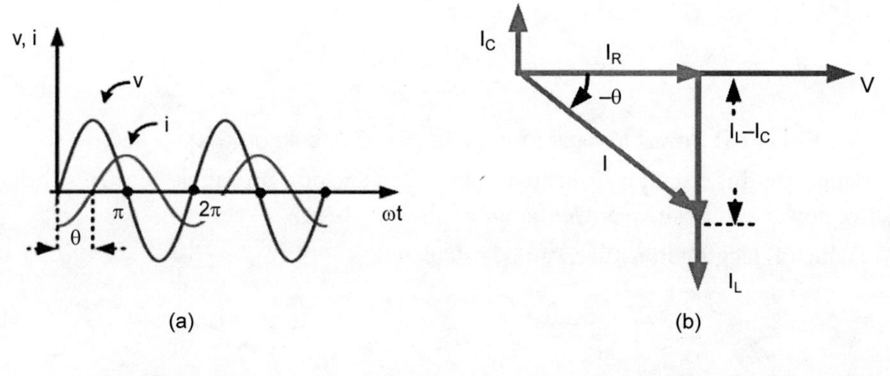

Fig. 1.60: Parallel *R-L-C* circuit for $X_L<X_C$ (a) waveforms; (b) phasor diagram

The relation between instantaneous value of supply voltage and input current as well as phasor form of supply voltage and input current are represented in Fig.1.60. Fig.1.60 and equations (1.187) and (1.210) clearly depict that the input current through a parallel *R-L-C* for $X_L<X_C$ is lagging the supply voltage by an angle of θ (impedance angle or power factor angle).

The active power or average power of this circuit is consumed by only *R* and it is given by,

$$P = \frac{1}{2}V_m I_m \cos\theta = |V_{rms}||I_{rms}|\cos\theta = |I_{R,rms}|^2 R = \frac{|V_{R,rms}|^2}{R} \tag{1.211}$$

$I_{R,rms}$ and $V_{R,rms}$ are the rms value of current through R and rms value of voltage across R. Similarly, the reactive power is only associated with reactive elements. Here the reactive power oscillates between the inductor, capacitor and source and its value is given by,

$$Q = \frac{1}{2} V_m I_m \sin\theta = |V_{rms}||I_{rms}| \sin\theta = |I_{X,rms}|^2 X = \frac{|V_{rms}|^2}{X} \quad (1.212)$$

Where X is the net reactance of the circuit which is nothing but the imaginary term of equation (1.193). The value of I_X is given by,

$$I_X = \frac{V}{X} \quad (1.213)$$

The expression of total power or apparent power for the parallel R-C circuit is mentioned as follows,

$$S = P + jQ \quad (1.214)$$

$$|S| = \sqrt{P^2 + Q^2} = \frac{1}{2} V_m I_m = |V_{rms}||I_{rms}| = |I_{rms}|^2 Z = \frac{|V_{rms}|^2}{Z} \quad (1.215)$$

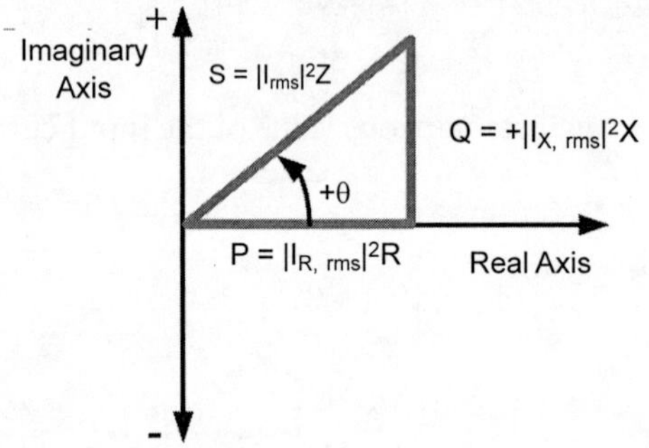

Fig. 1.61: Power triangle for parallel *R-L-C* circuit for $X_L < X_C$

The power triangle for this circuit is presented in Fig.1.61. The power triangle confirms the positive and lagging reactive power and positive power factor angle.

Power factor, which is lagging in nature, can be calculated as,

$$PF = \cos\theta = \frac{P}{S} = \frac{I_R}{I} \quad (1.216)$$

Example 1.22

A resistor of 12 ohm, an inductor of 0.05 H and a capacitor of 50μF are connected in parallel to each other. The supply voltage is 240V, 50 Hz single phase sinusoidal. Find the total current flowing through the circuit.

Solution:

Current through the resistor is given by,

$$I_R = \frac{240\sqrt{2}}{12} = 28.28A$$

Current through the capacitor is given by,

$$I_C = j\frac{240\sqrt{2}}{X_C} = j\frac{240\sqrt{2}}{63.69} = j5.32A[X_C = \frac{1}{\omega C}]$$

Current through the inductor is given by,

$$I_L = -j\frac{240\sqrt{2}}{X_L} = -j\frac{240\sqrt{2}}{15.7} = -j21.62A[X_L = \omega L]$$

Total current flowing through the circuit is given by,

$$I = I_R + I_C + I_L = 28.28 - j16.3 \text{ A} = 32.64\angle -29.95° \text{ A}$$

1.8 RESONANCE

Resonance in electrical circuits is a very significant phenomenon which is useful in communication applications, filter design etc.

Resonance in RLC circuits occur when the capacitive reactance is equal to the inductive reactance and circuit becomes resistive in nature.

1.8.1 Series Resonance

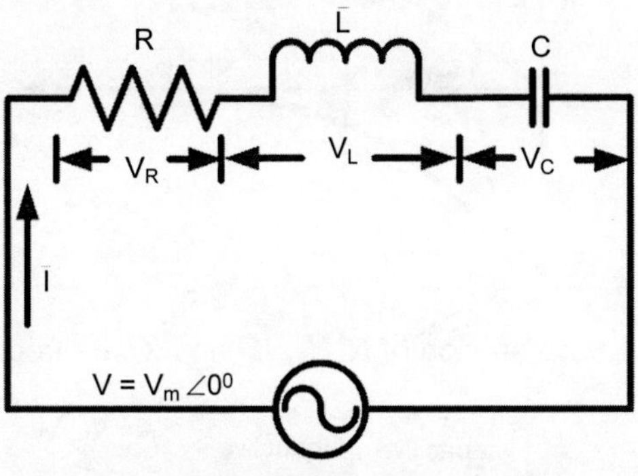

Fig. 1.62: Series *R-L-C* circuit

Fig. 1.62 shows a series *R-L-C* circuit indicating the all voltage and current values. In a series *R-L-C* if the inductive reactance becomes equal to the capacitive reactance then the impedance of the circuit is equal to the resistance of that circuit. The phenomenon during resonance is demonstrated as,

$$X_L = X_C \tag{1.217}$$

$$Z = R \pm j(X_L - X_C) = R \tag{1.218}$$

Hence it can be mentioned as,

$$X_L = X_C$$

$$\omega L = \frac{1}{\omega C}$$

$$\omega = \frac{1}{\sqrt{LC}} = \omega_r \tag{1.219}$$

$$f_r = \frac{1}{2\pi\sqrt{LC}} \quad (1.220)$$

Equations (1.219) and (1.220) represent the expression of resonant frequency (ω_r or f_r) in rad/sec or Hz respectively. This condition can be achieved by changing the frequency for a fixed value of L and C. It can also be achieved by varying L or C for a fixed frequency.

Variation of R, X_L, X_C and Z with frequency:

Fig. 1.63 shows the variations of R, X_L, X_C and Z with frequency. At zero frequency X_L is zero but X_C and Z are infinity. The value of R is independent of frequency. As frequency increases the value of X_L increases but the value of X_C decreases. At resonant frequency (f_r or ω_r), the $X_L = X_C$ and $Z = R$. At the frequencies below the resonant frequency, $X_C > X_L$ and Z decreases as X_C increases. At resonant frequency, Z has the lowest value. At the frequencies above the resonant frequency, $X_L > X_C$ and Z increases as X_C decreases or X_L increases. Fig. 1.64 demonstrates the variation of power factor angle with the frequency. At resonant frequency, the value of it is zero indicating the resistive nature of the circuit.

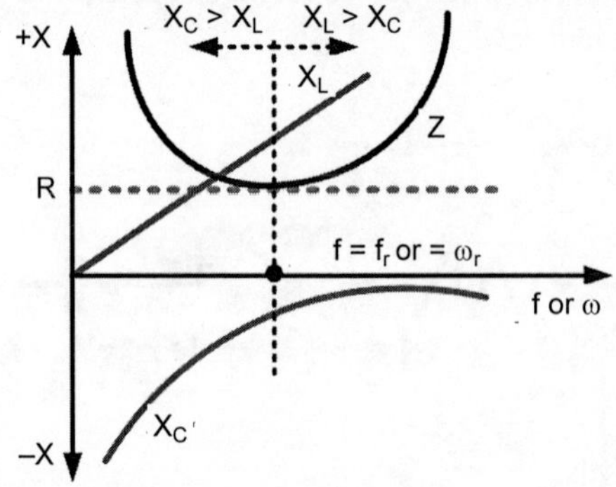

Fig.1.63: Variation of R, X_L, X_C and Z with frequency

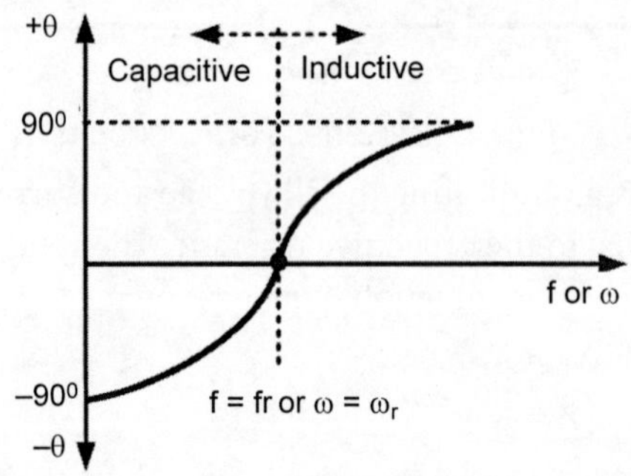

Fig.1.64: Variation of power factor angle (θ) with frequency

Current at resonance:

The current as resonant frequency is maximum as the impedance at that frequency is minimum. The R-L-C series circuit accepts maximum current at resonance, so under resonance it is known as acceptor circuit.

$$I_{max} = \frac{V}{R} \tag{1.221}$$

PF at resonance:

In a series R-L-C circuit, the power is the ratio of resistance and impedance of the circuit and impedance at resonant frequency is equal to the resistance. Hence the power factor at resonance is unity and the power factor angle is zero.

Bandwidth:

The active power dissipated across the resistor as resonance is given by,

$$\text{Presonance}(\omega) = I_{max}^2 R \tag{1.222}$$

This power is the highest power dissipated and it decreases as frequency increases or decreases.

Power dissipated at any other frequency for a series R-L-C circuit is given by,

$$P(\omega) = I^2 R \tag{1.223}$$

Where the value of I is given by,

$$I = \frac{V}{\sqrt{R^2 + \left(\omega L - \frac{1}{.\,\omega C}\right)^2}} \tag{1.224}$$

Using the equations (1.221) and (1.224), it can be written as,

$$\frac{I}{I_{\max}} = \frac{R}{\sqrt{R^2 + \left(\omega L - \frac{1}{\omega C}\right)^2}} \tag{1.225}$$

Let us assume at frequencies ω_1 and ω_2 the power dissipated in this circuit are equal and its value is half of the maximum power dissipated. These frequencies are known as half-power frequency. Then it can be written as,

$$P_1(\omega_1) = P_2(\omega_2) = I^2 R = \frac{P_{resonance}(\omega)}{2} = \frac{I_{\max}^2 R}{2} \tag{1.226}$$

It is clear from the above equation that the current at ω_1 and ω_2 is 0.707 times of I_{max}. Fig. (1.65) shows the variation of current through this circuit with the frequency. During this condition, the voltage across L and voltage across C are equal and opposite. Hence, the total voltage across the series combination of L-C is zero and the series combination of L-C is acting like a short circuit at resonance that is why the entire voltage is appeared across the resistor. The current through resistor is also maximum during this condition.

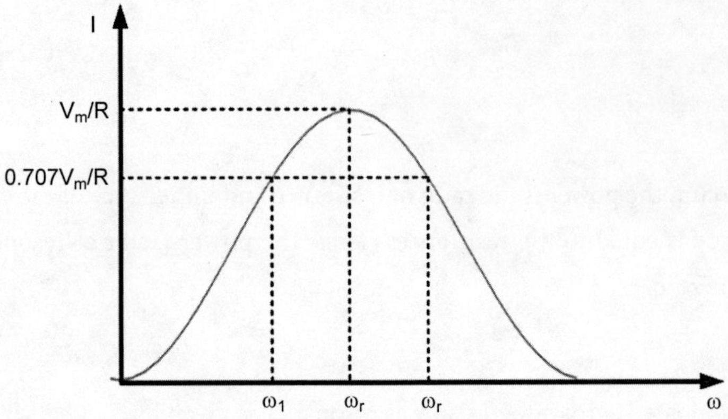

Fig. 1.65: Current variations with frequency for a series *R-L-C* circuit

At ω_1 and ω_2, the value of current is 0.707 times of I_{max}. Hence the equation (1.225) is modified for the frequencies ω_1 and ω_2 or at half-power frequencies as,

$$\frac{0.707I_{\max}}{I_{\max}} = \frac{R}{\sqrt{R^2 + \left(\omega L - \frac{1}{\omega C}\right)^2}} \tag{1.227}$$

Solving equation (1.227), the values of frequencies are obtained as,

$$\omega_1 = -\frac{R}{2L} + \sqrt{\left(\frac{R}{2L}\right)^2 + \frac{1}{LC}} \tag{1.228}$$

$$\omega_2 = \frac{R}{2L} + \sqrt{\left(\frac{R}{2L}\right)^2 + \frac{1}{LC}} \tag{1.229}$$

The relation between these half-power frequencies and resonant frequency is obtained by using equations (1.219), (1.228) and (1.229) as,

$$\omega_r = \sqrt{\omega_1 \omega_2} \tag{1.230}$$

ω_1 and ω_2 are known as lower cut-off frequency and upper cut-off frequency respectively. The difference between these two half-power frequencies is known as bandwidth. It is measured in rad or Hz. The expression for bandwidth is given by,

$$BW = (\omega_2 - \omega_1) = \frac{R}{L} \tag{1.231}$$

It can be also expressed as,

$$BW = (f_2 - f_1) = \frac{R}{2\pi L} \tag{1.232}$$

Quality factor:

The sharpness of the bandwidth is measured by quality factor. It is measured as the ratio of peak energy stored in inductor or capacitor to the energy dissipated in resistor in one period at resonance. It is expressed as,

$$Q\ Factor = 2\pi\frac{\frac{1}{2}LI^2}{\frac{1}{2}I^2R\left(\frac{1}{f_r}\right)} = 2\pi\frac{\frac{1}{2}C\left(\frac{I}{\omega_r C}\right)^2}{\frac{1}{2}I^2R\left(\frac{1}{f_r}\right)}$$

$$Q\ Factor = \frac{\omega_r L}{R} = \frac{1}{\omega_r CR} \tag{1.233}$$

Quality factor is dimensionless as it is the ratio of two quantities with the same dimensions. Quality factor can be expressed in terms of *BW* using equation (1.231) and (1.233) as,

$$Q\ Factor = \frac{\omega_r}{BW} \tag{1.234}$$

From equation (1.235), it is clear that higher the value of quality factor lower the bandwidth of the circuit. This means better the selectivity of the circuit. Selectivity of a circuit is the ability of the circuit to respond to a certain frequency and eliminate all other frequencies.

Voltage at Resonance:

At resonance the voltages across inductor and capacitor for a series *R-L-C* circuit are derived as,

$$\omega_r L = \frac{1}{\omega_r C}$$

$$I_{max}\omega_r L = \frac{V}{R}\omega_r L = I_{max}\frac{1}{\omega_r C} = \frac{V}{R}\frac{1}{\omega_r C}$$

$$V_{L,\ resonance} = V_{C,\ resosnance} = \text{V} \times (\text{Q Factor}) \tag{1.235}$$

The voltages across inductor and capacitor for a series *R-L-C* circuit at resonance are equal. As the current through the circuit under this condition is maximum, the voltage across resistance is maximum at resonance. The voltage across inductor and capacitor at resonance is magnified by *Q Factor* times than source voltage.

Example 1.23

A series R-L-C circuit has the following parameter: R = 10 ohm, L = 0.05 H and C = 50μF. Find out the resonant frequency, bandwidth, quality factor, lower cut-off frequency, upper cut-off frequency.

Solution:

Resonant frequency is calculated as,

$$\omega_r = \frac{1}{\sqrt{LC}} = \frac{1}{\sqrt{0.05 \times 50 \times 10^{-6}}} = 632.45 rad/\sec$$

Bandwidth is given by,

$$BW = \frac{R}{L} = 200 rad/\sec$$

Quality factor is calculated as,

$$Q\ Factor = \frac{\omega_r}{BW} = \frac{632.45}{200} = 3.16$$

Lower cut off frequency is obtained as,

$$\omega_1 = -\frac{R}{2L} + \sqrt{\left(\frac{R}{2L}\right)^2 + \frac{1}{LC}}$$

$$\omega_1 = -\frac{10}{2\times 0.05} + \sqrt{\left(\frac{10}{2\times 0.05}\right)^2 + \frac{1}{0.05\times 50\times 10^{-6}}} = 540.3 rad/\sec$$

Upper cut off frequency is obtained as,

$$\omega_2 = \frac{R}{2L} + \sqrt{\left(\frac{R}{2L}\right)^2 + \frac{1}{LC}}$$

$$\omega_1 = \frac{10}{2\times 0.05} + \sqrt{\left(\frac{10}{2\times 0.05}\right)^2 + \frac{1}{0.05\times 50\times 10^{-6}}} = 740.3 rad/\sec$$

1.8.2 Parallel Resonance

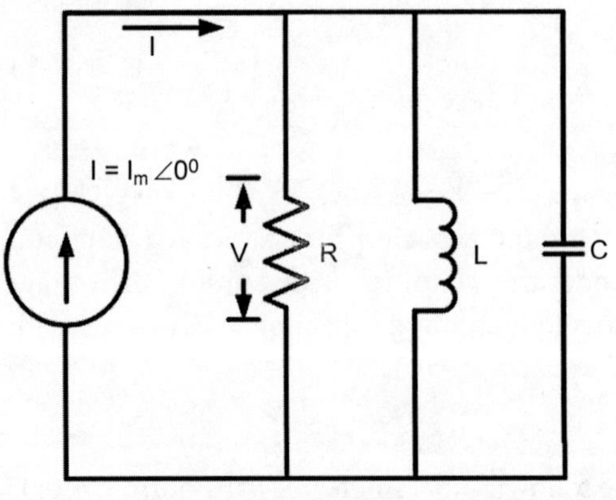

Fig. 1.66: Parallel *R-L-C* circuit

Fig. 1.66 shows a parallel *R-L-C* circuit indicating the all voltage and current values. In a parallel *R-L-C* if the inductive susceptance becomes equal to the capacitive susceptance then the impedance of the circuit is equal to the resistance of that circuit. The phenomenon is known as parallel resonance. Using the concept of equation (1.191), it can be stated that during parallel resonance the relation between inductive and capacitive reactance is given by,

$$\omega C = \frac{1}{\omega L} \tag{2.236}$$

The frequency at which parallel resonance occurs is given by,

$$\omega_r = \frac{1}{\sqrt{LC}} \tag{2.237}$$

The expression for parallel resonant frequency is same as of series resonant frequency. The variation in voltage in parallel *R-L-C* circuit with frequency is shown in Fig. 1.67. It is evident from the figure that the parallel combination of *L* and *C* is acting like open circuit at resonance that is why the entire current is flowing through the resistor.

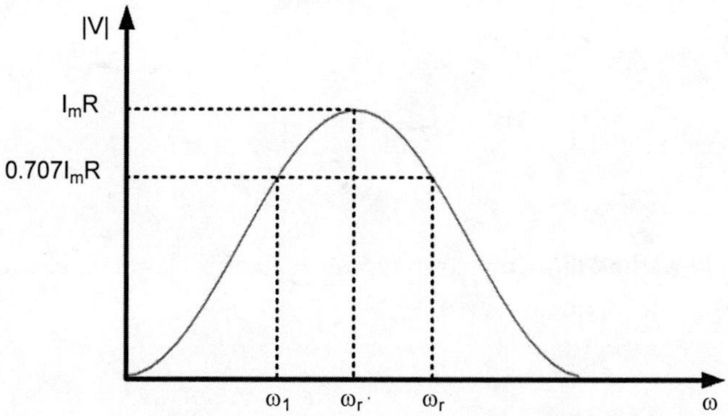

Fig. 1.67: Current variations with frequency for a series *R-L-C* circuit

Quality factor:

Quality factor for a parallel *R-L-C* is expressed as,

$$Q\ Factor = \frac{R}{\omega_r L} = \omega_r CR \tag{1.238}$$

Current at resonance:

The magnitude of current through *L* and *C* during at resonance are given by,

$$|I_L| = \frac{V}{\omega_r L} = \frac{I_m R}{\omega_r L} = Q\ Factor \times I_m \tag{1.239}$$

$$|I_C| = \omega_r CV = \omega_r CI_m R = Q\ Factor \times I_m \tag{1.240}$$

It is evident from the above two equations that the current through *L* and *C* during at resonance in a parallel *R-L-C* is magnified by Q Factor times as compared to source current.

Bandwidth:

By using the duality between series R-LC circuit and parallel *R-L-C* circuit and by comparing their impedance equations, the expressions for lower cut-off frequency and upper cut-off frequency are obtained as,

$$\omega_1 = -\frac{1}{2RC} + \sqrt{\left(\frac{1}{2RC}\right)^2 + \frac{1}{LC}} \tag{1.241}$$

$$\omega_2 = \frac{1}{2RC} + \sqrt{\left(\frac{1}{2RC}\right)^2 + \frac{1}{LC}} \tag{1.242}$$

The difference between these two half-power frequencies is known as bandwidth. It is measured in rad or Hz. The expression for bandwidth is given by,

$$BW = (\omega_2 - \omega_1) = \frac{1}{RC} \tag{1.243}$$

It can be also expressed as,

$$BW = (f_2 - f_1) = \frac{1}{2\pi RC} \tag{1.244}$$

Example 1.24

A parallel R-L-C circuit has the following parameter: R = 20 ohm, L = 0.02 H and C = 20μF. Find out the resonant frequency, bandwidth, quality factor.

Solution:

Resonant frequency is calculated as,

$$\omega_r = \frac{1}{\sqrt{LC}} = \frac{1}{\sqrt{0.02 \times 20 \times 10^{-6}}} = 1581.13 rad / \text{sec}$$

Bandwidth is given by,

$$BW = \frac{1}{RC} = \frac{1}{20 \times 20 \times 10^{-6}} = 2500 rad / \text{sec}$$

Quality factor is calculated as,

$$Q\ Factor = \frac{\omega_r}{BW} = \frac{1581.1}{2500} = 0.63$$

UNIT SUMMARY

1. Alternating voltage and current generated by alternators are sinusoidal in nature. Sinusoidal waveforms are periodic in nature. They are defined by their amplitude, frequency and phase.
2. Sinusoidal waveforms are represented in time domain as well as in phasor domain. It is convenient to perform mathematical operations in phasor domain.
3. RMS value and average values of symmetrical alternating waveforms are calculated by considering half of the total time period. Whereas, rms value and average values of asymmetrical alternating waveforms are calculated by considering the total time period. AC quantities are rated and specified by their rms values.

4. Impedance is the resistance offered by the ac circuit. It is measured in ohms. It has two components: resistance and reactance. Reactance can be inductive or capacitive in nature.
5. Current leads the supply voltage if the capacitance nature of the circuit is dominating. For inductance dominance circuit current lags the supply voltage. In case of purely resistive circuit, voltage and current are in same phase.
6. In ac circuit complex power is computed. Unit of complex power is VA. It has two components. First one is active power which is measured in Watts. Active power does the useful work. Another one is reactive power which is measured in VAr. Reactive power oscillates between reactance (inductor and capacitor) and source. Average power or active power consumes by pure inductor and pure capacitor is zero.
7. Relation between complex power, active power and reactive power is represented in terms of triangle in complex plane. It is known as power triangle.
8. Similarly, relation between impedance, resistance and reactance is represented in terms of triangle in complex plane. It is known as impedance triangle.
9. If the inductive reactance becomes equal to the capacitive reactance in a *R-L-C* circuit then the circuit becomes resistive and the phenomenon is known as resonance. Resonance can occur in series as well as parallel circuits. Resonance is an important phenomenon for communication systems, filter design etc.
10. Quality factor determines the sharpness of the bandwidth under resonance.

EXERCISES

Multiple Choice Questions

1. Unit of admittance is

(a) ohm (b) mho
(c) siemens (d) tesla

2. The current in a pure inductive circuit

(a) lags the supply voltage at any angle (b) leads the supply voltage at any angle
(c) lags the supply voltage by 90^0 (d) leads the supply voltage by 90^0

3. In a R-L-C series circuit, if the inductive reactance is larger than the capacitive reactance then the reactive power will be

(a) positive and lagging (b) positive and leading
(c) negative and lagging (d) negative and leading

4. The frequency of the reactive power is

(a) half of the supply frequency (b) one fourth of the supply frequency
(c) double the supply frequency (d) triple the supply frequency

5. The value of j operator

(a) 1 (b) $\sqrt{1}$
(c) $\sqrt{-1}$ (d) 0

6. If the impedance angle of a circuit is positive then the power factor angle will be
   (a) 0 (b) positive
   (c) negative (d) sign neutral.
7. If the current through a circuit leads the supply voltage by 30^0, then the power factor will be
   (a) 0.866 leading (b) 0.866 lagging
   (c) unity and lagging (d) unity and leading
8. If the current through a circuit is j and the supply voltage is 1V sinusoidal, then the impedance angle of the circuit will be
   (a) positive and 90° (b) negative and 90°
   (c) positive and 0° (d) negative and 0°
9. During resonance the circuit will become
   (a) purely inductive (b) purely capacitive
   (c) purely resistive (d) purely magnetic
10. If the Q Factor increases then for a fixed resonant frequency, bandwidth will
    (a) increase (b) decrease
    (c) remain same (d) either increase or decrease

ANSWERS															
1.	(c)	2.	(c)	3.	(d)	4.	(c)	5.	(c)	6.	(b)	7.	(a)	8.	(b)
9.	(c)	10.	(b)												

Short Answer Questions:

1. What is the meaning of alternating quantity?
2. Define frequency, time period, phase, amplitude, phase difference and cycle.
3. What is the positive phase sequence and negative phase sequence?
4. Give the mathematical definition of rms value and average value of a periodic function.
5. What do you mean by periodic function?
6. Draw the phasor relation between supply voltage and current for a pure resistive, pure inductive and pure capacitive circuit.
7. What do you mean by impedance? What is the unit of it?
8. What do you mean by power factor? Draw the power triangle of series R-L circuit.
9. Write down the condition of series resonance.
10. Mention the significance of quality factor under resonance.

Long Answer Questions:

1. Prove that the average power consumed by an inductor over a complete cycle is zero.
2. Show that the impedance angle of a series R-L-C circuit is positive if the capacitive reactance is dominant.
3. Derive the expression for instantaneous power, active power, reactive power and *PF* for a series R-C circuit excited by sinusoidal ac voltage source.

4. Discuss the series resonance phenomenon in a series R-L-C circuit and derive the expression of bandwidth considering an sinusoidal ac voltage excitation.
5. If the impedance angle of parallel R-L-C circuit is negative then draw the phasor relations between the three branch currents and derive the expression for active power, reactive power, apparent power and power factor. Consider the source as sinusoidal ac voltage.
6. Compare the series and parallel resonance.

Numerical Problems

1. A sinusoidalis expressed as $v(t) = 10\sin(100t + 20°)$. Find out the amplitude, phase, frequency, time period, angular frequency.
2. Calculate the phase angle between the two sinusoidals given by $v_1(t) = 10\sin(100t + 20°)$ and $v_2(t) = -20\cos(100t - 30°)$. Which sinusoidal is leading?
3. Find out the resultant of two phasors represented by $I_1 = 20\angle 10°$ and $I_2 = 30\angle 30°$.
4. Find out the resultant of two sinusoidals represented by $v_1(t) = 10\sin(100t - 20°)$ and $v_2(t) = 20\sin(100t - 30°)$.
5. Find out the rms and average value of a function given by

$$v(t) = 10\sin\omega t \quad 0 < \omega t < \frac{\pi}{2}$$

$$= 0\frac{\pi}{2} < \omega t < \pi$$

6. Determine the expressions of rms and average value of the voltage waveform shown in figure below considering both the curves (curve 1: 0 to $\pi/2$ and curve 2: $\pi/2$ to π) as sinusoidal. If V_m is 200 V, then calculate the RMS and average voltages of the waveform given below.

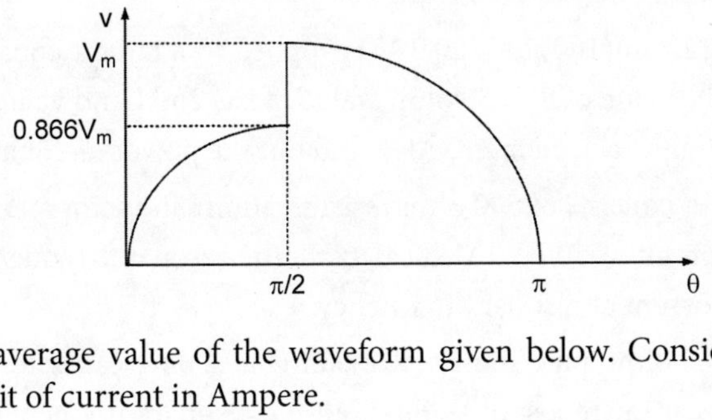

7. Find the rms and average value of the waveform given below. Consider the time period as 10 seconds and the unit of current in Ampere.

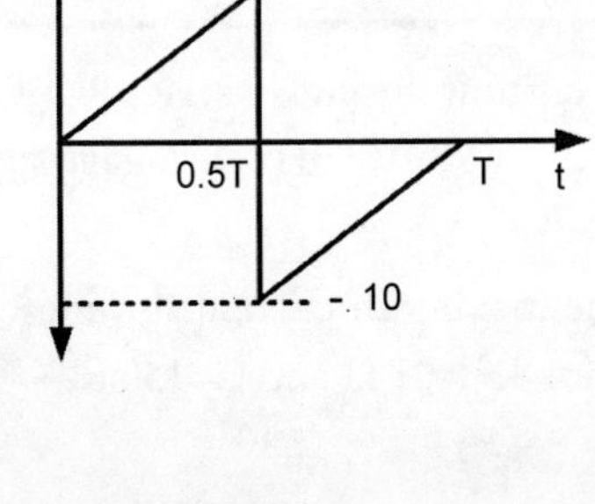

8. A single phase circuit consists of a pure resistance and a coil (combination of inductor and resistor in series) in series. Power dissipated in the resistance and in the coil are 1200W and 350W respectively. The voltage drops across the resistance and the coil are 250V and 400V respectively. Determine the (i) value of pure resistance (ii) resistance and reactance of the coil (iii) coil impedance (iv) combined resistance of the circuit.
9. The voltage applied to a series R-L circuit is $v(t) = 100 \sin \omega t$ and current flowing through it $i(t) = 10 \sin(\omega t - 45°)$. Find out the value of impedance, impedance angle, resistance, inductive reactance. Also calculate the active power, reactive power and total power drawn by the circuit.
10. In a series R-L circuit the value of applied voltage is 230 sin (ωt - 1500) Volts and current flow through it is 0.5 sin (ωt - 2100) A find the value of impedance, active power consumed, reactive power, power factor. Also draw the phasor diagram of voltage and current mentioning the values and angles.
11. A resistor of 50 ohm is connected in series with a 100μF capacitor to a 50Hz, 230V supply. Find the value of impedance, current through the circuit, voltage drop across each element, power factor.
12. In a series circuit containing resistance and capacitance, the current and voltage are expressed as $i(t) = 4 \sin(100\,\pi\, t + 10°)$ A and $v(t) = 120 \sin(100\,\pi\, t - 20°)$ V. Determine the (a) Impedance (b) Values of resistance and capacitance (c) Reactive power. Also draw the phasor relation between voltage and current.
13. A series circuit consumes 1000 Watt at 0.866 lagging power factor, when it is connected to a 220V, 50Hz single phase ac supply. Calculate reactive power, apparent power, current flowing through the circuit, impedance of the circuit, value of the resistor. Also draw the phasor diagram of voltage and current mentioning the values and angles.
14. A voltage of $v(t) = 120 \sin(100\,\pi\, t - 20°)$ V is applied to a circuit consisting of R, L and C are in parallel. Consider the value of R as 50 ohm, value of l as 2mH and value C as 50μF to calculate the current flowing through each element. Also find out the power factor angle.
15. A series R-L-C circuit consists of 500 ohm resistor, 50mH inductor and 0.05μF capacitor. Find the impedance of the circuit at 50 Hz. Also calculate the resonant frequency in rad/sec, bandwidth, quality factor and current at resonant frequency.
16. A pure resistor of 100 ohm, pure inductor of 60mH and pure capacitor of 0.5μF are connected in parallel across a 10A ac current source. Find out the resonant frequency and bandwidth.

PRACTICAL

1. Use LTspice to determine the current through a series R-L-C circuit with R = 100 ohm, L = 2 mH and C = 50 μF when connected to a 230V, 50Hz ac voltage supply. Also determine the power factor angle of this circuit.
2. Use LT spice to determine the maximum current flowing through a series R-L-C circuit during resonance. Consider R=10 ohm, L=0.01 H and C=150 μF.

KNOW MORE

1. A resistor of 50 ohm is connected in series with a 100μF capacitor to a 50Hz, 230V supply. Find the value of current through the circuit and power factor using LT spice.

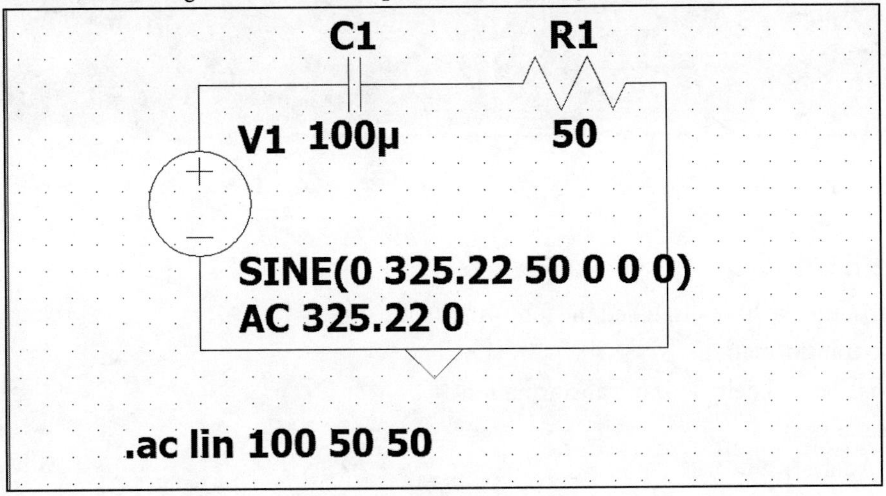

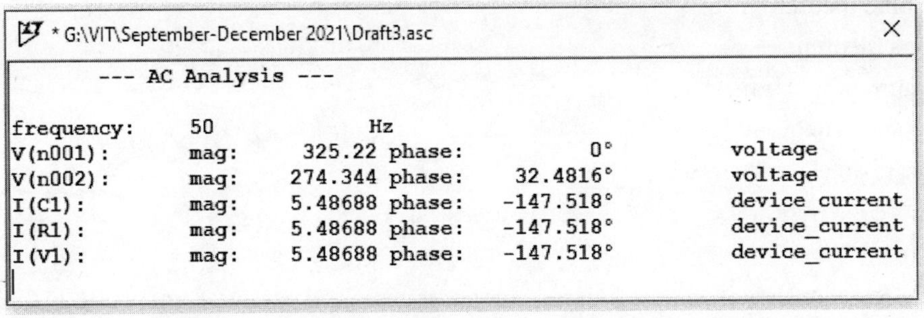


```
       --- AC Analysis ---

frequency:      50              Hz
V(n001):        mag:    325.22 phase:          0°       voltage
V(n002):        mag:   274.344 phase:    32.4816°       voltage
I(C1):          mag:   5.48688 phase:   -147.518°       device_current
I(R1):          mag:   5.48688 phase:   -147.518°       device_current
I(V1):          mag:   5.48688 phase:   -147.518°       device_current
```

Fig. 1. 68: Steady state response of series R-C circuit of the problem given using LTspice Current through the circuit is and power factor is leading.

REFERENCES AND SUGGESTED READINGS

1. Charles K. Alexander and Matthew N. O. Sadiku, "Fundamentals of Electric Circuits", McGraw Hill Publishers, 7th Edition, 2021.
2. D Roy Choudhury, "Networks and Systems", New Age International Publishers, Reprint 2005.

❑❑

2 Network Reduction and Theorems

UNIT SPECIFICS

Through this unit we have discussed the following aspects:

- Source transformation
- Star to delta and delta to star transformation
- Mesh analysis
- Node Analysis
- Superposition theorem
- Thevenin's theorem
- Norton's theorem
- Maximum power transfer theorem
- Reciprocity Theorem
- Duality of an electric circuit

RATIONALE

An electric circuit consists of voltage sources, current sources and passive elements (R, L, C) in series and parallel combinations. In order to analyse the circuit, the circuit is to be transformed for easy and fast analysis, hence both source transformation, star to delta transformation and vice versa are utilized. Based on the circuit formation, analysis could be done by mesh analysis or node analysis. In this chapter, few theorems are discussed to find the response of the circuit using the simplified method. The superposition theorem is applicable to the circuit consisting of more than one sources of the circuit. Both Thevenin's and Norton's theorem are used for solving any complicated circuit in effective manner. In most of the practical applications, the maximum power delivered by an electric circuit is calculated before the installation. Hence the maximum power transfer theorem helps to provide the load impedance value for the maximum power. The reciprocity theorem is applicable to the circuit which has only one source in it. The electric circuit have many dual pairs (voltage- current, resistance – conductance, inductance-capacitance) that has been discussed in this chapter.

PRE-REQUISITES

Vector calculus

Single-phase systems

UNIT OUTCOMES

List of outcomes of this unit is as follows:

U2-O1: Understand the concept of source transformations, star – delta and delta -star transformations

U2-O2: Apply the mesh and node analysis for a circuits

U2-O3: Comprehend and exploit the superposition theorem, Thevenin's and Norton's theorem

U2-O4: Realize the concepts of the maximum power transfer theorem, reciprocity theorem

U2-O5: Identify the dual pairs in a circuit

Unit-2 Outcome	EXPECTED MAPPING WITH COURSE OUTCOMES (1 - Weak Correlation; 2 - Medium Correlation; 3 - Strong Correlation)				
	CO-1	CO-2	CO-3	CO-4	CO-5
U2-O1	2	2	-	3	2
U2-O2	1	1	-	3	2
U2-O3	1	1	-	2	3
U2-O4	1	1	-	2	3
U2-O5	1	1	-	2	3

2.1 SOURCE TRANSFORMATION

The source transformation is one of the tools for analysing the circuits in a simple method. In dc circuits, the dc voltage source in series with the resistance is converted to the dc current source in parallel with the resistance which is represented in Fig. 2.1. The value of the current source is calculated as $I_s = \frac{V_s}{R}$.

Similarly, the current source in parallel with resistance is converted into voltage source in series with the resistance. The voltage source value is calculated using $V_s = Is * R$. The ac voltage source in series with the passive elements like R, L, C is converted into the ac current source in parallel with these passive elements as shown in Fig. 2.2 and vice versa. The conversion values are calculated as voltage source into current source using $I_s\angle\varphi = \frac{V_s\angle\theta}{Z}$ or current source into voltage source using $V_s \angle \theta = Is \angle \varphi * Z$.

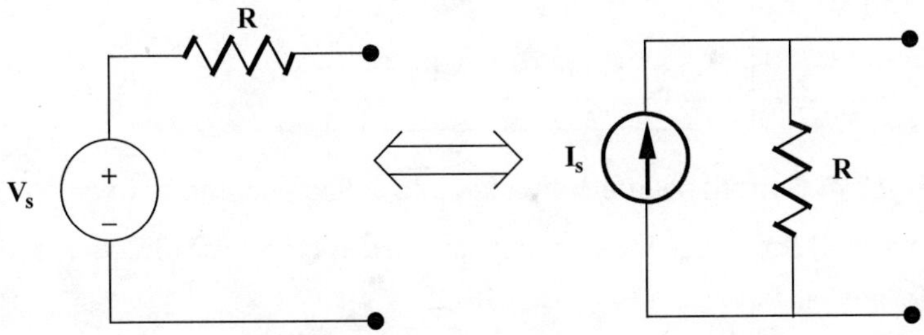

Fig. 2.1: Source transformation in dc circuits

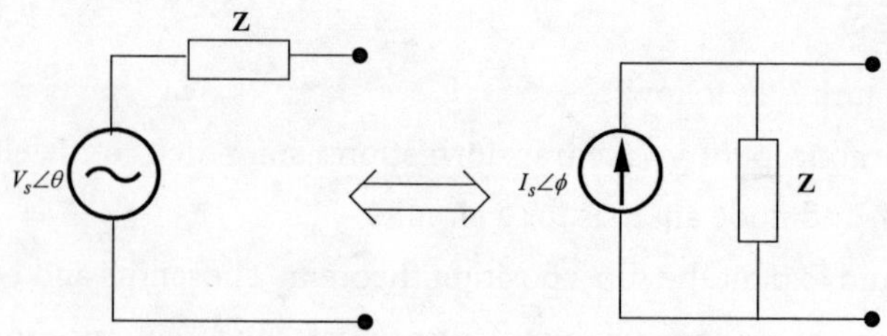

Fig. 2.2: Source transformation in ac circuits

Example 2.1

Find the voltage across 30Ω resistor shown in Fig.2.3 using source transformation method.

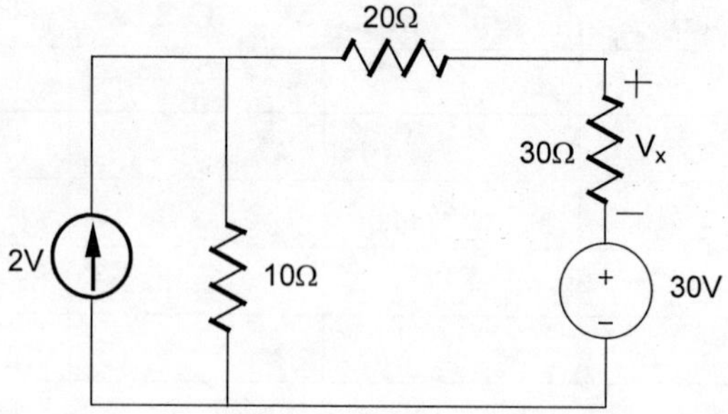

Fig. 2.3: Source transformation in dc circuits of Example 2.1

Solutions:

The circuit shown in Fig. 2.3 is the dc circuit. In this problem, current source and parallel 10Ω are converted into voltage source in series with resistance 10Ω which is shown in Fig.2.4.

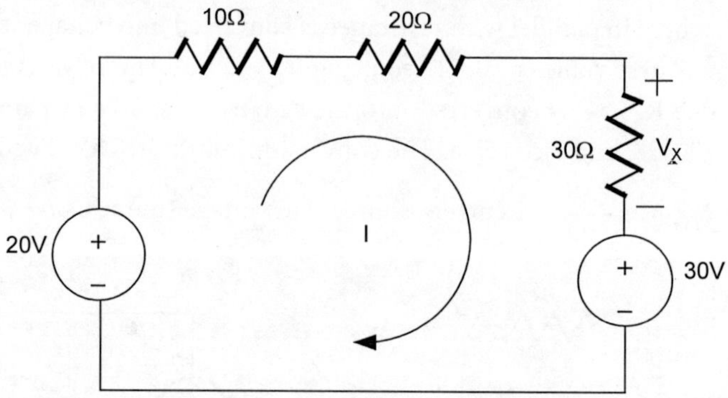

Fig. 2.4: Current source converted into voltage source of Example 2.1

Hence the current (I) flowing in the circuit is assumed as clockwise direction and is shown in Fig.2.3 that could be performed as follows.

$$+20 - 10I - 20I - 30I - 30 = 0$$

$$-60I - 10 = 0$$

$$60I = -10; \quad I = -\frac{10}{60} = -0.167A$$

Since the current I polarity is negative, it means that the assumed current direction is wrong and that to be reversed and the modified circuit is shown in Fig.2.5. Hence the current I flows through 30Ω as indicated by the arrow. Hence the voltage across 30Ω is calculated as follows.

$$V_{30\Omega} = I * 30\Omega = (0.167)(30) = 5.01V$$

Hence the voltage across 30Ω is needed to be found is named as V_x. Since the polarity of the both voltages V^x, $V_{30\Omega}$ are reversible. Hence $V_x = -V_{30\Omega} = -5.01V$.

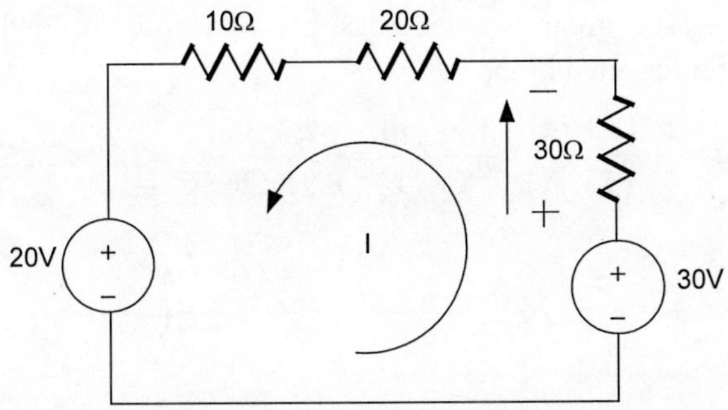

Fig. 2.5: Current through the circuit is reversed of Example 2.1

Example 2.2

Find the voltage across 10Ω resistor shown in Fig.2.6 using source transformation method

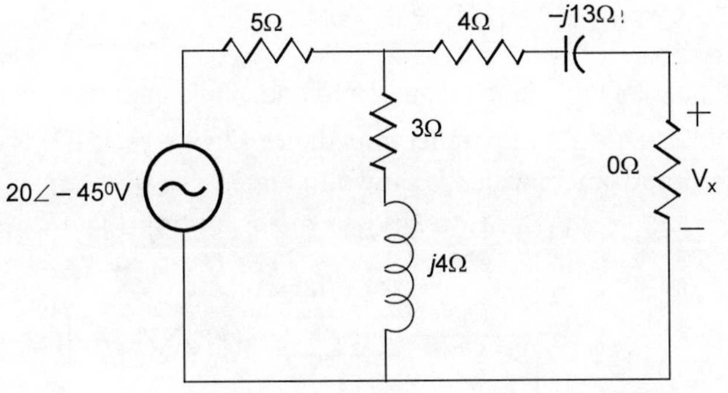

Fig. 2.6: Source transformation in ac circuits of Example 2.2

Solutions:

The circuit shown in Fig.2.6 is the ac circuit. In this problem, voltage source in series resistance 5Ω are converted into current source in parallel with resistance 5Ω which is shown in Fig.2.7.

$$I = \frac{V}{5} = \frac{20\angle -45^\circ}{5} = 4\angle -45^\circ A$$

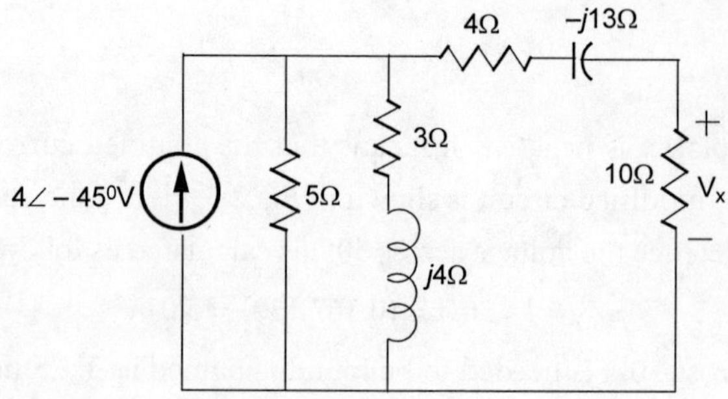

Fig. 2.7: Voltage source transformed into current source of Example 2.2

In Fig.2.7, 5Ω and (3 + *j*4) Ω are in parallel. These are made into the single impedance Z_{eq1} which is calculated as follows and represented in Fig.2.8.

$$Z_{eq1} = 5\Omega \,||\, (3+j4)\Omega = \frac{5*(3+j4)}{5+(3+j4)} = \frac{15+j20}{8+j4} = \frac{25\angle 53.1}{8.94\angle 26.56} = 2.79\angle 26.54 = (2.49+j1.25)$$

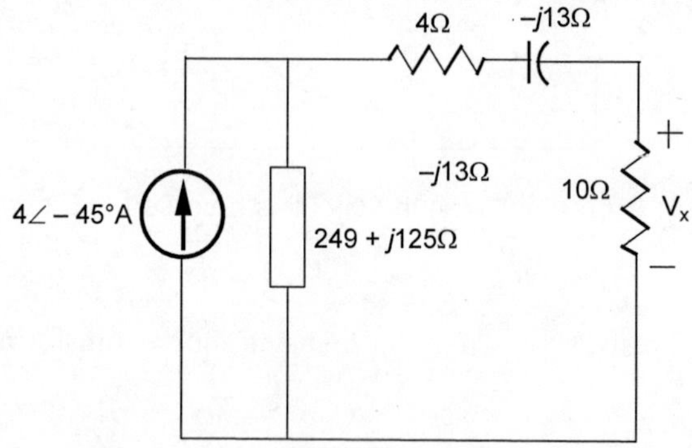

Fig. 2.8: Two parallel impedances made into one single impedances of Example 2.2

A current source (4∠ – 45° A) in parallel impedance (2.49 + *j*1.25)Ω) is converted into a voltage source in series with the impedances which is shown in Fig.2.9.

$$V_{SS} = (4\angle -45°\ A) * (2.49 + j1.25) = 11.16\angle -18.44V$$

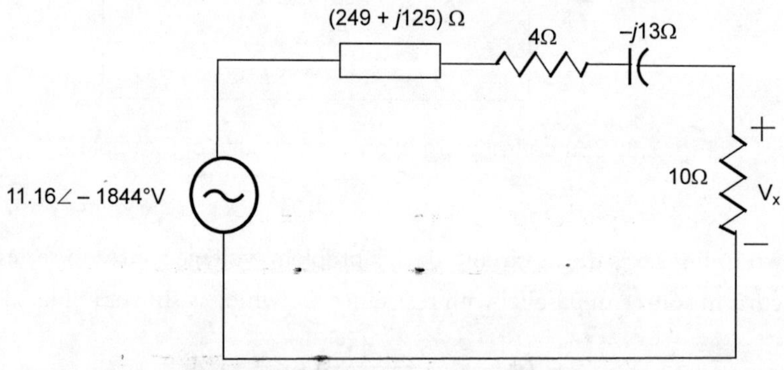

Fig. 2.8: Two parallel impedances made into one single impedances of Example 2.2

A total impedance of the circuit shown in Fig.2.8 is (16.49-j11.75) Ω. Hence the current flowing that circuit is $I = \frac{11.16\angle -18.44^\circ}{(16.49 - j11.75)} = \frac{11.16\angle -18.44}{20.25\angle -35.47} = 0.55\angle 17^\circ A$

Hence the voltage across 10Ω resistance is $V_{10\Omega} = (0.55\angle 17) * 10 = 5.5 \angle 17^\circ$ V

2.2 STAR/DELTA AND DELTA/STAR TRANSFORMATION

As like source transformation, star/delta transformation is also applicable to both dc and ac circuits. The conversion of star connected impedances is converted into delta connected impedances which is pictorially represented in Fig.2.9.

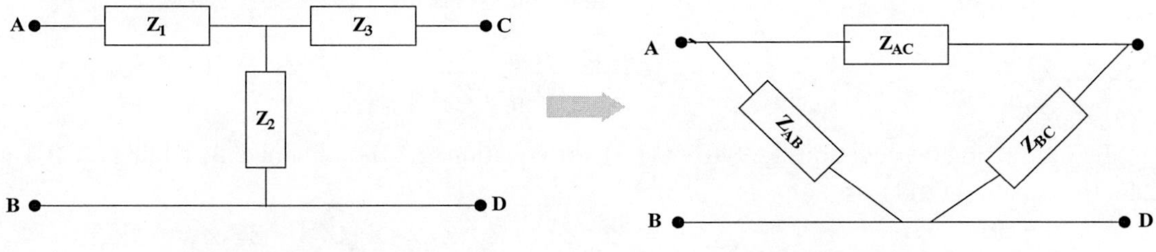

(a) Star Connection (b) Delta Connection

Fig. 2.9: Star connected load transformed into delta connected load

Based on the Fig.2.9a, the equivalent value of the impedances by looking from terminals A and B, that could be written as equation (2.1).

$$Z_{AB} = Z_1 + Z_2 \tag{2.1}$$

Similarly, the equivalent impedances are obtained from terminals A and C, terminals C and D and they are presented in equations (2.2) and (2.3) respectively. Both B and D are the same potential. Hence D could be replaced with (b)

$$Z_{AC} = Z_1 + Z_3 \tag{2.2}$$

$$Z_{CD} = Z_{CB} = Z_2 + Z_3 \tag{2.3}$$

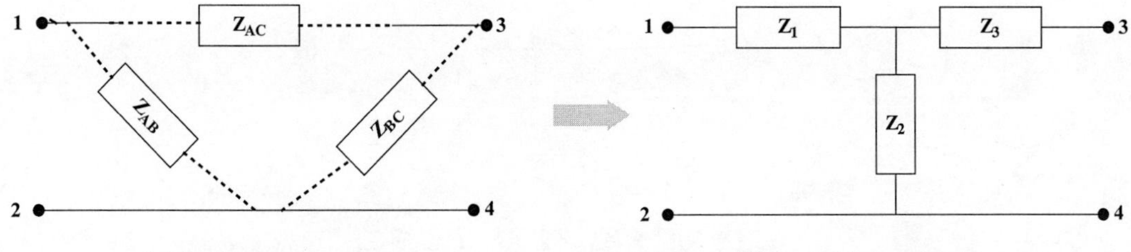

(a) Delta Connection (b) Star Connection

Fig. 2.10: Delta connected load transformed into star connected load

Based on the Fig.2.10a, the equivalent value of the impedances by looking from terminals 1 and 2, that could be written as equation (2.4).

$$Z_{12} = Z_{AB} \,||\, Z_{AC} + Z_{BC} \tag{2.4}$$

Similarly, the equivalent impedances are obtained from terminals 1 and 3, terminals 3 and 4 and they are presented in equations (2.5) and (2.6) respectively. Both terminals 2 and 4 are the same potential. Hence 4 could be replaced with 2.

$$Z_{13} = Z_{Ac} \,||\, Z_{AB} + Z_{BC} \tag{2.5}$$

$$Z_{23} = Z_{BC} \,||\, Z_{AB} + Z_{BC} \tag{2.6}$$

By comparing Fig.2.9 and Fig.2.10, the potential A, potential B are equal to potential at point 1 and 2 respectively. Hence, $Z_{AB} = Z_{12}$. Similarly, $Z_{Ac} = Z_{13}$ and $Z_{CB} = Z_{23}$. The following equations (2.7) to (2.9) are written based on the above statements.

$$Z_1 + Z_2 = \frac{Z_{AB} * (Z_{AC} + Z_{BC})}{Z_{AB} + Z_{AC} + Z_{BC}} \tag{2.7}$$

$$Z_1 + Z_3 = \frac{Z_{Ac} * (Z_{AB} + Z_{BC})}{Z_{AB} + Z_{AC} + Z_{BC}} \tag{2.8}$$

$$Z_1 + Z_3 = \frac{Z_{Ac} * (Z_{AB} + Z_{BC})}{Z_{AB} + Z_{AC} + Z_{BC}} \tag{2.9}$$

By subtracting the equations (2.9) from (2.7), the equation (2.10) is obtained. By adding (2.10a) and (2.8), the equation (2.11) is attained.

$$Z_1 - Z_3 = \frac{Z_{AB}Z_{AC} + Z_{AB}Z_{BC} - Z_{BC}Z_{AB} - Z_{BC}Z_{AC}}{Z_{AB} + Z_{AC} + Z_{BC}} \tag{2.10}$$

$$Z_1 - Z_3 = \frac{Z_{AB}Z_{AC} - Z_{BC}Z_{AC}}{Z_{AB} + Z_{AC} + Z_{BC}} \tag{2.10a}$$

$$2Z_1 = \frac{Z_{Ac}Z_{AB} + Z_{Ac}Z_{BC} + Z_{AB}Z_{AC} - Z_{BC}Z_{AC}}{Z_{AB} + Z_{AC} + Z_{BC}} \tag{2.11}$$

$$= \frac{2Z_{AC}Z_{AB}}{Z_{AB} + Z_{AC} + Z_{BC}}$$

Hence,

$$Z_1 = \frac{Z_{AC}Z_{AB}}{Z_{AB} + Z_{AC} + Z_{BC}} \tag{2.12}$$

$$Z_2 = \frac{Z_{AB}Z_{BC}}{Z_{AB} + Z_{AC} + Z_{BC}} \tag{2.13}$$

$$Z_3 = \frac{Z_{AC}Z_{BC}}{Z_{AB} + Z_{AC} + Z_{BC}} \tag{2.14}$$

For Star to Delta conversion, use the equations (2.12) to (2.14) and perform $Z_1Z_2 + Z_2Z_3 + Z_1Z_3$,

$$Z_1Z_2 + Z_2Z_3 + Z_1Z_3 = \frac{Z_{AB}^2 Z_{AC}Z_{BC} + Z_{BC}^2 Z_{AB}Z_{AC} + Z_{AC}^2 Z_{AB}Z_{BC}}{(Z_{AB} + Z_{AC} + Z_{BC})^2}$$

$$= \frac{Z_{AB}Z_{AC}Z_{BC}(Z_{AB} + Z_{AC} + Z_{BC})}{(Z_{AB} + Z_{AC} + Z_{BC})^2}$$

$$= \frac{Z_{AB} Z_{AC} Z_{BC}}{Z_{AB} + Z_{AC} + Z_{BC}} = Z_{AB} * \frac{Z_{AC} Z_{BC}}{Z_{AB} + Z_{AC} + Z_{BC}}$$

$$Z_1 Z_2 + Z_2 Z_3 + Z_1 Z_3 = Z_{AB} * \frac{Z_{AC} Z_{BC}}{Z_{AB} + Z_{AC} + Z_{BC}} \quad (2.15)$$

By using (2.14), the equation (2.15) becomes as follows.

$$Z_1 Z_2 + Z_2 Z_3 + Z_1 Z_3 = Z_{AB} * Z_3$$

Hence,

$$Z_{AB} = \frac{Z_1 Z_2 + Z_2 Z_3 + Z_1 Z_3}{Z_3} \quad (2.16)$$

Similarly,

$$Z_{AC} = \frac{Z_1 Z_2 + Z_2 Z_3 + Z_1 Z_3}{Z_2} \quad (2.17)$$

$$Z_{BC} = \frac{Z_1 Z_2 + Z_2 Z_3 + Z_1 Z_3}{Z_1} \quad (2.18)$$

In general, $Z_1 Z_2 + Z_2 Z_3 + Z_1 Z_3$ could be represented as $\Sigma Z_1 Z_2$. Hence equations (2.16) to (2.18) are written as equations (2.19) to (2.21) respectively. These equations are used for the conversion of star (Y) into delta (Δ) connection.

$$Z_{AB} = \frac{\sum Z_1 Z_2}{Z_3} \quad (2.19)$$

$$Z_{AC} = \frac{\sum Z_1 Z_2}{Z_2} \quad (2.20)$$

$$Z_{BC} = \frac{\sum Z_1 Z_2}{Z_1} \quad (2.21)$$

Similarly, $Z_{AB} + Z_{AC} + Z_{BC}$ could be represented as ΣZ_{AB}. Hence equations (2.12) to (2.14) are written as equations (2.22) to (2.24) respectively. These equations are used for the conversion of delta (Δ) into star (Y) connection.

$$Z_1 = \frac{Z_{AC} Z_{AB}}{\sum Z_{AB}} \quad (2.22)$$

$$Z_2 = \frac{Z_{AB} Z_{BC}}{\sum Z_{AB}} \quad (2.23)$$

$$Z_3 = \frac{Z_{AC} Z_{BC}}{\sum Z_{AB}} \quad (2.24)$$

When all the impedances are equal, $Z_{AB} = Z_{AC} = Z_{BC} = Z_\Delta$ & $Z_1 = Z_2 = Z_3 = Z_Y$, the equations (2.22) could be mentioned as follows,

$$Z_Y = \frac{Z_\Delta^2}{3Z_\Delta}$$

$$Z_Y = \frac{Z_\Delta}{3} \tag{2.25}$$

From the equation (2.25), it is understood that the delta impedances are converted into star impedances by dividing with 3. In other words, obtaining delta impedances from the star impedances by multiplying with 3.

Example 2.3

Perform the star connected impedance into delta connected impedances in Fig.2.11 using star/delta transformation method. Consider terminals 1, 2 and 3 are the terminals of star connection. $Z_1 = 5\Omega$; $Z_2 = (3 + j4)\Omega$; $Z_3 = (4 - j13)\Omega$

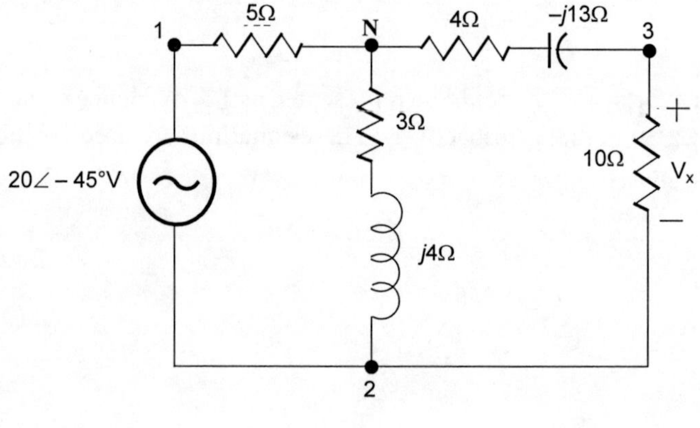

Fig. 2.11: Star/delta transformation in ac circuits of Example 2.3

Solutions:

The circuit shown in Fig.2.11 is the ac circuit. In this problem, the star impedances are converted into delta connected impedances which is represented in Fig.2.12. The Fig.2.12 is modified for the convenient to make the loops as like in Fig.2.13.

$$Z_1 = 5\Omega;\ Z_2 = (3 + j4)\Omega = 5\angle 53.13°;\ Z_3 = (4 - j13)\Omega = 13.6\angle -72.9°$$

Hence delta connected impedances are $Z_{12} = \dfrac{\sum Z_1 Z_2}{Z_3}$; $Z_{13} = \dfrac{\sum Z_1 Z_2}{Z_2}$; $Z_{23} = \dfrac{\sum Z_1 Z_2}{Z_1}$;

$$\Sigma Z_1 Z_2 = (5)(3 + j4) + (3 + j4)(4 - j13) + (5)(4 - j13)$$

$$= 99 - j68 = 120.1\angle -34.48°$$

$$Z_{12} = \frac{120.1\angle -34.48}{13.6\angle -72.9} = 8.83\angle 38.42 = (6.92 + j5.48)\ \Omega$$

$$Z_{13} = \frac{120.1\angle -34.48}{5\angle 53.13} = 24\angle -87.61 = (1 - j23.97)\,\Omega$$

$$Z_{23} = \frac{120.1\angle -34.48}{5} = 24\angle -34.48 = (19.78 - j13.58)\,\Omega$$

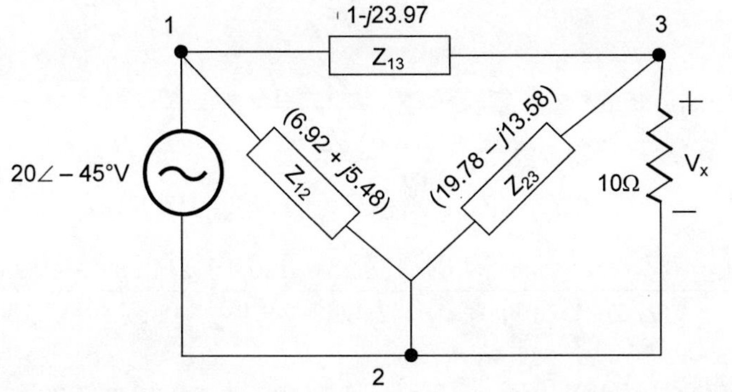

Fig. 2.12: Star to delta transformation in ac circuits from Fig.2.11

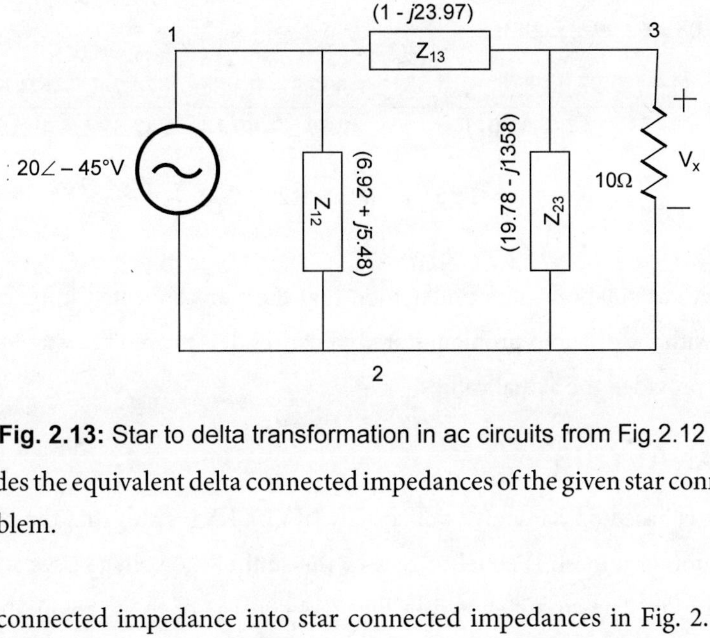

Fig. 2.13: Star to delta transformation in ac circuits from Fig.2.12

Fig. 2.13 provides the equivalent delta connected impedances of the given star connected impedances as given in this problem.

Example 2.4

Perform the delta connected impedance into star connected impedances in Fig. 2.13 using star/delta transformation method. $Z_{12} = (6.92 + j5.48)\ \Omega$; $Z_{13} = (1 - j23.97)\ \Omega$; $Z_{23} = (19.78 - j13.58)\Omega$

Solutions:

The circuit shown in Fig.2.13 is the ac circuit. In this problem, the delta impedances are converted into star connected impedances.

Z12 = $(6.92 + j5.48)\ \Omega$; $Z_{13} = (1 - j23.97)\ \Omega$; $Z_{23} = (19.78 - j13.58)\ \Omega$

Hence star connected impedances are $Z_1 = \dfrac{Z_{AC}Z_{AB}}{\sum Z_{AB}}; Z_2 = \dfrac{Z_{AB}Z_{BC}}{\sum Z_{AB}};\ Z_3 = \dfrac{Z_{AC}Z_{BC}}{\sum Z_{AB}}$ according to the equations (2.22) to (2.24). These equations are replaced according to the problem. Since terminals are mentioned as 1, 2 and 3. Hence the suffixes A,B and C are replaced with 1, 2, 3 respectively in the above equations.

The obtained equations are $Z_1 = \dfrac{Z_{13}Z_{12}}{\sum Z_{12}}; Z_2 = \dfrac{Z_{12}Z_{23}}{\sum Z_{12}};\ Z_3 = \dfrac{Z_{13}Z_{23}}{\sum Z_{12}}$

$$\Sigma Z_{12} = Z_{12} + Z_{13} + Z_{23} = (27.7 - j32.07)\ \Omega$$

$$Z_1 = \frac{Z_{13}Z_{12}}{\sum Z_{12}} = \frac{(1 - j23.97)(6.92 + j5.48)}{(27.7 - j32.07)} = \frac{138.28 - j160.4}{42.37\angle -49.18} = \frac{211.78\angle -49.23}{42.37\angle -49.18} = 4.998\,\Omega$$

$$Z_2 = \frac{Z_{12}Z_{23}}{\sum Z_{12}} = \frac{(6.92 + j5.48)(19.78 - j13.58)}{(27.7 - j32.07)} = \frac{211.29 + j14.42}{42.37\angle -49.18} = \frac{211.78\angle 3.9}{42.37\angle -49.18}$$

$$= 4.99\angle 53.08 = (2.99 + j3.98)\Omega$$

$$Z_3 = \frac{Z_{13}Z_{23}}{\sum Z_{12}} = \frac{(1 - j23.97)(19.78 - j13.58)}{(27.7 - j32.07)} = \frac{-300.3 - j13.56}{42.37\angle -49.18} = \frac{572.74\angle -121.67}{42.37\angle -49.18}$$

$$= 13.51 = (4.06 - j12.8)\ \Omega$$

$$Z_3 = (4.06 - j12.8)\ \Omega$$

From the above calculations, it is understood that the star connected impedances Z_1, Z_2 and Z_3 values are similar with the previous problem star impedances. It is proved that star to delta and then delta to star conversion provided the actual values.

2.3 MESH ANALYSIS

The mesh analysis is based on Kirchoff's voltage law (KVL). KVL states that the phasor summation of the voltages in a loop or a mesh is equal to zero or the sum of the voltage rises is equal to the sum of voltage drops in a loop. The circuit shown in Fig. 2.14 consists of the three meshes namely ABGHA, BCFGB and CDEFC respectively. In this circuit, the mesh currents I_1, I_2 & I_3 are to be found. For that, the three loops will be formed and the corresponding equations (2.26) to (2.28) are written respectively.

$$+V_1\angle\Theta - I_1 Z1 - Z_2 (I_1 - I_2) = 0 \qquad (2.26)$$

$$-I_2 Z_3 - Z_4 (I_2 - I_3) - Z_2 (I_2 - I_1) = 0 \qquad (2.27)$$

$$-I_3 Z_5 - Z_4 (I_3 - I_2) - Z_6 I_3 = 0 \qquad (2.28)$$

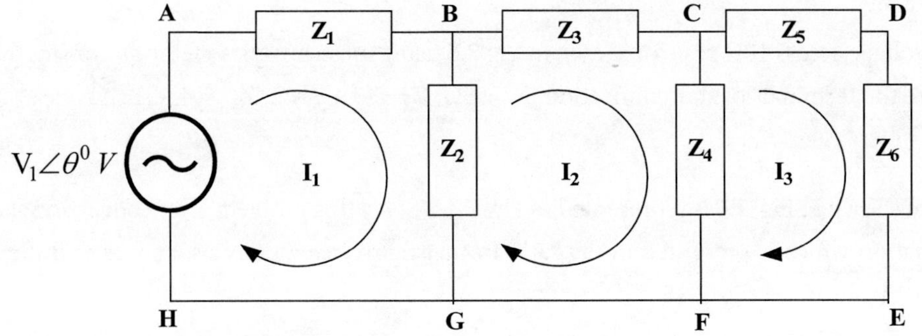

Fig. 2.14: Mesh analysis in ac circuits

Eq. (2.26) could be modified as, $V_1\angle\Theta = I_1 Z_1 + Z_2 (I_1 - I_2)$

$$V_1\angle\Theta = I_1(Z_1 + Z_2) - I_2 Z_2 \quad (2.29a)$$

$$I_1 (Z_1 + Z_2) - I_2 Z_2 = V_1\angle\Theta \quad (2.29b)$$

Eq. (2.27) and Eq. (2.28) could be modified as

$$-I_2 Z_3 - Z_4 I_2 - Z_4 I_3 - Z_2 I_2 + Z_2 I_1 = 0$$

$$-I_1 Z_2 + I_2 (Z_2 + Z_3 + Z_4) - I_3 Z_4 = 0 \quad (2.30)$$

$$-I_2 Z_4 + I_3 (Z_4 + Z_5 + Z_6) = 0 \quad (2.31)$$

These equations (2.29) to (2.31) are formed as matrix format which is represented in equations (2.32).

$$\begin{bmatrix} (Z_1+Z_2) & -Z_2 & 0 \\ -Z_2 & (Z_2+Z_3+Z_4) & -Z_4 \\ 0 & -Z_4 & (Z_2+Z_3+Z_4) \end{bmatrix} \begin{bmatrix} I_1 \\ I_2 \\ I_3 \end{bmatrix} = \begin{bmatrix} V_1\angle \\ 0 \\ 0 \end{bmatrix} \quad (2.32)$$

$$[Z][I] = [V] \quad (2.32a)$$

The equation (2.32) will be solved by Cramer's rule to find the loop currents. Δ_Z is the determinant of Z matrix. Δ_1, Δ_2 and Δ_3 are the determinant of the matrix, where those matrices are formed by replacing the 1st, 2nd and 3rd column by [V] matrix. The values of the loop currents are found by equation (2.33).

$$I_1 = \frac{\Delta_1}{\Delta_Z}; I_2 = \frac{\Delta_2}{\Delta_Z}; I_3 = \frac{\Delta_3}{\Delta_Z} \quad (2.33)$$

Steps for Mesh Analysis:

1. Mark the node terminals and name them.
2. Identify the number of closed loops and mark the current flow direction and also name these mesh currents as I_1, I_2, I_3 and etc.
3. Apply the KVL and write the equations.
4. Simplify the equations and form the matrix format as like $[Z][I] = [V]$.
5. Solve the matrix by Cramer's rule and find the mesh current values.

Example 2.5:

Find the voltage across 10Ω resistor shown in Fig.2.6 using mesh analysis method. Consider terminals 1, 2 and 3 are the terminals of star connection. $Z_1 = 5\Omega$; $Z_2 = (3 + j4)\Omega$; $Z_3 = (4 - j13)\Omega$

Solutions:

The circuit shown in Fig. 2.6 is represented as Fig. 2.11. Solve this problem after converting star into delta transformation which is presented in Fig.2.13. In that figure, the nodes are named and the meshes are formed that is shown in Fig. 2.15.

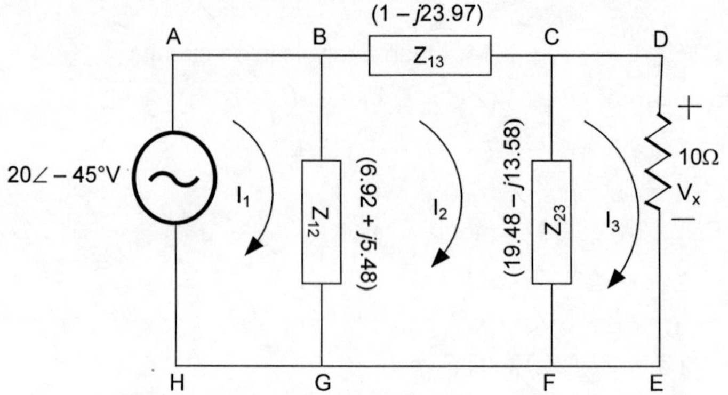

Fig. 2.15: Mesh analysis of Example 2.5

The loop equations (2.34) to (2.36) are formed according to the KVL and then simplified equations are presented as equations (2.37) to (2.39) respectively.

$$+20\angle -45 - Z_{12}(I_1 - I_2) = 0 \quad (2.34)$$

$$-I_2 Z_{13} - Z_{23}(I_2 - I_3) - Z_{12}(I_2 - I_1) = 0 \quad (2.35)$$

$$-Z_{23}(I_3 - I_2) - 10 I_3 = 0 \quad (2.36)$$

After simplification,

$$Z_{12} I_1 - Z_{12} I_2 = 20\angle -45 \quad (2.37)$$

$$Z_{12} I_1 - (Z_{12} + Z_{23} + Z_{13})I^2 + Z_{23} I_3 = 0 \quad (2.38)$$

$$Z_{23} I_2 - (10 + Z_{23})I_3 = 0 \quad (2.39)$$

The simplified equation is formed as matrix which is shown below.

$$\begin{bmatrix} Z_{12} & -Z_{12} & 0 \\ Z_{12} & -(Z_{12}+Z_{23}+Z_{13}) & Z_{23} \\ 0 & Z_{23} & -(10+Z_{23}) \end{bmatrix} \begin{bmatrix} I_1 \\ I_2 \\ I_3 \end{bmatrix} = \begin{bmatrix} 20\angle -45^0 \\ 0 \\ 0 \end{bmatrix}$$

$= 6.92 + j5.98 + 1 - j23.97 + 19.78 - j13.58 = 27.7 - j32.07;$

$= 10 + 19.78 - j13.58 = 29.78 - j13.58;$

$$\begin{bmatrix} 6.92+j5.48 & -6.92-j5.48 & 0 \\ 6.92+j5.48 & (-27.7+j32.07) & 19.78-j13.58 \\ 0 & 19.78-j13.58 & (-29.78+j13.58) \end{bmatrix} \begin{bmatrix} I_1 \\ I_2 \\ I_3 \end{bmatrix} = \begin{bmatrix} 20\angle-45^0 \\ 0 \\ 0 \end{bmatrix}$$

The determinant values are calculated as per the description given in section 4.3.

$$\Delta z = 4052.7 - j6510.1; \Delta_3 = 3191.6 - j2783.8$$

simplified equation is formed as matrix which is shown below.

$$\Delta_Z = 7668.49\angle-58.09°$$

$$\Delta_3 = 4235.07\angle-41.09°$$

$$I_3 = \frac{\Delta_3}{\Delta_Z} = 0.552\angle17°A$$

$$V_{10\Omega} = 10 * I_3 = 5.52\angle17° \text{ V}$$

Example 2.6

Find the current (I) flowing through the inductor using mesh analysis for the circuit shown in Fig.2.16.

Solutions:

The circuit shown in Fig.2.16 has the two voltage sources and its value is represented in time domain which needs to be converted into polar form. The conversion is described in the next page. Based on the description, 50 sin $5t$ = 50$\angle$ – 90°, V; 50 cos $5t$ = 50$\angle$0°, V

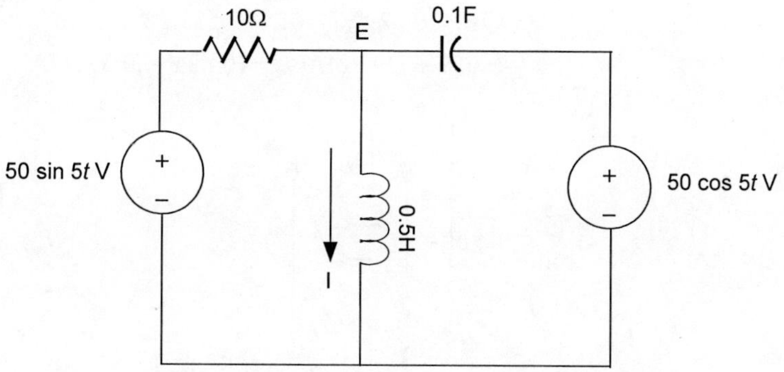

Fig. 2.16: Mesh analysis of Example 2.6

Inductive reactance $X_L = (2\pi f L) = (2\pi * 50 * 0.5) = 5\pi = 15.71\Omega$

Capacitive reactance $X_C = \frac{1}{(2\pi fC)} = \frac{1}{(2\pi * 50 * 0.1)} = 0.318\Omega$

The loop current direction is marked and the impedances are represented in ohms which is represented in Fig.2.17. The KVL is applied and the corresponding equations are presented in equation (2.40) and (2.41).

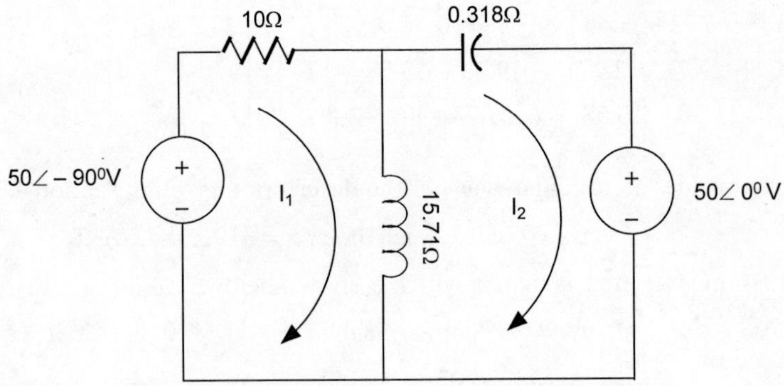

Fig. 2.17: Mesh analysis of Example 2.6 with the loop current direction

$$50\angle -90^\circ - 10I_1 - j15.71(I_1 - I_2) = 0 \quad (2.40)$$

$$-j15.71(I_2 - I_1) - 0.318I_2 - 50\angle 0^\circ = 0 \quad (2.41)$$

$$(10 + j15.71)I_1 - j15.71I_2 = 50\angle -90^\circ \quad (2.41a)$$

$$j15.71I_1 - (0.318 + j15.71)I_2 = 50\angle 0^\circ \quad (2.41b)$$

$$\begin{bmatrix} (10+j15.71) & -15.71 \\ 15.71 & -(0.318+j15.71) \end{bmatrix}\begin{bmatrix} I_1 \\ I_2 \end{bmatrix} = \begin{bmatrix} 50\angle -90^0 \\ 50\angle 0^0 \end{bmatrix} = \begin{bmatrix} -50j \\ 50 \end{bmatrix}$$

The determinant values are calculated as per the description given in section 4.3.

$$\Delta z = 490.4 - j162.1;\ \Delta_1 = j15.9;\ \Delta_2 = 500 + j157$$

$$\Delta_z = 516.49\angle -18.29^\circ;\ \Delta_1 = 15.9\angle 90^\circ;\ \Delta_2 = 524.06\angle 17.43^\circ$$

$$I_1 = 30.78\angle 108.29^\circ;\ I_2 = 1.015\angle 35.72^\circ$$

$$I_1 = (-9.659 + j29.22)\text{A};\ I_2 = (0.824 + j0.5925)\text{A}$$

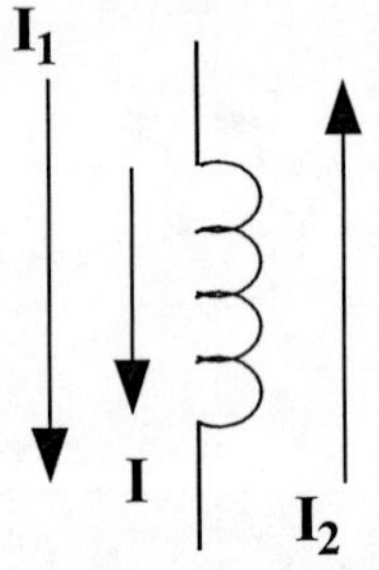

Fig. 2.18: Mesh analysis for Example 2.6 to find the current I

The current flowing through the inductor is I. In order to find the current I, the loop currents' direction is necessary to know and those are presented in Fig.2.18. The current I is in the same direction of I_1 and in the opposite direction of I_2. Hence the current I is found as $(I_1 - I_2)$ (a)

$$I = (I_1 - I_2) = (-9.659 + j29.22 - 0.824 - j0.5925)$$

$$= (-10.483 - j28.6275)\text{A}$$

$$I = 30.48\angle -110.112^\circ \text{ A}$$

2.4 NODE ANALYSIS

The node analysis is performed based on Kirchoff's current law (KCL). KCL states that the algebraic sum of the currents in a node is equal to zero or the sum of the incoming currents is equal to the sum of outgoing currents in a node. Node is a point where one or more elements are joined together. The circuit shown in Fig. 2.19 is taken for explaining the node analysis.

Steps for Node Analysis:

1. Identify the nodes and name them as A, B, C and etc.
2. Identify the different potential nodes and mark them as E_1, E_2, E_3 and etc. Where E_1, E_2, E_3 are the node potential.
3. Mark the current flow direction in the elements of the circuit (called as branch currents) like I_1, I_2, I_3 and etc.
4. Apply the KCL at each node potential points and write the equations.
5. Simplify the equations and form the matrix format as like [Y][E] =[I].
6. Solve the matrix by Cramer's rule and find the node voltage values.

Time domain into phasor form

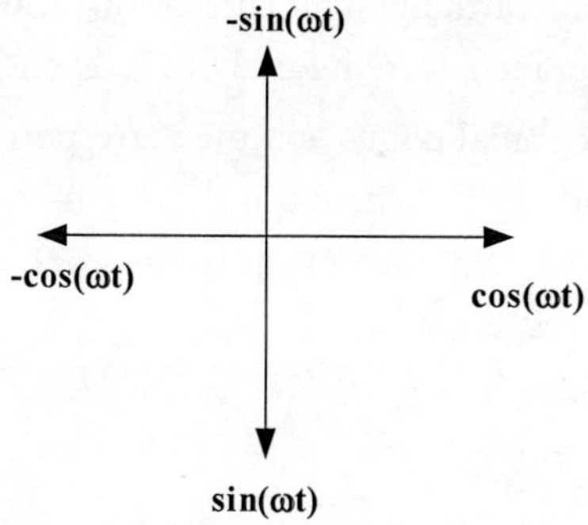

Fig. 2. S Phasor Reference

The voltage is represented in time domain as like given in the example 2.7. $v_1\ (t) = 50 \sin 5t$,V; $v_2\ (t) = 50 \cos 5t$, V

Cosine is taken as the reference: $v_2\ (t) = 50 \cos 5t$,V is represented in phasor form with zero degree.

$$V_2 = 50\angle 0°, \text{V}$$

Another voltage $v_1\ (t) = 50 \sin 5t$,V is represented as follows.

$$v_1\ (\text{t}) = 50 \sin 5\text{t} = 50 \cos(5t - 90), \text{V}$$

Hence, $V_1 = 50\angle - 90°$, V

Sine is taken as the reference: $v_1\ (t) = 50 \sin 5t$, V is represented in phasor form with zero degree.

$$V_1 = 50\angle 0°, \text{V}$$

Another voltage $v_2\ (t) = 50 \cos 5t$,V is represented as follows.

$$v_2\ (t) = 50 \cos 5t = 50 \sin(5t + 90), V$$

Hence, $V_2 = 50\angle + 90°$, V

Hint:

As "cos" reference:

$$A\cos(\omega t + \phi) = A\angle\phi$$

$$A\sin(\omega t + \phi) = A\cos(\omega t + \phi - 90) = A\angle(\phi - 90)$$

$$-A\cos(\omega t + \phi) = A\cos(\omega t + \phi \pm 180) = A\angle(\phi \pm 180)$$

$$-A\sin(\omega t+\phi) = A\cos(\omega t + \phi + 90) = A\angle(\phi + 90)$$

As "sin" reference:

$$A\sin(\omega t+\phi) = A\angle\phi$$

$$A\cos(\omega t + \phi) = A\sin(\omega t + \phi + 90) = A\angle(\phi + 90)$$

$$-A\cos(\omega t + \phi) = A\sin(\omega t + \phi - 90) = A\angle(\phi - 90)$$

$$-A\sin(\omega t + \phi) = A\sin(\omega t + \phi \pm 180) = A\angle(\phi \pm 180)$$

$$I_1 = \frac{V_1\angle\theta_1 - E_1}{Z_1} \tag{2.42}$$

Nodes are identified in the circuit shown in Fig.2.19, namely A, B, C, D, E, F, G and H. Now mark the different potential nodes and mark the branch current direction which is shown in Fig.2.20. Two potential nodes E_1 and E_2 and the branch currents are I_1, I_2, I_3, I_4 and I_5.

Apply the KCL at the node potential points and the corresponding equations are formed which is mentioned in equations (2.42a) and (2.43). Where I_1 is flowing from source $V_1\angle\theta_1$ to node potential E_1. Hence current I_1 could be written as mentioned in equation (2.42). Hence, equations (2.42a) and (2.43) are elaborated as (2.44) and (2.45) respectively.

$$I_1 = I_2 + I_3 \tag{2.42a}$$

$$I_3 + I_5 = I_4 \tag{2.43}$$

$$\frac{V_1\angle\theta_1 - E_1}{Z_1} = \frac{E_1 - 0}{Z_2} + \frac{E_1 - E_2}{Z_3} \tag{2.44}$$

$$(V_1\angle\theta_1 - E_1)\,Y_1 = E_1\,Y_2 + (E_1 - E_2)\,Y_3$$

$$\left(\frac{Y_1 + Y_2 + Y_3}{Y_1}\right)E_1 - \left(\frac{Y_3}{Y_1}\right)E_2 = V_1\angle\theta_1 \tag{2.44a}$$

$$\frac{E_1 - E_2}{Z_3} + \frac{V_2\angle\theta_2 - E_2}{Z_5} = \frac{E_2 - 0}{Z_4} \tag{2.45}$$

$$-\left(\frac{Y_3}{Y_5}\right)E_1 + \left(\frac{Y_3 + Y_4 + Y_5}{Y_5}\right)E_2 = V_2\angle\theta_2 \tag{2.45a}$$

These equations (2.44a) and (2.45a) are formed as matrix format which is represented in equation (2.46) and this can be written as equation (2.46a).

$$\begin{bmatrix} \left(\dfrac{Y_1+Y_2+Y_3}{Y_1}\right) & -\left(\dfrac{Y_3}{Y_1}\right) \\ -\left(\dfrac{Y_3}{Y_5}\right) & \left(\dfrac{Y_3+Y_4+Y_5}{Y_5}\right) \end{bmatrix} \begin{bmatrix} E_1 \\ E_2 \end{bmatrix} = \begin{bmatrix} V_1\angle\theta_1 \\ V_2\angle\theta_2 \end{bmatrix} \tag{2.46}$$

OR

$$\begin{bmatrix} Y_1+Y_2+Y_3 & -Y_3 \\ -Y_3 & Y_3+Y_4+Y_5 \end{bmatrix} \begin{bmatrix} E_1 \\ E_2 \end{bmatrix} = \begin{bmatrix} V_1\angle\theta_1 * Y_1 \\ V_2\angle\theta_2 * Y_5 \end{bmatrix} \tag{2.46a}$$

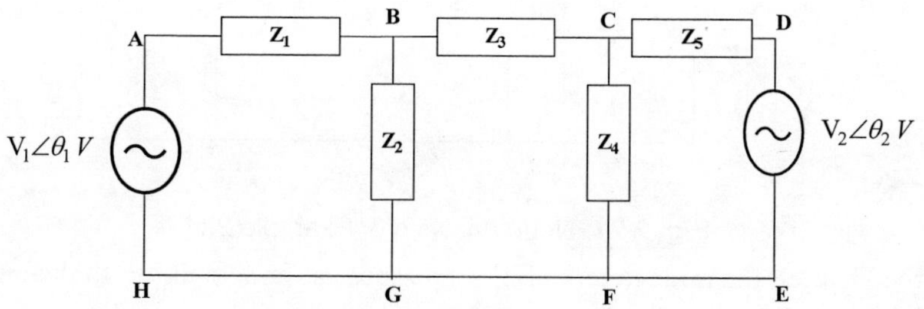

Fig. 2.19: Node analysis in ac circuits

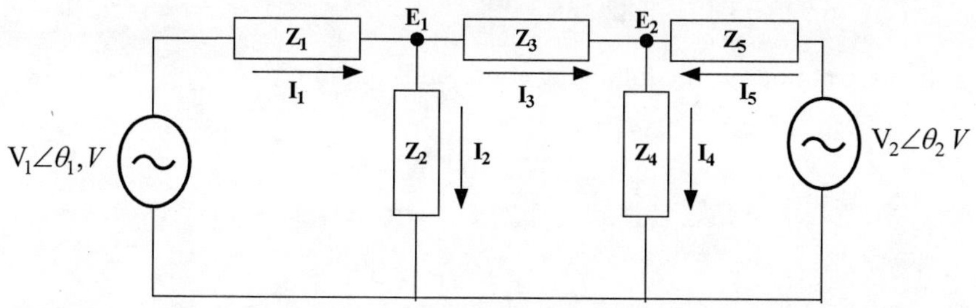

Fig. 2.20: Node analysis in ac circuits with node potential and branch currents

Solve the equation (2.46a) using Cramer's rule and find the node potential E_1 and E_2. Δ_Y is the determinant of the Y matrix. Δ_1 and Δ_2 are the determinant of the matrix, where those matrices are formed by replacing the 1st and 2rd column by [V*Y] or [I] matrix. The values of the node potentials are found by equation (2.47).

$$E_1 = \frac{\Delta_1}{\Delta_Y}; E_2 = \frac{\Delta_2}{\Delta_Y} \tag{2.47}$$

Example 2.6

Find the voltage across 10Ω resistor shown in Fig.2.6 using node analysis method. Consider terminals 1, 2 and 3 are the terminals of star connection. $Z_1 = 5\Omega$; $Z_2 = (3 + j4)\Omega$; $Z_3 = (4 - j13)\Omega$

Solutions:

The circuit shown in Fig.2.6 is redrawn as Fig.2.11. Solve this problem after converting star into delta transformation which is presented in Fig.2.13. In that figure, the nodes are named and the loops are formed that is shown in Fig.2.15. In order to solve that the Fig.2.15 is modified as Fig.2.21 with node potential markings and current flow direction.

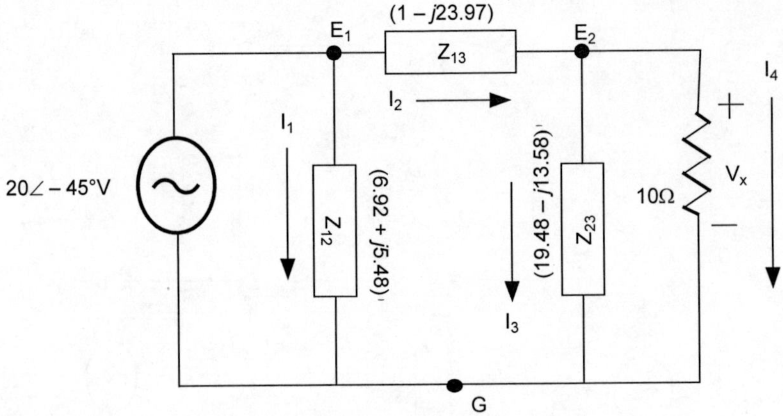

Fig. 2.21: Node analysis of Example 2.6

Based on the Fig.2.21, the node potential E_1 is measured at node point E_1 to the reference potential G. As per the definition, E_1 is equal to $20\angle-45°$V.

Hence, $I_1 = \dfrac{E_1}{Z_{10}} = \dfrac{20\angle-45}{6.92+j5.48} = \dfrac{20\angle-45}{8.82\angle38.3} = 2.26\angle-83.3°A$

Applying the KCL at node 2, the following equation is obtained.

$$I_2 = I_3 + I_4$$

$$\frac{E_1 - E_2}{Z_{12}} = \frac{E_2 - 0}{Z_{20}} + \frac{E_2 - 0}{10}$$

$$\frac{E_1}{Z_{12}} = E_2\left(\frac{1}{Z_{12}} + \frac{1}{Z_{20}} + \frac{1}{10}\right)$$

$$\frac{20\angle-45}{1-j23.97} = E_2\left(\frac{1}{1-j23.97} + \frac{1}{19.78-j13.58} + \frac{1}{10}\right)$$

$$\frac{20\angle-45}{23.99\angle-87.6} = E_2\left(0.0416\angle87.6 + 0.0416\angle34.4 + 0.1\right)$$

$$0.833\angle42.6 = E_2\ (0.135 + j0.0645)$$

$$0.833\angle42.6 = E_2\ (0.15\angle25.5)$$

$$E_2 = \frac{0.833\angle42.6}{0.15\angle25.5} = 5.55\angle17.1°V$$

The node potential E2 is measured at node point E2 to the ground potential G. It is equivalent to the voltage across the impedance Z_{20} and also the voltage across the resistance 10Ω. Since they are connected in parallel.

Hence the voltage across the 10Ω resistance is $5.55\angle17.1°$ V.

Example 2.7

Find the node potential (E) and the current (I) flowing through the inductor using Node analysis for the circuit shown in Fig.2.22.

Solutions:

The circuit shown in Fig.2.7 has the two voltage sources and its value is represented in time domain which needs to be converted into polar form. The conversion is described in the next page. Based on the description, 50 sin 5t = 50∠ – 90°, V; 50 cos 5t = 50°,V

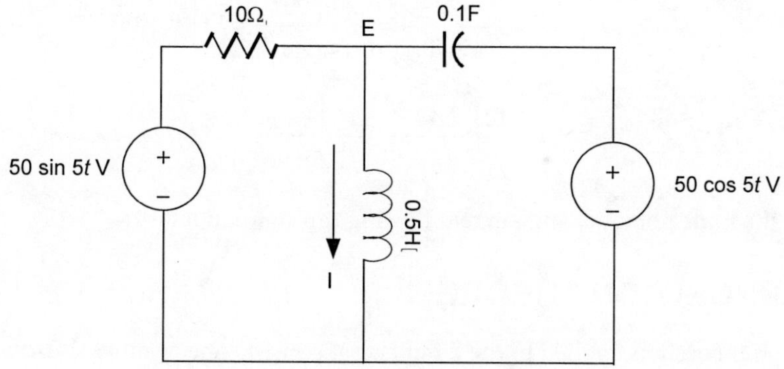

Fig. 2.22: Node analysis of Example 2.7

Inductive reactance $XL = j(2\pi fL) = j(\omega L) = j(5 * 0.5) = j2.5\Omega$

Capacitive reactance $X_C = \frac{1}{j(2\pi fC)} = \frac{1}{j(5*0.1)} = -j2\Omega$

The branch current direction is marked and the impedances are represented in ohms which is represented in Fig.2.23. The node potential is marked as E. The KCL is applied at that node point and the corresponding equation is presented in equation (2.48).

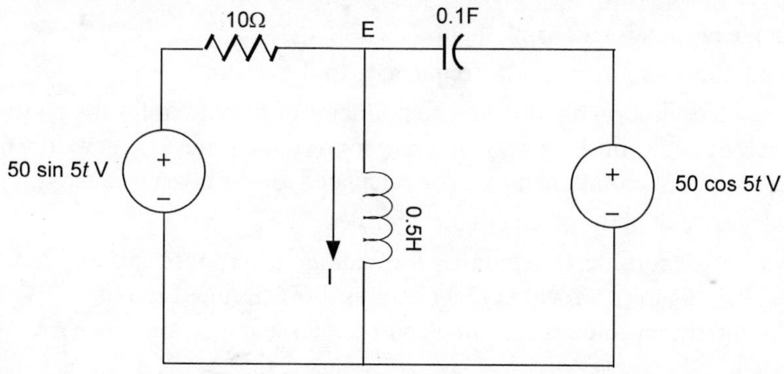

Fig. 2.23: Node analysis of Example 2.7 with designated nodes and branch current directions

$$I_1 = I_2 + I_3 \tag{2.48}$$

$$\frac{50\angle-90^0-E}{10}=\frac{E}{j2.5}+\frac{E-50\angle 0^0}{-j2}$$

$$\frac{50\angle-90^0}{10}+\frac{50\angle 0^0}{-j2}=\frac{E}{10}+\frac{E}{j2.5}+\frac{E}{-j2}$$

$$5\angle-90° + 25\angle 90° = E(0.1 + 0.4\angle-90° + 0.5\angle 90°)$$

$$-5j + j25 = E(0.1 - j0.4 + j0.5) = E(0.1 + j0.1)$$

$$E=\frac{j20}{0.1+j0.1}=\frac{20\angle 90^0}{0.141\angle 45^0}=141.8\angle 45° V$$

$$I=\frac{141.8\angle 45^0}{j2.5}=\frac{141.8\angle 45^0}{2.5\angle 90^0}=56.72\angle-45^0 A$$

E and I are the node potential and current flowing through inductor.

2.5 SUPERPOSITION THEOREM

The superposition theorem is applied to the circuits that consists one or more sources. This theorem is applicable to both dc and ac circuits or in other words, applicable to linear circuits. For analysing those ac circuits, superposition theorem is very useful for finding the responses.

Superposition theorem states that in a circuit of linear elements containing more than one sources with single frequency, the current flowing through an element of the circuit when all sources considered at a time is the sum of all the currents which would flow through that element of a circuit when individual sources considered separately with that frequency and all other sources are replaced by their internal impedances.

Steps for Superposition Theorem:

1. Mark the node terminals and name them.
2. Check the number of sources and its frequencies in the circuit.
3. Find the impedances of each individual components of the circuit for the given frequency. Let us assume that the circuit consists of two different sources and some RLC components. The equivalent impedance of each LC components will be calculated for the given frequency.
4. The equivalent circuit of the given circuit is formed,
5. Now, each source is considered separately for finding the response of the circuit. During that time, any other voltage sources present in that circuit is short circuited and the current sources are open circuited. Similarly, all sources are to be considered separately and the corresponding responses are calculated.
6. All calculated responses (O_1, O_2, O_3, ...O_N) of each individual circuits obtained by considering the independent sources are presented in time domain. O_i is the responses of the circuit, where i

= 1, 2, 3,…., N. N is the number of sources in the circuit. Let us assume $O_1(t)$ = Asinωt is due to the first source. Similarly, $O_2(t) = B\sin(\omega t + \theta)$, $O_3(t) = C\sin(\omega t - \theta)$, ….$O_N(t) = D\sin(\omega t + \varphi)$. Hence the overall response is $O(t) = O_1(t) + O_2(t) + ... + O_N(t) = A\sin\omega t + B\sin(\omega t + \theta) + ... + D\sin(\omega t + \varphi)$.

Superposition theorem for the circuits having multiple sources with single frequency:

The circuit shown in Fig.2.24 consists of two sources with a frequency of ω_1. By applying the superposition theorem, find the current flowing through the inductor. First step is to find the impedances of inductor and capacitor for the given frequency ω_1.

$$X_L = \omega_1 L, \text{ in } \Omega$$

$$X_C = 1/(\omega_1 C), \text{ in } \Omega$$

The source voltages are converted into the phasor form as discussed in Example 2.7, $V_1 \sin\omega_1 t = V_1\angle -90°$; $V_2 \cos\omega_1 t = V_2\angle 0°$; Hence the equivalent circuit in phasor form of sources is shown in Fig.2.25. Consider E as the node potential. Apply node analysis to find the current flowing through the inductor (Ref. Sec. 4.4). Let us assume that the current flowing through the inductor is I for the circuit shown in Fig.2.25.

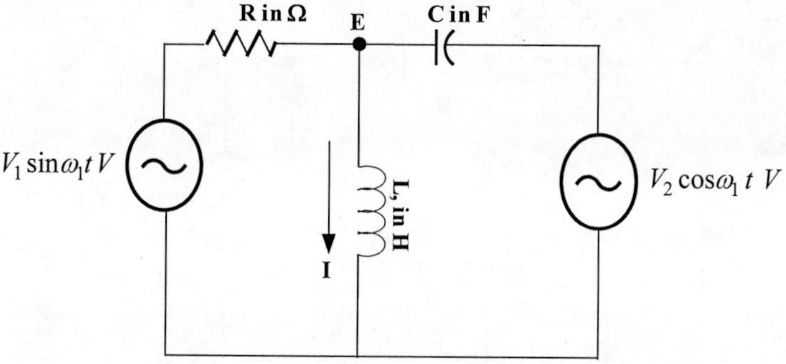

Fig. 2.24: Finding current in the circuit with two sources having single frequency using superposition theorem

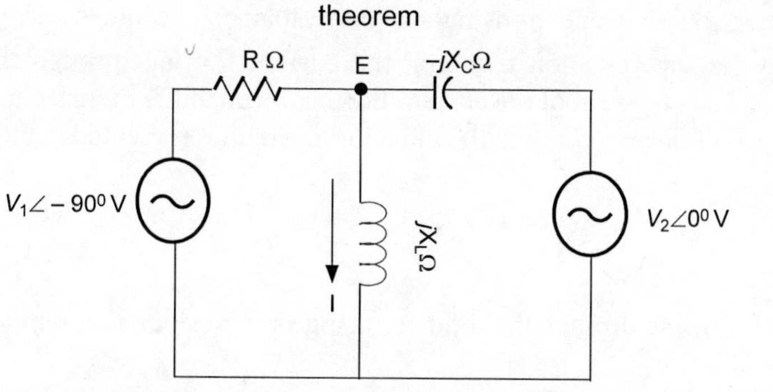

Fig. 2.25: Finding current in the circuit with two sources having single frequency in phasor form

Second step is to consider only one source (V_1) acting on the circuit, another voltage source is short circuited and the current flowing through the inductor is assumed to be I_1.The diagram is shown in Fig.2.26. Find the current I_1 by applying node analysis or by simplifying the impedances and apply current division technique or voltage division technique.

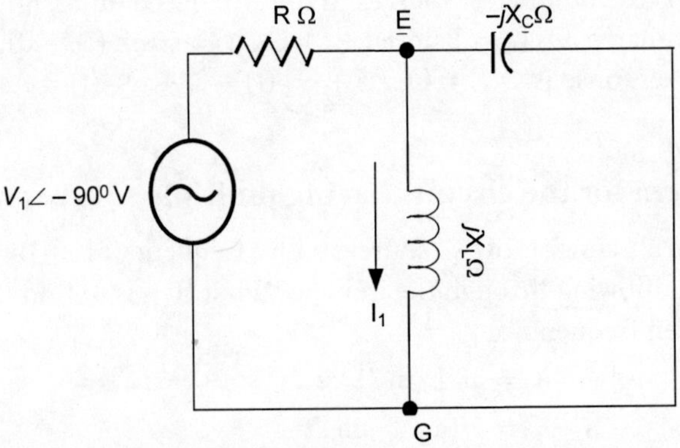

Fig. 2.26: Finding current in the circuit with first sources acting alone

Third step is to consider second source (V_2) acting alone on the circuit, another voltage source is short circuited and the current flowing through the inductor is assumed to be I_2. The concern diagram is shown in Fig.2.27. Find the current I_2 by applying node analysis or by simplifying the impedances and apply current division technique or voltage division technique.

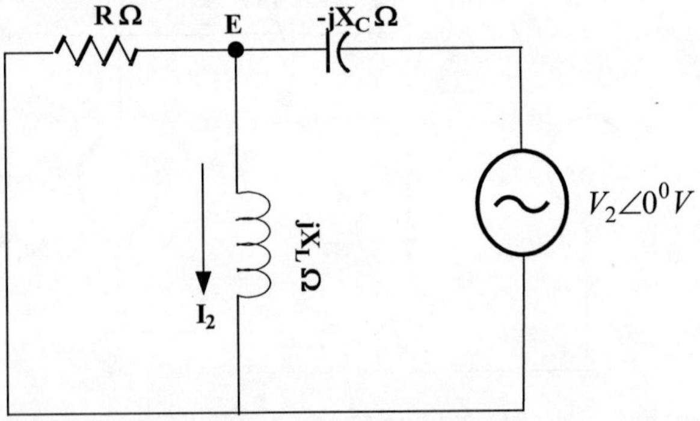

Fig. 2.27: Finding current in the circuit with second sources acting alone

According to the superposition theorem, the current flowing through the inductor in circuit represented in Fig.2.25 is the sum of the current flowing through the inductor in circuit represented in Fig.2.26 and the current flowing through the inductor in circuit is presented in Fig.2.27 and the equation (2.49).

$$I = I_1 + I_2 \tag{2.49}$$

Example 2.8

Find the current (I) flowing through the inductor using superposition theorem for the circuit shown in Fig.2.22.

Solutions:

The circuit shown in Fig.2.22 has the two voltage sources having same frequency of ω = 5Hz and its value is represented in time domain which needs to be converted into polar form. 50 sin 5t = $50\angle -90°$,V; 50 cos5t = $50\angle 0^0$,V. Also, the inductive impedance and capacitive impedance are to be calculated (Ref. Example 2.7). The circuit of the incorporated phasor values and calculated impedances are shown in

Fig. 2.28.

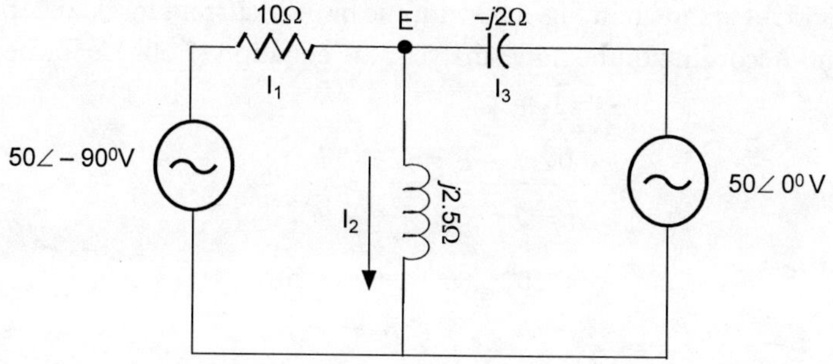

Fig. 2.28: Finding current in the circuit with two sources having same frequency

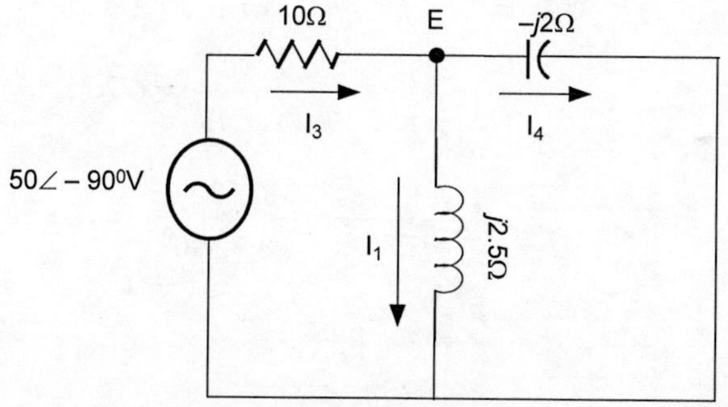

Fig. 2.29: Finding current in the inductor when only 1st voltage source acting alone

Step 1: Both sources are acting together, the current flowing through the inductor is calculated (Ref. Example 2.7) and the value is $I = 56.72\angle -45^0$ A, which is represented in time domain $i(t) = 56.72 \cos(5t - 45^0)$A.

Step 2: First source is acting alone, the current flowing through the inductor ($i_1(t)$) is calculated and the equivalent circuit is shown in Fig. 2.29 with the branch current direction. According to the node analysis, the equation (2.50a) is formed.

$$I_3 = I_1 + I_4 \tag{2.50a}$$

$$\frac{50\angle -90^0 - E}{10} = \frac{E}{j2.5} + \frac{E}{-j2}$$

$$\frac{50\angle -90^0}{10} = \frac{E}{10} + \frac{E}{j2.5} + \frac{E}{-j2}$$

$$25\angle 90^0 = E(0.1 + 0.4\angle -90^0 + 0.5\angle 90^0)$$

$$25j = E(0.1 - j0.4 + j0.5) = E(0.1 + j0.1)$$

$$E = \frac{}{0.1 \quad 0.1} = \frac{5\angle -90}{0.141\angle 45} = 35.46\angle -135 \ V$$

$$I_1 = \frac{35.46\angle -135^0}{2.5\angle 90^0} = 14.18\angle -225^0 \ A$$

Step 3: Second source is acting alone, the current flowing through the inductor ($i_2(t)$) is calculated and the equivalent circuit is shown in Fig. 2.30 with the branch current marked. Apply the node analysis and find the current. According to the node analysis, the equation (2.50b) is formed.

$$I_5 = I_2 + I_6 \tag{2.50b}$$

$$\frac{50\angle 0^0 - E}{-j2} = \frac{E}{j2.5} + \frac{E}{10}$$

$$\frac{50\angle 0^0}{-j2} = \frac{E}{10} + \frac{E}{j2.5} + \frac{E}{-j2}$$

$$25\angle 90^0 = E(0.1 + 0.4\angle -90^0 + 0.5\angle 90^0)$$

$$25j = E(0.1 - j0.4 + j0.5) = E(0.1 + j0.1)$$

$$E = \frac{25j}{0.1 + j0.1} = \frac{25\angle 90^0}{0.141\angle 45^0} = 177.3\angle 45^0\,V$$

$$I_2 = \frac{177.3\angle 45^0}{j2.5} = \frac{3177.3\angle 45^0}{2.5\angle 90^0} = 70.92\angle -45^0\,A$$

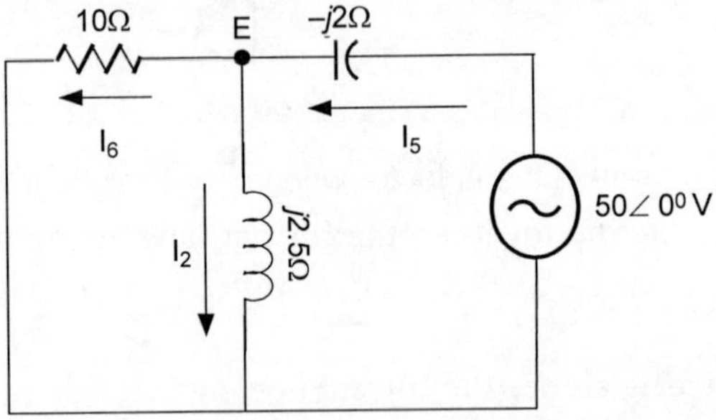

Fig. 2.30: Finding current in the inductor when only 2nd voltage source acting alone with branch current direction and node potential marking

Step 4: Finding the sum of the current due to the two sources are acted alone, the calculation is done as follows.

$$I_1 + I_2 = 14.18\angle -225^0 + 70.92\angle -45^0$$

$$I_1 + I_2 = -10.026 + j10.026 + 50.148 - j50.148$$

$$I_1 + I_2 = 40.122 - j40.122 = 56.74\angle -45^0$$

The total current of the individual sources could be written in time domain as $i_1(t) + i_2(t) = 56.74 \cos(5t - 45^0)$A. It is equal to the current flowing through the inductor when both sources acting together that is given in step 1. Hence the superposition theorem is proved.

Example 2.9

Find the current (I) flowing through the resistor using superposition theorem for the circuit shown in Fig.2.31.

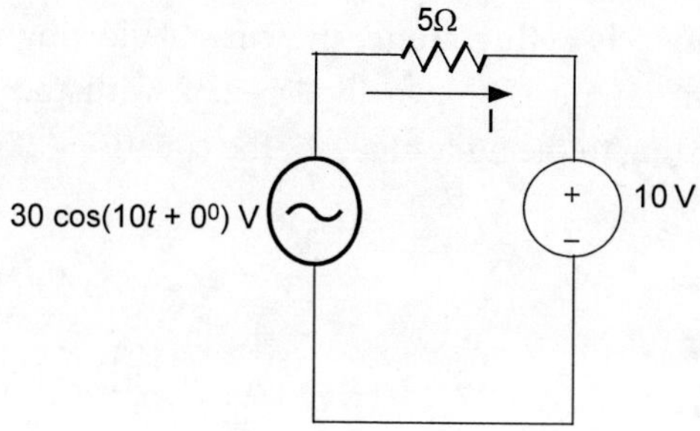

Fig. 2.31: Finding current in the resistor using superposition theorem for Example 2.10

Solutions:

The circuit shown in Fig.2.31 has the two voltage sources, one is ac voltage source and another is dc source. The ac source is represented in phasor form $30\cos(10t + 0^0) = 30\angle 0^0$.

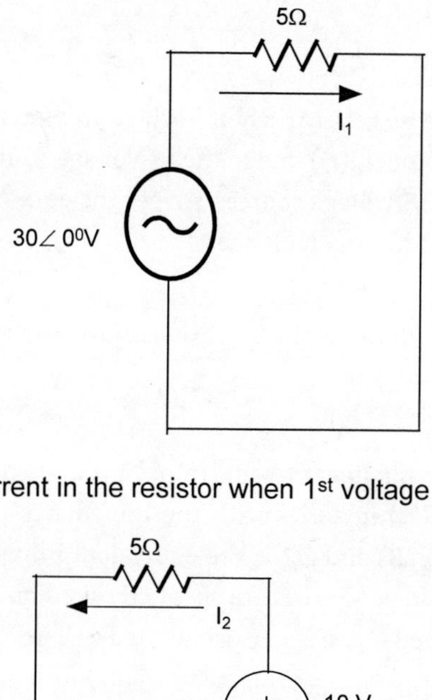

Fig. 2.32: Finding current in the resistor when 1st voltage source acting alone

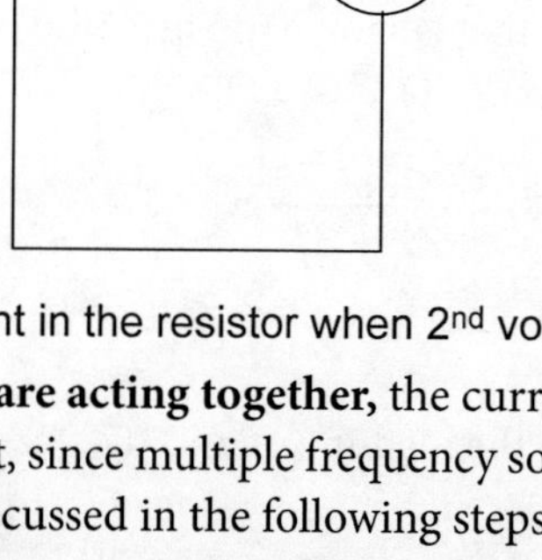

Fig. 2.33: Finding current in the resistor when 2nd voltage source acting alone

Step 1: Both ac and dc sources are acting together, the current flowing through the resistor is not possible to calculate as a single circuit, since multiple frequency sources present in the circuit. Hence the method of finding the response is discussed in the following steps.

Step 2: First voltage source is acting alone, the current flowing through the resistor (i_1 (t)) is calculated and the equivalent circuit is shown in Fig. 2.32 with the branch current direction and equivalent impedances. According to the node analysis, the equation (2.51) is formed.

$$I_1 = \frac{V_1}{5}$$

$$I_1 = \frac{30\angle 0^0}{5} = 6\angle 0^0\, A$$

Step 3: Second voltage source is acting alone, the current flowing through the inductor (i_2 (t)) is calculated and the equivalent circuit is shown in Fig. 2.33 with the branch current direction and equivalent impedances.

$$I_2 = \frac{V_2}{5}$$

$$I_2 = \frac{10}{5} = 2A$$

Step 4: Current flowing in the resistor when both sources acting together is equal to the sum of current flowing through the resistor (i_1 (t)) due to the 1st voltage source and the current flowing through the resistor (i_2 (t)) due to the 2nd st voltage source. As per the equivalent circuit, I_1 is in the direction of I and the I_2 is in the opposite direction of I. Hence

$$I = I_1 - I_2$$

$$i(t) = i_1\,(t) - i_2\,(t) = 6\cos(10t + 0^0) - 2\text{ A}$$

2.6 THEVENIN'S THEOREM

The Thevenin's theorem states that a linear two terminal circuit is replaced by an equivalent circuit that consists of the voltage source (V_{th}) in series with the impedance (Z_{th}). Where V_{th} is the open circuit voltage at the open-ended terminals and Z_{th} is the equivalent impedance of the circuit from the open-ended terminals when all the voltage sources are short circuited and current sources are open circuited or replaced by their internal impedances. Z_{th} is also named as Thevenin's impedance.

The statement is described in Fig.2.34. The linear circuit is represented as the Thevenin's circuit which is equal to the equivalent voltage source (V_{th}) in series with the equivalent impedance (Z_{th}). The load impedance is now replaced as Z_L that is shown in Fig.2.35. The current through the load impedance is calculated that is shown in equation (2.51).

$$I_L = \frac{V_{th}}{Z_{th} + Z_L} \tag{2.51}$$

Example 2.10

Find the current (I) flowing through the load impedance (Z_L) using Thevenin's theorem for the circuit shown in Fig.2.36.

Solutions:

The circuit shown in Fig.2.36 is represented with its impedance values as shown in Fig. 2.37. The angular frequency ω = 50Hz. The impedances and source voltage in phasor form are calculated as follows.

$$X_L = j\omega\ L, \text{ in } \Omega = j(50)(0.4) = j20\Omega$$

$$X_C = \frac{1}{j\omega C}, in\Omega = 1/\ j(50)(0.667) = -j30\Omega$$

$$150\ \sin(50t + 30^0) = 150\angle(-90^0 + 30^0) = 150\angle - 60^0$$

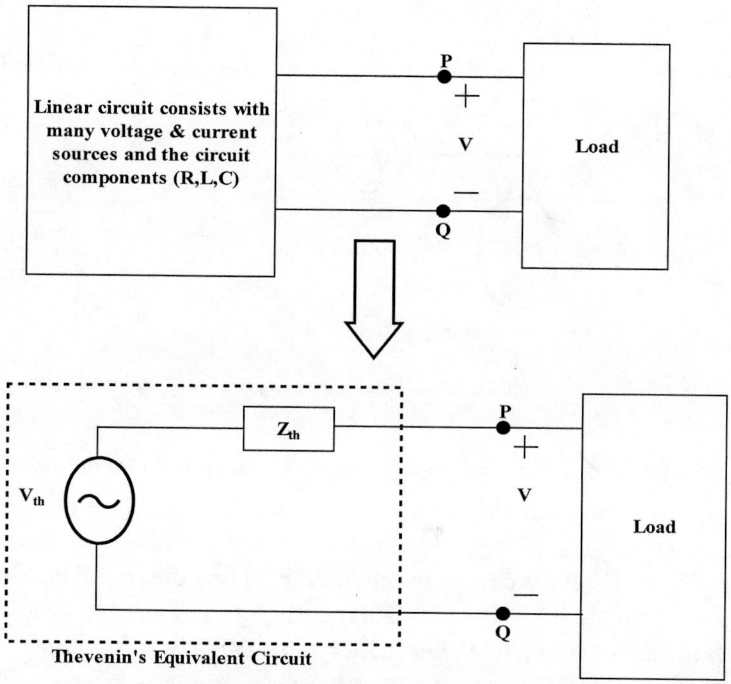

Fig. 2.34: The transformation of the linear two open-ended circuit into the Thevenin's equivalent circuit

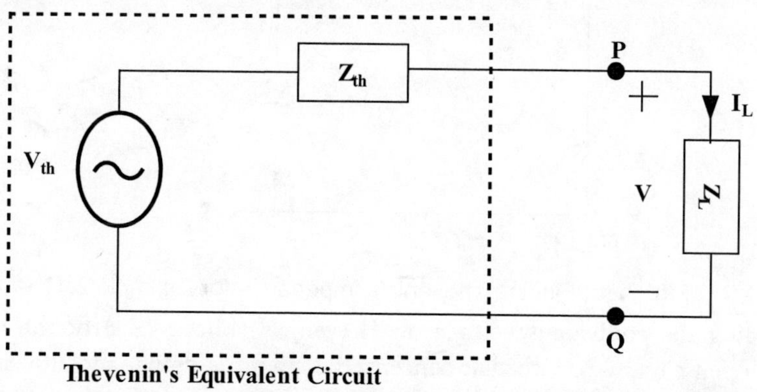

Fig. 2.35: Thevenin's equivalent circuit with load impedance to find the load current

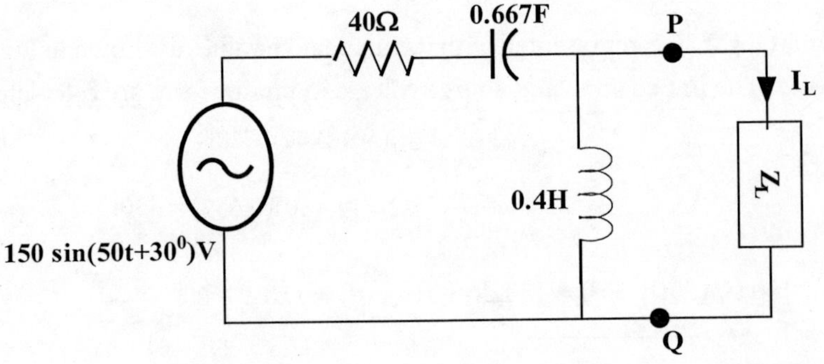

Fig. 2.36: Finding current in the load impedance using Thevenin's theorem for Example 2.11

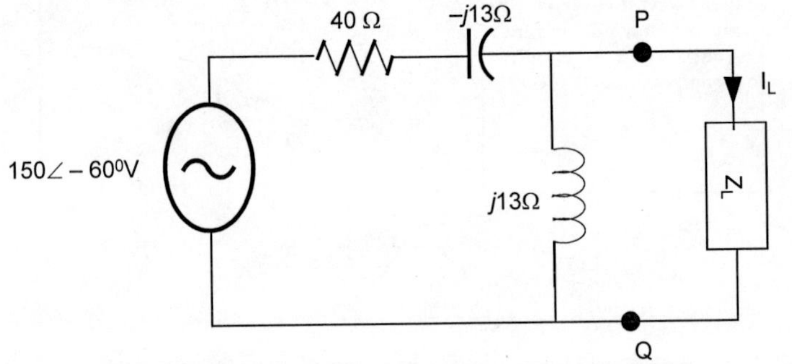

Fig. 2.37: Finding current in the load impedance using Thevenin's theorem for Example 2.11

Step1: Finding the equivalent impedance or Thevenin's impedance (Z_{th}) of the circuit – the load impedance is removed. The voltage sources are short circuited and the current sources are open circuited. After implementing these points, the circuit for finding the impedance is drawn and shown in Fig.2.38. In the given circuit, the 40Ω and (–j30Ω) are in series. Hence (40 – j30) Ω is obtained. That impedance value is in parallel with (j20Ω), hence $Z_{th} = (40 - j30)||(j20)$.

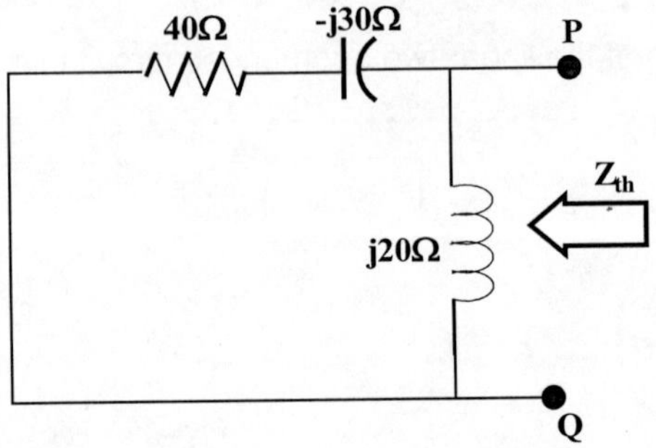

Fig. 2.38: Finding Thevenin's impedance for Example 2.11

Step2: Finding the equivalent voltage or Thevenin's voltage (Z_{th}) of the circuit – the load impedances are open circuited. With that concept, the circuit is drawn and shown in Fig.2.39. The Thevenin's voltage is also known as open circuited voltage which means that the voltage across the open-ended load terminals.

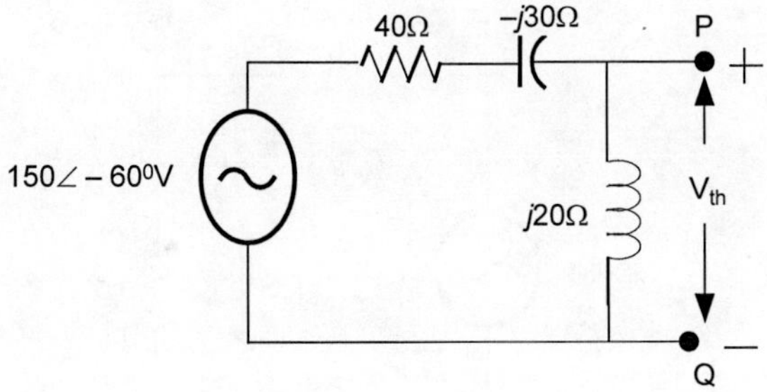

Fig. 2.39: Finding Thevenin's voltage for Example 2.11

By applying voltage division concept, the voltage across the inductor is calculated. The voltage across the inductor is equal to the potential of the open-ended load terminals which is the Thevenins's voltage.

$$V_{th} = \frac{150\angle -60^0 \times (j20)}{40 - j30 + j20} = \frac{150\angle -60^0 \times 20\angle 90^0}{40 - j10} = \frac{3000\angle 30^0}{41.231\angle -14.03^0}$$

$V_{th} = 72.76\angle 44.03^0$ V

Step3: Finding the load current of the circuit – the circuit is replaced with the Thevenin's equivalent circuit and the load impedance is connected in series with the Thevenin's equivalent circuit and the circuit diagram is shown Fig.2.40.

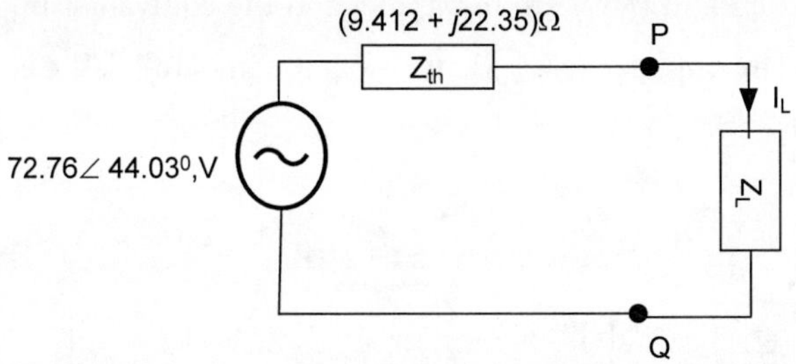

Fig. 2.40: Thevenin's equivalent circuit with load impedance for Example 2.11

Let us assume $Z_L = 10\Omega$; the load current is calculated as follows.

$$I_L = \frac{V_{th}}{Z_{th} + Z_L} = \frac{72.76\angle 44.03^0}{(9.412 + j22.35) + 10} = \frac{72.76\angle 44.03^0}{29.6\angle 49.02^0} = 2.45\angle -4.99^0\,A$$

If $Z_L = (10 + j10)\Omega$, the load current is

$$I_L = \frac{V_{th}}{Z_{th} + Z_L} = \frac{72.76\angle 44.03^0}{(9.412 + j22.35) + (10 + j10)} = \frac{72.76\angle 44.03^0}{19.412 + j32.35} = \frac{72.76\angle 44.03^0}{37.73\angle 59.03^0} = 1.93\angle -15^0\,A$$

Example 2.11

Find the current (I_L) flowing through the load impedance (Z_L) using Thevenin's theorem for the circuit shown in Fig.2.41.

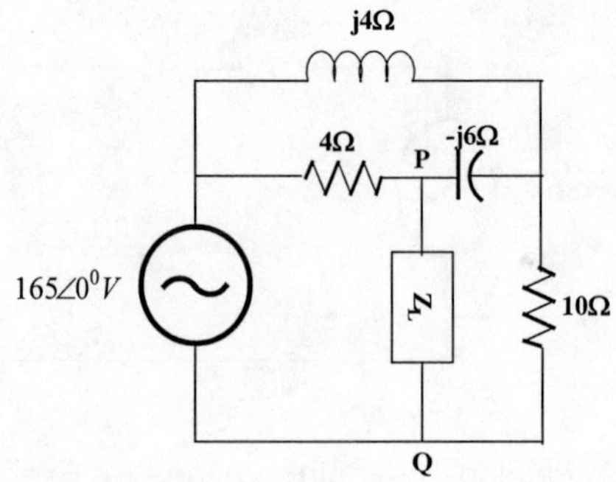

Fig. 2.41: Find the load current of the circuit using the Thevenin's theorem of Example 2.12

Solutions:

The circuit shown in Fig.2.49, find the Thevenin's impedance and Thevenin's voltage in the following steps.

Step1: Finding the equivalent impedance or Thevenin's impedance (Z_{th}) of the circuit – the load impedance is removed. The voltage sources are short circuited and the current sources are open circuited. After implementing these points, the circuit for finding the impedance is drawn and shown in Fig.2.42a. In the given circuit, the 10Ω and (*j*4Ω) are in parallel and the equivalent impedance (Z_{eq1}) is obtained that shown in Fig.2.42b. The impedance (Z_{eq1}) and –*j*6Ω are in parallel, the equivalent impedance (Z_{eq2}) is calculated and mentioned in Fig. 2.42c. Hence Z_{th} = (Z_{eq2}) || (4Ω), which is equal to Z_{th} = 1.956∠ – 36.23⁰ Ω.

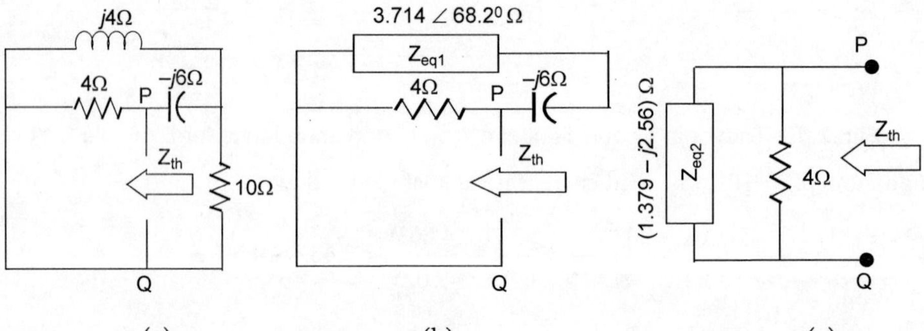

(a) (b) (c)

Fig. 2.42: The Thevenin's impedance of Example 2.12

Step2: Finding the equivalent voltage or Thevenin's voltage (V_{th}) of the circuit – the load impedances are open circuited. With that concept, the circuit is drawn and shown in Fig.2.43a. The Thevenin's voltage is also known as open circuited voltage which means that the voltage across the open-ended load terminals which is shown in Fig.2.43b.

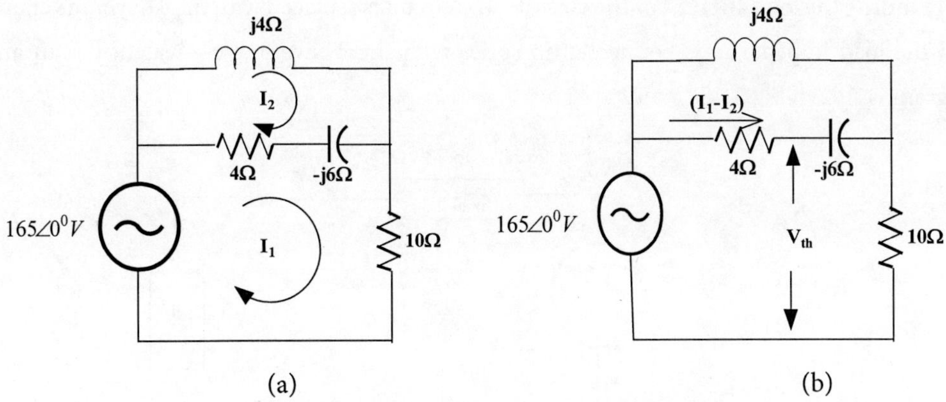

Fig. 2.43: Finding the Thevenin's voltage of Example 2.12

By applying mesh analysis, two equations are formed as follows.

$$+165\angle 0^0 - (4 - j6)(I_1 - I_2) - 10I_1 = 0$$

$$-(4 - j6)(I_2 - I_1) - j4I_2 = 0$$

After simplification and in matrix form,

$$\begin{bmatrix} 14 - j6 & -(4 - j6) \\ (4 - j6) & -(4 - j2) \end{bmatrix} \begin{bmatrix} I_1 \\ I_2 \end{bmatrix} = \begin{bmatrix} 165\angle 0^0 \\ 0 \end{bmatrix}$$

Apply Kramer's rule, to find the loop currents I_1 and I_2.

$$I_1 = (10.59 - j4.5)\text{A}$$

$$I_2 = (11.24 - j14.76)\text{A}$$

To find V_{th}, the circuit shown Fig.2.51b is used. The open-ended terminal voltage or the Thevenin's voltage is equal to either by subtracting voltage across 4Ω from the voltage source or the addition of voltage across (-j6)Ω and (10Ω). The first choice is chosen to solve this problem. Hence the equivalent diagram is shown in Fig.2.44.

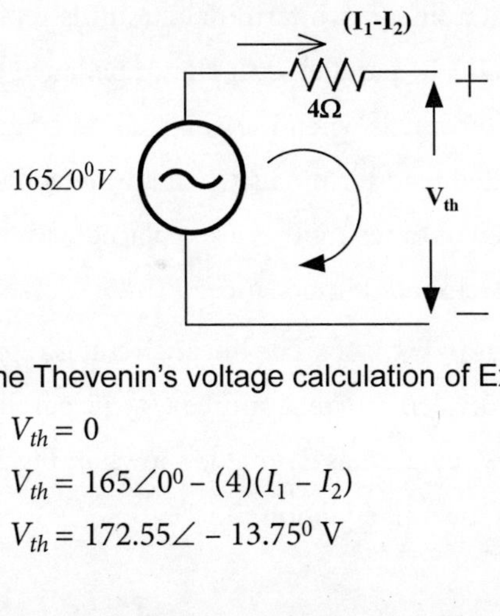

Fig. 2.44: The Thevenin's voltage calculation of Example 2.12

$$+165\angle 0^0 - (4)(I_1 - I_2) - V_{th} = 0$$

$$V_{th} = 165\angle 0^0 - (4)(I_1 - I_2)$$

$$V_{th} = 172.55\angle - 13.75^0 \text{ V}$$

Step3: Finding the load current of the circuit – the circuit is replaced with the Thevenin's equivalent circuit and the load impedance is connected in series with the Thevenin's equivalent circuit and the circuit diagram is shown Fig.2.45.

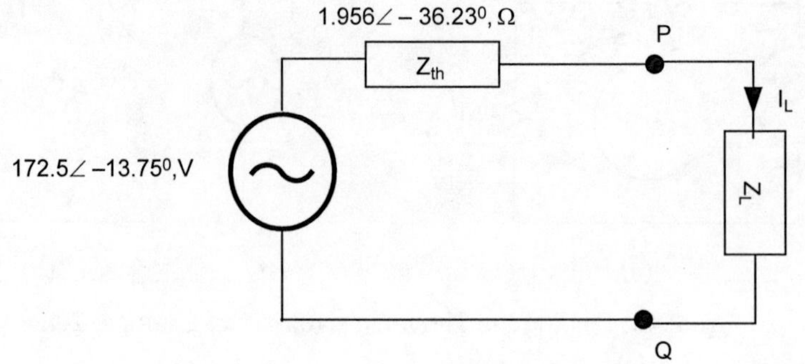

Fig. 2.45: Find the load current of the circuit using the Thevenin's theorem of Example 2.12

Let us assume $Z_L = 10\Omega$; the load current is calculated as follows.

$$I_L = \frac{V_{th}}{Z_{th}+Z_L} = \frac{172.55\angle -13.75^0}{(1.577 - j1.156)+10} = \frac{172.55\angle -13.75^0}{11.63\angle -5.7^0} = 14.83\angle -8.05^0\,A$$

If $Z_L = (10 + j10)\Omega$, the load current is

$$I_L = \frac{V_{th}}{Z_{th}+Z_L} = \frac{172.55\angle -13.75^0}{(1.577 - j1.156)+(10+j10)} = \frac{172.55\angle -13.75^0}{11.77 - j8.844} = \frac{172.55\angle -13.75^0}{14.56\angle -37.36^0}$$

$$= 11.85\angle 23.61^0\ A$$

2.7 NORTON'S THEOREM

The Norton's theorem states that a linear two terminal circuit is replaced by an equivalent circuit that consists of the current source (I_N) in parallel with the impedance (Z_N)Where I_N is the short circuit current of the open-ended load terminals when it is short circuited and Z_N is the equivalent impedance of the circuit from the open-ended load terminals when all the voltage sources are short circuited and current sources are open circuited or other sources are replaced with their internal resistances. The Z_N is the Norton's impedance or short-circuited impedance.

The statement is described as in Fig.2,46. The linear circuit is represented as the Norton's equivalent circuit which is equal to the equivalent current source (I_N) in parallel with the equivalent impedance (Z_N). The load impedance is now replaced as Z_L that is shown in Fig.2.47. The current through the load impedance is calculated that is shown in equation (2.52).

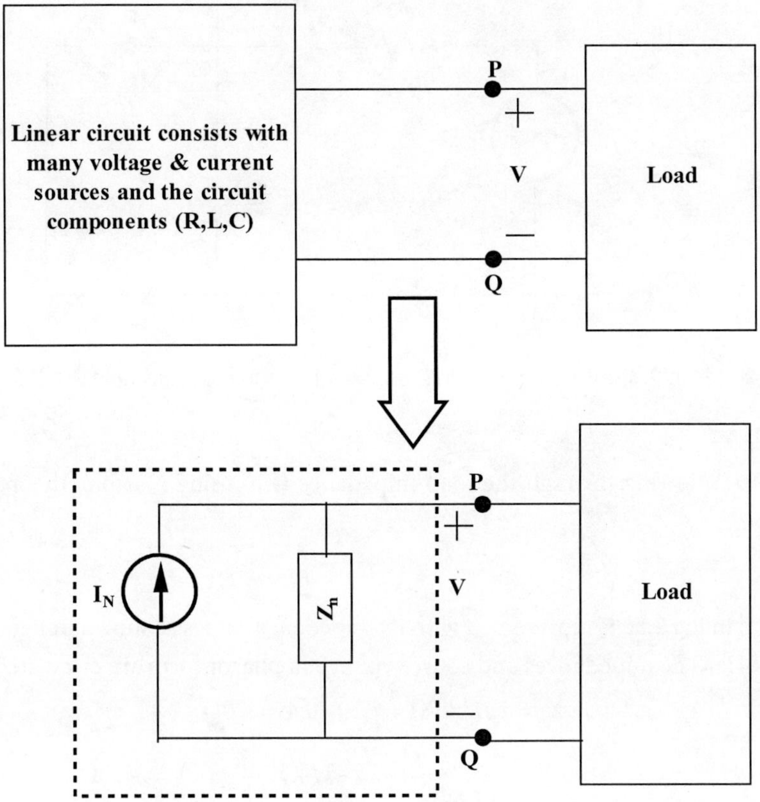

Norton's Equivalent Circuit

Fig. 2.46: The transformation of the linear two open-ended circuit into the Norton's equivalent circuit

$$I_L \quad I_N \frac{}{Z_N \quad Z_L} \tag{2.52}$$

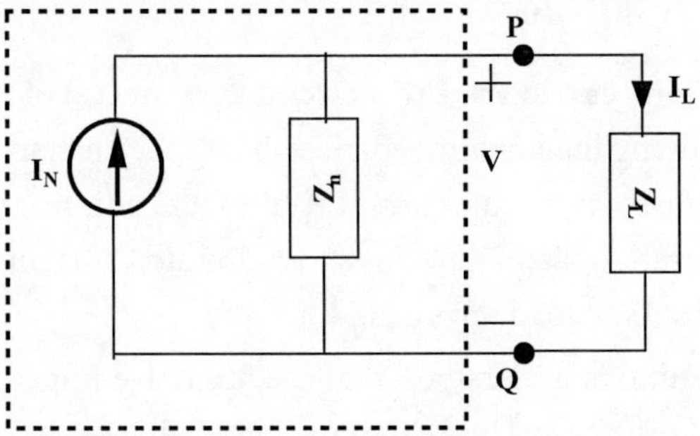

Norton's Equivalent Circuit

Fig. 2.47: Norton's equivalent circuit with load impedance to find the load current

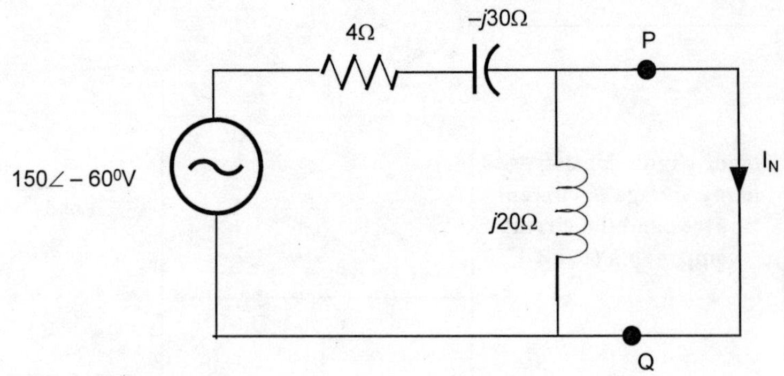

Fig. 2.48: Finding Norton's equivalent circuit for Example 2.13

Example 2.12

Find the current (I) flowing through the load impedance (Z_L) using Norton's theorem for the circuit shown in Fig.2.36.

Solutions:

The circuit shown in Fig.2.36 is represented with its impedance values as shown in Fig. 2.48. The angular frequency ω = 50Hz. The impedances and source voltage in phasor form are calculated as follows.

$$X_L = j\omega L,\ in\ \Omega = j(50)(0.4) = j20\Omega$$

$$X_C = \frac{1}{j\omega C}, in\ \Omega = 1/j(50)(0.667) = -j30\Omega V$$

$150 \sin(50t + 30^0) = 150\angle(-90^0 + 30^0) = 150\angle - 60^0$, V

Step1: Finding the equivalent impedance or Thevenin's impedance (Z_{th}) of the circuit – is explained (Ref. Example 2.11).

$$Z_{th} = \frac{(40 - j30)\times(j20)}{40 - j30 + j20} = \frac{600 + j800}{40 - j10} = 24.25\angle 67.16^0\,\Omega = (9.412 + j22.35)\Omega$$

Step2: Finding the Norton's current or short circuit current (I_N) of the circuit – the load terminals are short circuited after removing the load impedances. Then the current flowing in the short-circuited load terminals is known as the short circuit current. With that concept, the circuit is drawn and shown in Fig.2.56. The Norton's current is also known as short circuited current which means that the current flowing through the short-circuited load terminals.

According to the diagram, it is understood that the inductive impedance is negligible since it is in parallel to the short-circuited terminals. Hence the current flowing through the short-circuited terminals is calculated as follows.

$$I_N = \frac{150\angle -60^0}{40 - j30} = \frac{150\angle -60^0}{50\angle -36.86^0} = 3\angle -23.14^0, A$$

Step3: Finding the load current of the circuit – the circuit is replaced with the Norton's equivalent circuit and the load impedance is connected in parallel with the Thevenin's equivalent circuit and the circuit diagram is shown Fig.2.49.

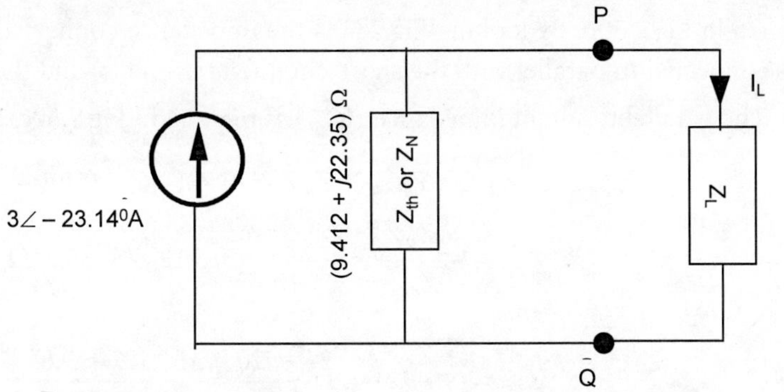

Fig. 2.49: Norton's equivalent circuit with load impedance for Example 2.13

Let us assume $Z_L = 10\Omega$; the load current is calculated by applying current division techniques as follows.

$$I_L = I_N \times \frac{Z_N}{Z_N + Z_L} = 3\angle -23.14^0 \times \frac{(9.412 + j22.35)}{(9.412 + j22.35) + 10}$$

$$= 3\angle -23.14^0 \times \frac{24.25\angle 67.163^0}{29.6\angle 49.02^0} = 2.457\angle -4.99^0$$

If $Z_L = (10 + j10)\Omega$, the load current is

$$I_L = I_N \times \frac{Z_N}{Z_N + Z_L} = 3\angle -23.14^0 \times \frac{(9.412 + j22.35)}{(9.412 + j22.35) + (10 + j10)}$$

$$= 3\angle -23.14^0 \times \frac{24.25\angle 67.163^0}{37.73\angle 59.03^0} = 1.928\angle -15^0\, A$$

Example 2.13

Find the current (I_L) flowing through the load impedance (Z_L) using Norton's theorem for the circuit shown in Fig.2.41.

Solutions:

The circuit shown in Fig.2.42, find the Norton's impedance and Norton current in the following steps.

Step1: Finding the equivalent impedance or Norton's impedance (Z_N) of the circuit – the calculation part is explained (Ref. Example 2.11). The Norton's impedance is $Z_N = 1.956\angle -36.23^0\ \Omega = (1.577 - j1.156)\Omega$.

Step2: Finding the short-circuited current or Norton's current (I_N) of the circuit – the load impedances is removed and it is short circuited. With that concept, the circuit is drawn and shown in Fig.2.50a. The impedances $Z_R = (j\,4\Omega)$, $Z_P = (-j6\Omega)$ and $Z_Q = (10\Omega)$ are connected in star connection. The equivalent delta connected impedances are calculated as follows and the incorporated delta connected impedances is shown in Fig.2.50b. By looking Fig.2.50b, the impedance connected between P and Q is invalid (since it is connected in parallel with the short-circuited terminals) and the impedance Z_{RP} and 4Ω are in parallel. The parallel resultant impedance ($Z_{RP}^{||}$) is marked in Fig.2.50c.

$$\Sigma Z_R Z_P = Z_R Z_P + Z_P Z_Q + Z_R Z_Q = ((j4)(-j6) + (10)(j4) + (10)(-j6) = 24 - j20$$

$$Z_{RP} = \frac{\sum Z_R Z_P}{Z_{Qt}} = \frac{24 - j20}{10} = 3.124\angle -39.8^0\Omega = (2.4 - j2)\Omega$$

$$Z_{PQ} = \frac{\sum Z_R Z_P}{Z_R} = \frac{24 - j20}{j4} = \frac{31.24\angle -39.8^0}{4\angle 90^0} = 7.81\angle -129.8^0$$

$$Z_{RQ} = \frac{\sum Z_R Z_P}{Z_P} = \frac{24 - j20}{-j6} = \frac{31.24\angle -39.8^0}{6\angle -90^0} = 5.2\angle 50.2^0\Omega = (3.33 + j4)\Omega$$

$$Z_{RP}^{||} = Z_{RP} \,||\, (4\Omega) = \frac{(2.4 - j2)\times 4}{(2.4 - j2) + 4} = \frac{9.6 - j8}{6.4 - j2} = \frac{12.5\angle -39.8^0}{6.7\angle -17.35^0}$$

$$= 1.865\angle -22.45^0\ \Omega = (1.723 - j0.712)\Omega$$

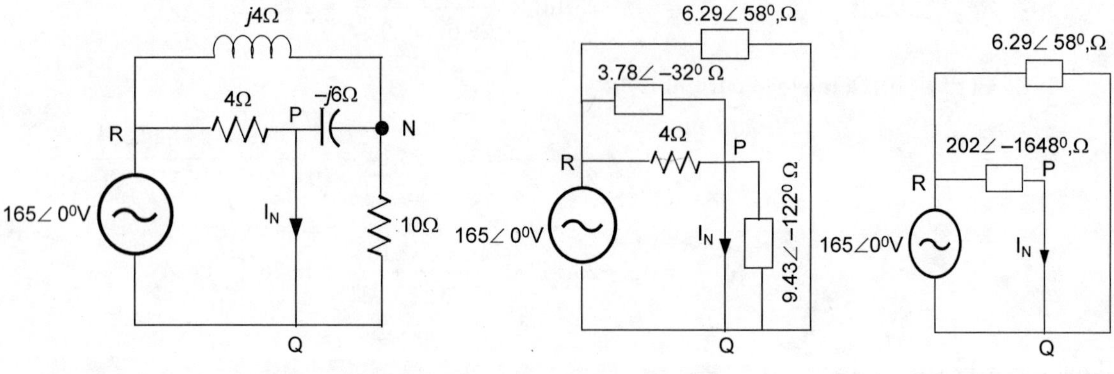

(a) (b) (c)

Fig. 2.50 Finding the Norton's current of Example 2.14

The Norton's current is calculated as follows.

$$I_N = \frac{165\angle 0^0}{1.865\angle -22.45^0} = 88.471\angle 22.45^0\,A$$

Step3: Finding the load current of the circuit – the circuit is replaced with the Norton's equivalent circuit and the load impedance is connected in parallel with the Thevenin's equivalent circuit and the circuit diagram is shown Fig.2.51.

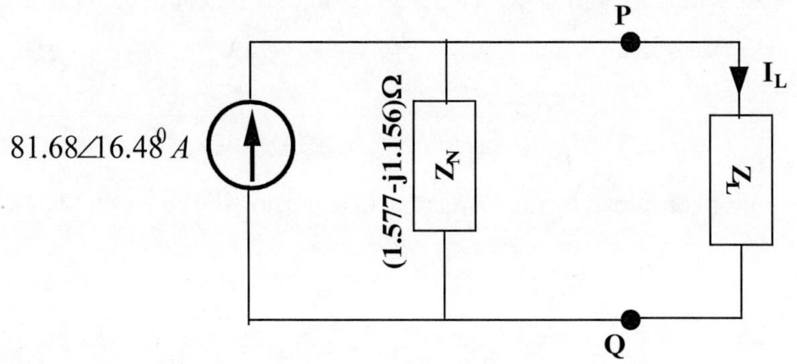

Fig. 2.51: Find the load current of the circuit using the Norton's theorem of Example 2.14

Let us assume $Z_L = 10\Omega$; the load current is calculated by applying current division techniques as follows.

$$I_L = I_N \times \frac{Z_N}{Z_N + Z_L} = 88.471\angle 22.45^0 \times \frac{(1.577 - j1.156)}{(1.577 - j1.156) + 10}$$

$$= 88.471\angle 22.45^0 \times \frac{1.956\angle -36.24^0}{11.63\angle -5.7^0} = 14.8\angle -8.096^0\, A$$

If $Z_L = (10 + j10)\Omega$, the load current is

$$I_L = I_N \times \frac{Z_N}{Z_N + Z_L} = 88.471\angle 22.45^0 \times \frac{(1.577 - j1.156)}{(1.577 - j1.156) + (10 + j10)}$$

$$= 88.471\angle 22.45^0 \times \frac{1.956\angle -36.24^0}{14.56\angle -37.36^0} = 11.88\angle 23.57^0\, A$$

2.8 MAXIMUM POWER TRANSFER THEOREM

The maximum power transfer theorem is applicable to both ac and dc circuits. The maximum power is transferred from circuits to the load when the load resistance is equal to the Thevenin's resistance of the circuit in a dc circuit. The power components in an ac circuit are the real power, the reactive power and apparent power. There is another term called "average power". Average power is defined as the integral of the instantaneous power over a time period. The maximum power transferred from ac circuits to the load is the average power only. Hence this theorem is also denoted as maximum average power transfer theorem for ac circuits.

The linear ac circuit is connected to the load impedance which is shown in Fig.2.52. The first step is to find the Thevenin's equivalent model of the linear ac circuit that has to be connected to the load impedance. The maximum power is transferred to the load impedance only when the load impedance is the conjugate of the Thevenin's impedance. The concept of the maximum power transfer theorem is shown in Fig. 2.53. In order to prove the theorem, the Thevenin's equivalent circuit is connected to the load impedance as shown in Fig.2.43 (Ref. Sec.2.6). The current flowing through the load impedance is expressed in equation (2.51) (Ref. Sec.2.6).

Where, $Z_{th} = R_{th} + jX_{th}$; $Z_L = R_L + jX_L$. Hence the load current could be represented as in equation (2.53).

$$I_L = \frac{V_{th}}{R_{th} + jX_{th} + R_L + jX_L} = \frac{V_{th}}{(R_{th} + R_L) + j(X_{th} + X_L)} \tag{2.53}$$

The average power consumed by the load is calculated using the equation (2.54).

$$P_L = \frac{1}{2}|I_L|^2 R_L \tag{2.54}$$

$$P_L = \frac{1}{2}\frac{V_{th}^2}{(R_{th} + R_L)^2 + (X_{th} + X_L)^2} R_L$$

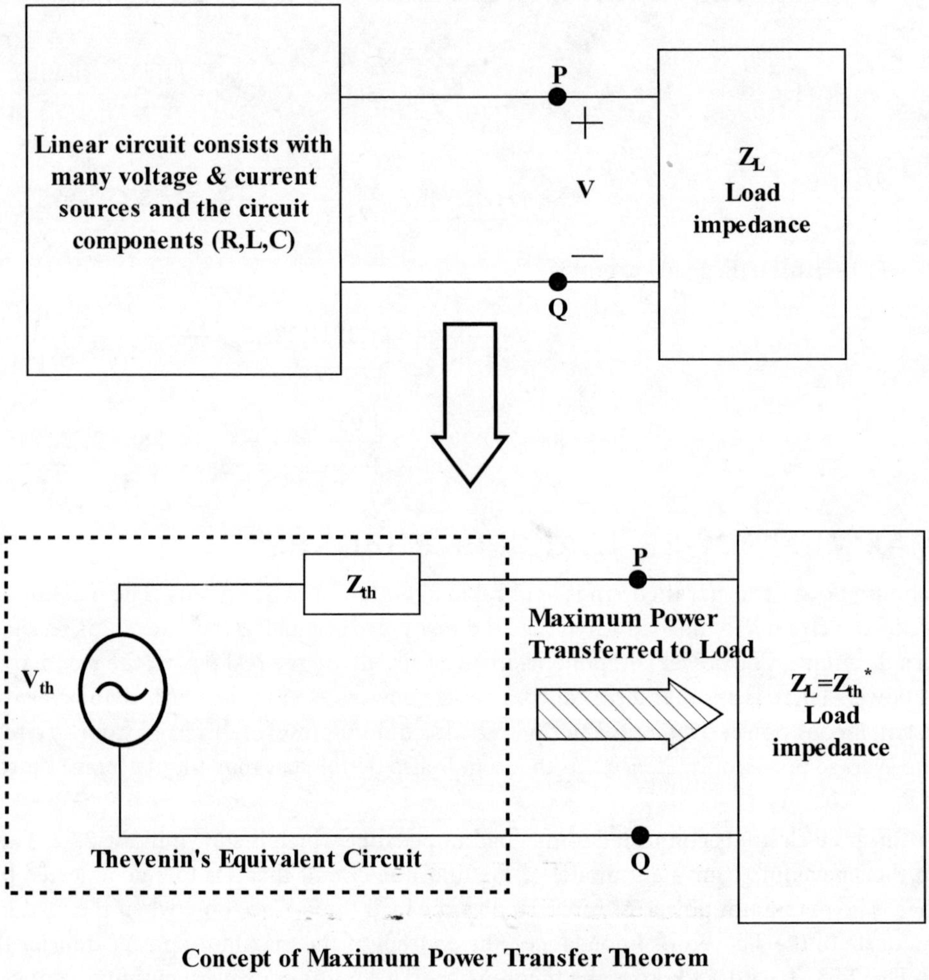

Fig. 2.52: The concept of the maximum power transfer theorem for ac circuits

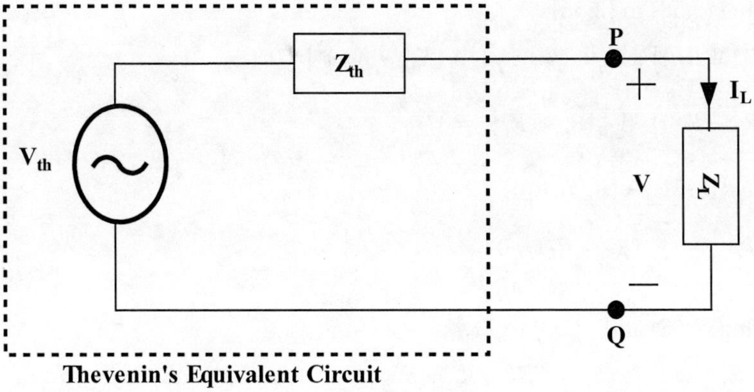

Fig. 2.53: Thevenin's equivalent circuit with load impedance to find the maximum power transferred to the load circuit

For finding the maximum power consumed by the load resistance, the first order derivative of load power w.r.t the load impedance is set to zero and the second order derivative should be less than zero. Hence the first order derivative w.r.t load resistance and w.r.t. load reactance is expressed in equations (2.55) and (2.57) respectively.

$$P_L = \frac{1}{2}\frac{V_{th}^2}{(R_{th}+R_L)^2+(X_{th}+X_L)^2}R_L = \frac{V_{th}^2 R_L}{2}[(R_{th}+R_L)^2+(X_{th}+X_L)^2]^{-1}$$

$$\frac{\partial P_L}{\partial R_L} = \frac{V_{th}^2 R_L}{2}(-1)\left[(R_{th}+R_L)^2+(X_{th}+X_L)^2\right]^{-2}\left(2\left(R_{th}+R_L\right)\right)+\frac{V_{th}^2}{2}\left[(R_{th}+R_L)^2+(X_{th}+X_L)^2\right]^{-1}=0 \tag{2.55}$$

$$\frac{\partial P_L}{\partial R_L} = \frac{1}{2}\frac{-V_{th}^2}{\left[(R_{th}+R_L)^2+(X_{th}+X_L)^2\right]^2}R_L\left(2\left(R_{th}+R_L\right)\right)+\frac{1}{2}\frac{V_{th}^2}{(R_{th}+R_L)^2+(X_{th}+X_L)^2}=0$$

$$\frac{\partial P_L}{\partial R_L} = \frac{1}{2}\frac{-V_{th}^2}{\left[(R_{th}+R_L)^2+(X_{th}+X_L)^2\right]^2}R_L\left(2\left(R_{th}+R_L\right)\right)+\frac{1}{2}\frac{V_{th}^2[(R_{th}+R_L)^2+(X_{th}+X_L)^2]}{\left[(R_{th}+R_L)^2+(X_{th}+X_L)^2\right]^2}=0$$

$$-V_{th}^2 R_L\left(2\left(R_{th}+R_L\right)\right)+V_{th}^2[(R_{th}+R_L)^2+(X_{th}+X_L)^2]=0 \tag{2.56}$$

$$\frac{\partial P_L}{\partial X_L} = \frac{V_{th}^2 R_L}{2}(-1)\left[(R_{th}+R_L)^2+(X_{th}+X_L)^2\right]^{-2}\left(2\left(X_{th}+X_L\right)\right)=0 \tag{2.57}$$

$$\frac{\partial P_L}{\partial X_L} = \frac{1}{2}\frac{-V_{th}^2}{\left[(R_{th}+R_L)^2+(X_{th}+X_L)^2\right]^2}R_L\left(2\left(X_{th}+X_L\right)\right)=0$$

$$X_{th} + X_L = 0$$

$$X_L = -X_{th} \tag{2.58}$$

By substituting (2.58) in (2.56),

$$-V_{th}^2 R_L\left(2\left(R_{th}+R_L\right)\right)+V_{th}^2[(R_{th}+R_L)^2+(X_{th}-X_{th})^2]=0$$

$$-V_{th}^2 R_L\left(2\left(R_{th}+R_L\right)\right)+V_{th}^2(R_{th}+R_L)^2=0$$

$$V_{th}^2\left(\left(R_{th}+R_L\right)\right)(-2R_L+(R_{th}+R_L))=0$$

$$(R_{th}-R_L)=0$$

$$R_L=R_{th} \tag{2.59}$$

By substituting (2.58) and (2.59) in Z_L,

$$Z_L=R_L+jX_L=R_{th}+j\left(-X_{th}\right)=R_{th}-jX_{th}=Z_{th}^*$$

$$Z_L=Z_{th}^*$$

The maximum power is transferred to the load impedance only when the load impedance is the conjugate of the Thevenin's impedance. In order to calculate the maximum power, the condition for the maximum power is substituted in equation (2.54).

$$P_{L_{\max}}=\frac{1}{2}\frac{V_{th}^2}{(R_{th}+R_{th})^2+(X_{th}-X_{th})^2}R_{th}$$

$$P_{L_{\max}}=\frac{1}{2}\frac{V_{th}^2}{(R_{th}+R_{th})^2}R_{th}=\frac{1}{2}\frac{V_{th}^2}{(2R_{th})^2}R_{th}=\frac{V_{th}^2}{8R_{th}}$$

*The equation (2.60) produces the maximum power consumed by the load. In this equation, the obtained V_{th} represents the maximum value of the voltage. In this book, all phasor is represented as maximum voltage only.

If the readers want to convert this as the rms value, $V_{\max}=\frac{V_{rms}}{\sqrt{2}}$. Hence the equation (2.63) based on the rms value could be represented as follows.

$$V_{th_{\max}}=\frac{V_{th_{rms}}}{\sqrt{2}}$$

$$P_{L_{\max}}=\frac{\left|V_{th_{\max}}\right|^2}{8R_{th}}=\left\{\frac{\left|V_{th_{\max}}\right|^2}{2}\right\}\left\{\frac{1}{4R_{th}}\right\}=\frac{\left|V_{th_{rms}}\right|^2}{4R_{th}}$$

$$P_{L_{\max}}=\frac{\left|V_{th}\right|^{2^*}}{8R_{th}} \tag{2.60}$$

From (2.56),

$$-V_{th}^2 R_L\left(2\left(R_{th}+R_L\right)\right)+V_{th}^2\left[(R_{th}+R_L)^2+(X_{th}+X_L)^2\right]=0$$

$$-R_L\left(2\left(R_{th}+R_L\right)\right)+[(R_{th}+R_L)^2+(X_{th}+X_L)^2]=0$$

$$-2R_LR_{th}-2R_L^2+(R_{th}+R_L)^2+(X_{th}+X_L)^2=0$$

$$-2R_LR_{th}-2R_L^2+R_{th}^2+R_L^2+2R_LR_{th}+(X_{th}+X_L)^2=0$$

$$R_{th}^2-R_L^2+(X_{th}+X_L)^2=0$$

$$R_L^2=R_{th}^2+(X_{th}+X_L)^2$$

$$R_L=\sqrt{R_{th}^2+(X_{th}+X_L)^2}$$

When the load is purely resistive load, then, ; Hence,

$$R_L=\sqrt{R_{th}^2+\left(X_{th}\right)^2}$$

$$R_L=\left|Z_{th}\right|$$

It is understood that the maximum power is transferred to the load when the load resistance is equal to the magnitude of the Thevenin's impedance for the resistive loads.

Example 2.14

Find the load impedance of the circuit shown in Fig.2.36 that absorbs the maximum power from the circuit. Also calculate the maximum average power consumed by the load.

Solutions:

The circuit shown in Fig 2.36 is represented with its impedance values as shown in Fig. 2.37. The angular frequency ω=50Hz. The impedances and source voltage in phasor form are calculated as follows.

$$X_L=j\omega L, in\ \Omega=\text{j}(50)(0.4)=\text{j}20\Omega$$

$$X_C=\frac{1}{j\omega C}, in\ \Omega=1/\text{j}(50)(0.667)=-\text{j}30\Omega$$

$$150\sin\left(50t+30^0\right)=150\angle\left(-90^0+30^0\right)=150\angle-60^0, V$$

Step1: **Finding the equivalent impedance or Thevenin's impedance (of the circuit**

$$Z_{th}=24.25\angle 67.16^0\Omega=(9.412+j22.35)\Omega$$

(Ref. Sec 4.6 – Example 2.10)

Step2: Finding the Thevenin's voltage of the circuit

$$V_{th}=72.76\angle 44.03^0\,\text{V}$$

(Ref. Sec 4.6 – Example 2.10)

Step3: Obtaining the Thevenin's equivalent circuit with Thevenin's voltage and Thevenin's impedance

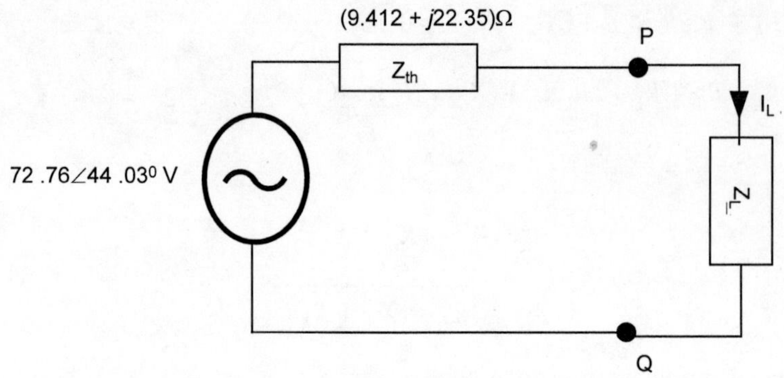

Fig. 2.54: Thevenin's equivalent circuit with load impedance of Example 2.10 (Repetition of Fig. 2.40)

(Ref. Sec 4.6 – Example 2.10)

Step 4: Calculate the load impedance

According to the maximum power transfer theorem, the load impedance is the conjugate of the Thevenin's impedance.

$$Z_L = Z_{th}^* = 9.412 - j22.35$$

Step5: Calculate the maximum power to the load impedance

According to the maximum power transfer theorem,

$$P_{\max} = \frac{|V_{th}|^2}{8R_{th}} = \frac{72.76^2}{8 \times 9.412} = 70.31W$$

Example 2.15:

Find the load impedance of the circuit shown in Fig.2.41 that absorbs the maximum power from the circuit. Also calculate the maximum average power consumed by the load.

Solutions:

The circuit shown in Fig.2.41, find the Thevenin's impedance and Thevenin's voltage in the following steps.

Step1: Finding the equivalent impedance or Thevenin's impedance (Z_{th}) of the circuit

$$Z_{th} = 1.956\angle -36.23^0\,\Omega = 1.577 - j1.156\Omega$$

(Ref. Sec 4.6 – Example 2.11)

Step2: Finding the Thevenin's voltage of the circuit

$$V_{th} = 172.55\angle -13.75^0\,V$$

(Ref. Sec 4.6 – Example 2.11)

Step3: Obtaining the Thevenin's equivalent circuit with Thevenin's voltage and Thevenin's impedance

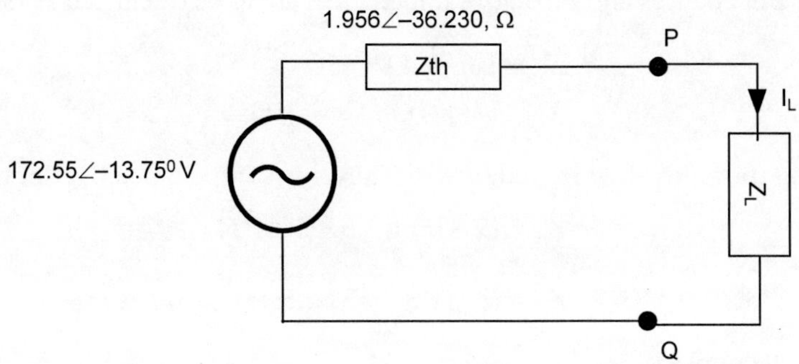

Fig. 2.55: Thevenin's equivalent circuit with load impedance for Example 2.11 (Repetition of Fig. 2.53)

(Ref. Sec 4.6 – Example 2.11)

Step4: Calculate the load impedance

According to the maximum power transfer theorem, the load impedance is the conjugate of the Thevenin's impedance.

$$Z_L = Z_{th}^* = (1.577 + j1.157)\Omega$$

Step5: Calculate the maximum power to the load impedance

According to the maximum power transfer theorem,

$$P_{max} = \frac{|V_{th}|^2}{8R_{th}} = \frac{172.55^2}{8 \times 1.577} = 2359.9W = 2.4kW$$

Example 2.16:

Find the load resistance of the circuit shown in Fig. 2.56 using maximum power transfer theorem. Also calculate the maximum power absorbed by the load.

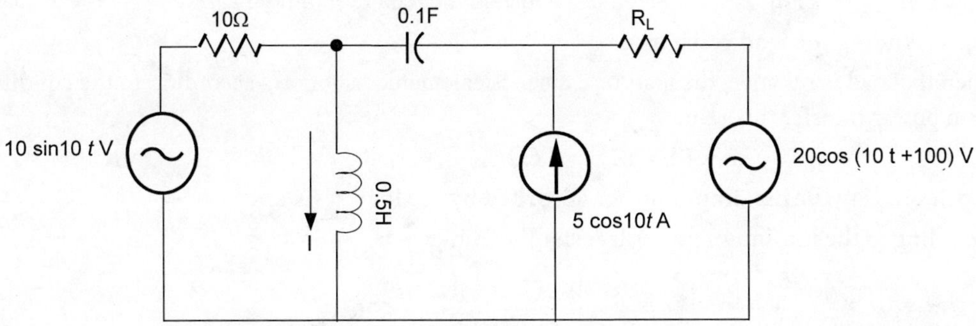

Fig. 2.56: Find the load resistance and maximum power absorbed by the load of Example 2.18 (Similar to Example 2.15)

Solutions:

The circuit shown in Fig.2.56, find the Thevenin's impedance and Thevenin's voltage.

$$10\sin 10t = 10\angle -90^0, V; 20\cos\cos\left(10t + 10^0\right) = 20\angle 10^0, V; 5\cos\cos\left(10t\right) = 5\angle 0^0, A;$$

The inductive impedances and capacitive impedances are to be calculated as follows.

$$X_L = j\omega L, in\ \Omega = \text{j}5\Omega$$
$$X_C = 1/(\omega C), in\ \Omega = -\text{j}1\Omega$$

Step1: Finding the equivalent impedance or Thevenin's impedance (Z_{th}) of the circuit

$$Z_{th} = (2 + \text{j}3) = 3.6\angle 56.31^0\Omega$$

$$Z_{th} = |Z_{th}|\angle\theta_{th} = 3.6\angle 56.31^0\Omega$$

(Ref. Sec 4.7 – Example 2.15)

Step2: Finding the equivalent voltage or Thevenin's voltage (V_{th}) of the circuit

$$V_{th} = -5.71 + j9.49 = 11.07\angle 121.03^0 V$$

(Ref. Sec 4.7 – Example 2.15)

Step3: Finding the Thevenin's Equivalent circuit

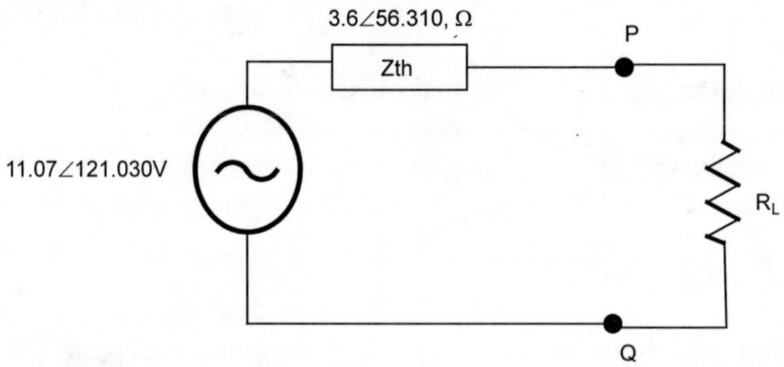

Fig. 2.57: Thevenin's' equivalent circuit of Example 2.18

Step 4: Finding the load resistance

When the load is resistive, the load resistance is calculated as follows according to the condition of maximum power transfer theorem.

$$\text{RL} = |Z_{th}| = 3.6\Omega$$

Step 5: Finding the maximum power absorbed by the load resistance

According to the maximum power transfer theorem,

$$P_{\max} = \frac{|V_{th}|^2}{8R_{th}} = \frac{11.07^2}{8 \times 3.6} = 4.255W$$

2.9 RECIPROCITY THEOREM

The reciprocity theorem states that the current flowing through any branch (first branch) of the circuit due to a voltage source or a current source is equal to the current flowing through the branch (source branch) when a voltage source or a current source positioned at the first branch.

This statement is explained in the following section with the example circuit shown in Fig.2.58. The current flowing through the branch of resistance R_2 is calculated by either mesh or node analysis.

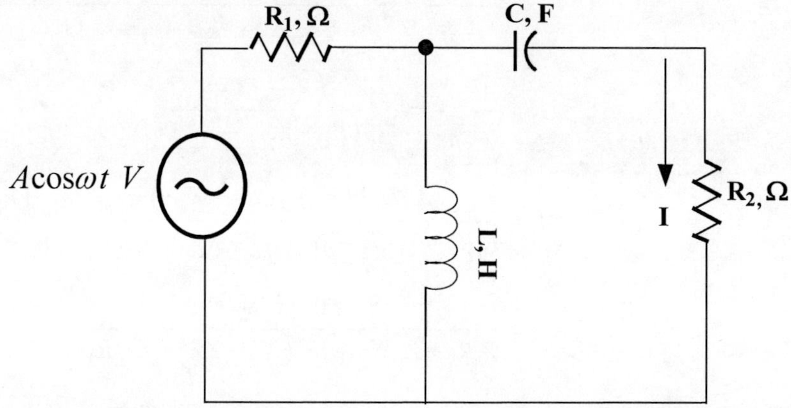

Fig. 2.58: Find current I using reciprocity theorem

First step is to convert the source into phasor form $A\angle 0^0$,V and the impedances are found in Ω. Hence it is found asjX_L Ω, – jX_C Ω. By applying these values, the circuit is shown in Fig.2.74.

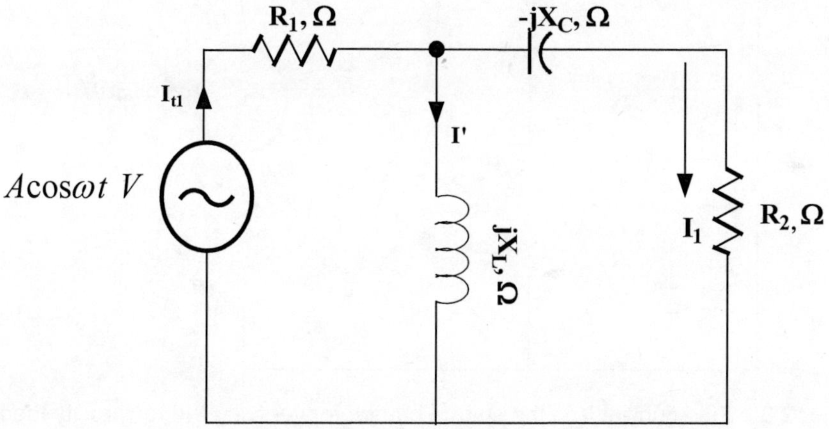

Fig. 2.59: Find current I_1 for verifying the reciprocity theorem

Total impedance of the circuit in Fig. 2.58 is found in equation (2.61). R_2 and $(-jX_C)$ are in series, it is $(R_2 - jX_C)$. This impedance is in parallel with (jX_L) and then in series with R_1.

$$Z_{eq1} = R_1 + \frac{jX_L R_2 + X_C X_L}{(R_2 + j(X_L - X_C))}$$

$$Z_{eq1} = \frac{R_1(R_2 + j(X_L - X_C)) + jX_L R_2 + X_C X_L}{(R_2 + j(X_L - X_C))}$$

$$Z_{eq1} = \frac{R_1 R_2 + X_C X_L + jR_1(X_L - X_C) + jX_L R_2}{(R_2 + j(X_L - X_C))} \tag{2.61}$$

The total current in the

$$I_{t1} = \frac{A\angle 0^0}{Z_{eq1}} = \frac{A\angle 0^0 * \left(R_2 + j\left(X_L - X_C\right)\right)}{R_1R_2 + X_CX_L + jR_1\left(X_L - X_C\right) + jX_LR_2}$$

$$I_1 = \frac{A\angle 0^0}{Z_{eq1}} = I_{t1} * \frac{\left(R_2 - jX_C\right)}{\left(R_2 + j(X_L - X_C\right)}$$

$$I_1 = \frac{A\angle 0^0 * \left(R_2 + j\left(X_L - X_C\right)\right)}{R_1R_2 + X_CX_L + jR_1\left(X_L - X_C\right) + jX_LR_2} * \frac{\left(R_2 - jX_C\right)}{\left(R_2 + j(X_L - X_C\right)}$$

$$I_1 = \frac{A\angle 0^0 * \left(jX_L\right)}{R_1R_2 + X_CX_L + jR_1\left(X_L - X_C\right) + jX_LR_2} \tag{2.62}$$

Second step is the voltage source is placed to location where the current I_1 is found and the existing source branch is short circuited and the current flowing through that short-circuited branch is calculated and corresponding circuit is shown in Fig.2.60.

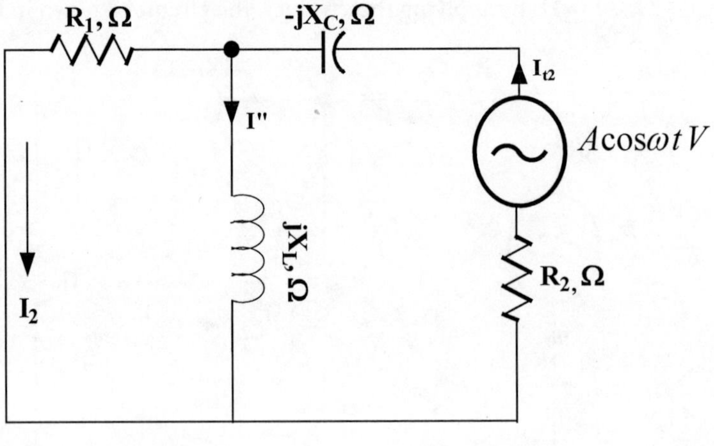

Fig. 2.60: Find current I_2 in the source branch for verifying the reciprocity theorem

Total impedance of the circuit in Fig. 2.75 is found in equation (2.63). R_1 and (jX_L) are in parallel, it is in series with (R2 – jX_C).

$$Z_{eq2} = (R_2 - jX_C) + (R_1 \,||\, (jX_L))$$

$$Z_{eq2} = \frac{R_1R_2 + X_CX_L + jR_1\left(X_L - X_C\right)) + jX_LR_2}{(R_1 + jX_L)} \tag{2.63}$$

The total current in the

$$I_{t2} = \frac{A\angle 0^0}{Z_{eq2}} = \frac{A\angle 0^0 * \left(R_1 - jX_C\right)}{R_1R_2 + X_CX_L + jR_1\left(X_L - X_C\right) + jX_LR_2}$$

$$I_2 = \frac{A\angle 0^0}{Z_{eq2}} = I_{t2} * \frac{(jX_L)}{(R_1 + jX_L)}$$

$$I_2 = \frac{A\angle 0^0 * (R_1 - jX_C)}{R_1R_2 + X_CX_L + jR_1(X_L - X_C) + jX_LR_2} * \frac{(jX_L)}{(R_1 + jX_L)}$$

$$I_2 = \frac{A\angle 0^0 * (jX_L)}{R_1R_2 + X_CX_L + jR_1(X_L - X_C) + jX_LR_2} \tag{2.64}$$

Hence the current obtained from the circuit shown in Fig. 2.74 and Fig.2.75 is shown in equation (2.62) and (2.64) respectively. It is understood that both currents are equal. The reciprocity theorem is proved.

Example 2.17

Prove the reciprocity theorem and find the current in the circuit shown in Fig. 2.61.

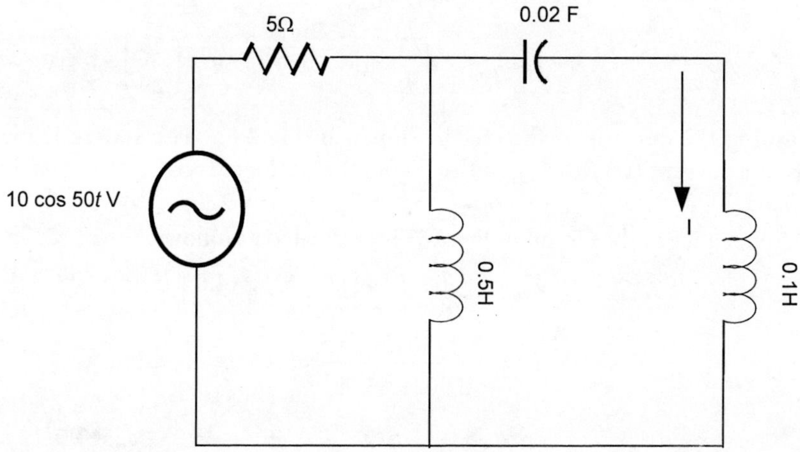

Fig. 2.61: Find the current I and prove the reciprocity theorem of Example 2.19

Solutions:

The circuit shown in Fig.2.76,

The inductive impedances and capacitive impedances are to be calculated as follows. By incorporating these values and the circuit is shown in Fig.2.62.

$$X_{L1} = j\omega L_1, \text{ in } \Omega = j25\Omega$$

$$X_{L2} = j\omega L_2, \text{ in } \Omega = j5\Omega$$

$$X_C = 1/(j\omega C), \text{ in } \Omega = -j1\Omega$$

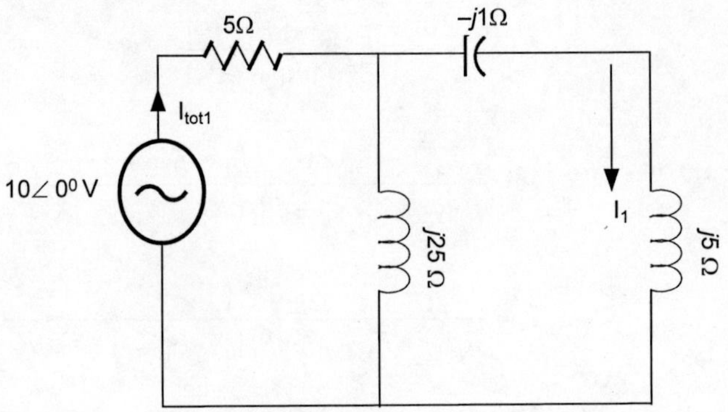

Fig. 2.62: Find the current I_1

Step1: Finding the current of the circuit shown in Fig.2.62.

The total impedance of the circuit in Fig.2.62 is calculated as follows.

$$Z_{tot1} = 5 + \left(j25 \,||\, (j5 - j1)\right) \cong j3.448\Omega = 3.448\angle 90^0\,\Omega.$$

$$I_{tot1} = \frac{10\angle 0^0}{3.448\angle 90^0} = 1.647\angle -34.59^0\,A$$

$$I_1 = I_{tot1} * \frac{j25}{j29} = 1.647\angle -34.59^0 * \left(\frac{25}{29}\right) = 1.419\angle -34.59^0\,A$$

Step2: Finding the current of the circuit shown in Fig.2.63. The source is placed to the first branch (where the current (I_1) flowing in the Fig2.62) and the current is found at the source branch shown in Fig.2.63.

The total impedance of the circuit in Fig.2.63 is calculated as follows.

$$Z_{tot2} = (j5 - j1) + \left(j25 \,||\, 5\right) \cong (4.8 + j4.96)\Omega = 6.902\angle 45.93^0\,\Omega$$

$$I_{tot2} = \frac{10\angle 0^0}{6.902\angle 45.93^0} = 1.45\angle -45.93^0\,A$$

$$I_2 = I_{tot2} * \frac{j25}{(5 + j25)} = \frac{1.45\angle -45.93^0 \times 25\angle 90^0}{25.49\angle 78.69^0} = 1.422\angle -34.62^0\,A$$

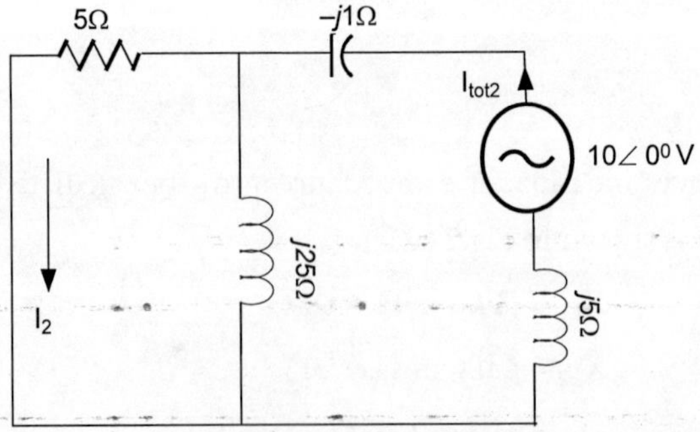

Fig. 2.63: Find the current I_2

Step3: Prove of the reciprocity theorem

The current found in Fig.2.62 and 2.63 is remain the same. Hence the theorem is proved.

2.10 DUALITY IN ELECTRIC CIRCUITS

Duality concepts present in electric circuit for the certain parameters and theorems which helps to solve problems faster. Hence it is a time saving and effective tool for solving an electric circuit. Some of the dual pairs in an electric circuit is listed as below. During the circuit analysis, two different circuits have similar equations expect the certain duality elements are interchanged. This characteristic of interchangeability is known as duality principle. In Table 2.1, the parameter power is not having the duality parameter. Since power is a non-linear parameter. It is understood that duality concept is applicable only to the linear elements.

The process of formulating the duality circuit is explained as follows.

- Find the number of loops in the circuit. Mark the nodes inside that closed loop and number it. These nodes are named as non-reference nodes.
- Also mark the ground node outside the circuit.
- Draw a line between nodes that has to pass through an element.
- For the transformation of voltage and current sources, the polarity of voltage sources and the direction of current sources are found by following method.
    - ➢ If the mesh current in the voltage source is in clockwise direction, then the duality has the current source with the current direction from ground node to non-reference nodes.

Table 2.1: Dual Pairs in an electric circuit

Resistance	Conductance
Inductance	Capacitance
Voltage	Current
Voltage Source	Current Source
Mesh	Node
Series circuit	Parallel circuit
Open circuit	Short circuit
KVL	KCL
Thevenin	Norton
Voltage division technique	Current division technique

Example 2.18:

Apply the duality concept and obtain the dual circuit of the circuit shown in Fig.2.64

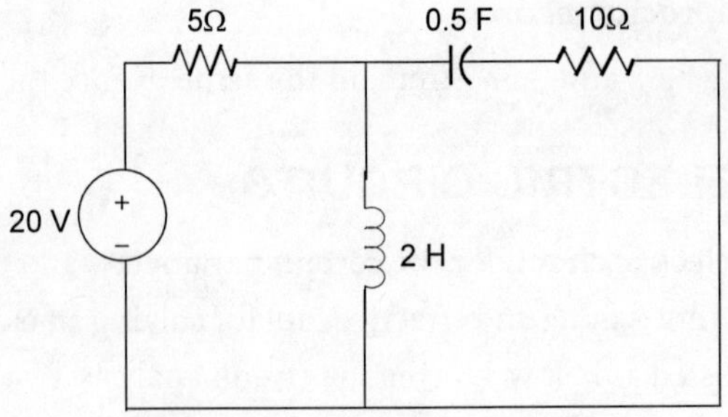

Fig. 2.64: The circuit taken for explaining the concept of duality

Solutions:

The circuit shown in Fig.2.64, the dual circuit is obtained with the following steps.

Step1: Mark the node and the concern circuit is in Fig.2.65.

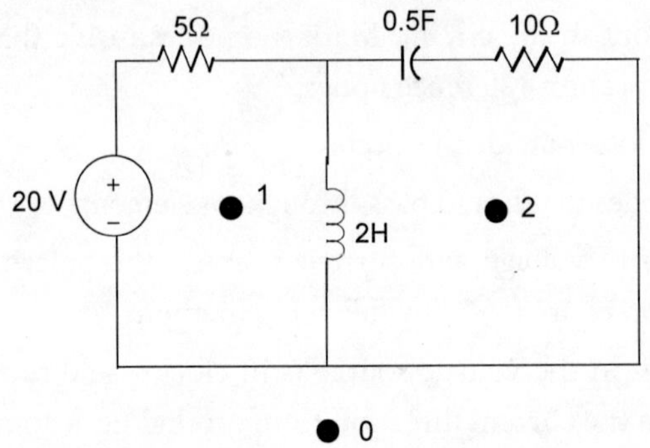

Fig. 2.64: The circuit with marked nodes for explaining the concept of duality

Step2: Draw a line between nodes that has to pass through an element and the circuit is shown in Fig.2.65. Totally 5 connecting lines are passing through an element in the circuit.

Voltage source into current source, inductance into capacitance, capacitance into inductance, resistance R into resistance (no change in the image, but value is 1/R, Ω - Series become parallel). Hence the duality circuit is obtained and the circuit drawn is shown in Fig.2.66.

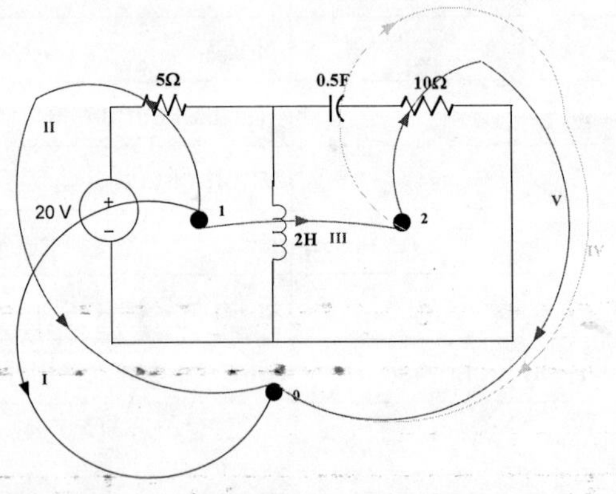

Fig. 2.65: The connected lines among the nodes for explaining the concept of duality

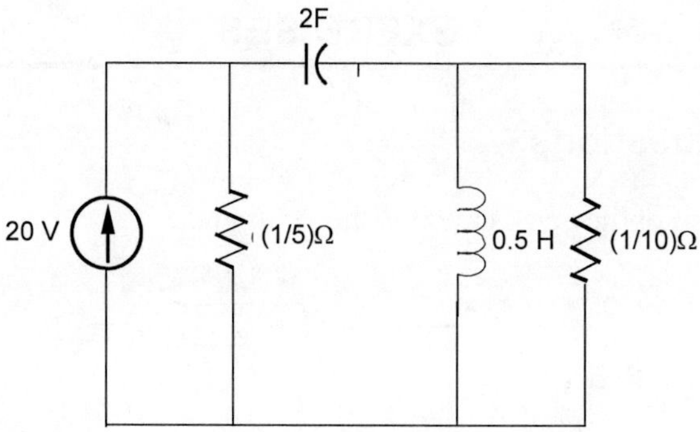

Fig. 2.66: The dual circuit of the circuit shown in Fig.2.64 (with the concept of duality)

UNIT SUMMARY

1. The source transformation is helpful to simplify the circuit analysis by transforming voltage source into current source and vice versa.
2. Both the star to delta transformation and the delta to star transformation are discussed and the examples are proved with transformations.
3. The mesh/node analysis is formed based on the Kirchoff's voltage/current law respectively. The formed voltage/current equations are solved by the Cramer's rule to find the loop currents/node potentials respectively. The detailed steps for both analyses are discussed.
4. The superposition theorem is discussed for the circuit that consists of the sources having single frequency. The simple problems are solved to understand the concepts.
5. The Thevenin's and Norton's theorem are applicable to both ac and dc circuits. In this chapter, the theorem is explained for ac circuits. The Thevenin's equivalent circuit is the Thevenin's voltage source in series with the Thevenin's impedance. The Thevenin's voltage is the open circuit voltage observed at the load terminals and the Thevenin's impedance is the equivalent impedance of the original circuit which is looked from the open-ended load terminals when the voltage sources are short circuited and the current sources are open circuited.
6. The Norton's equivalent circuit is the Norton's current source in parallel with the Norton's impedance. The Norton's current is the short circuit current that flows through the short circuited at the load terminals. Norton's impedance is the impedance of the original circuit which is looked from the open-ended load terminals when the voltage sources are short circuited and the current sources are open circuited.
7. The maximum power is transferred from the circuit to the load when the load impedance is the conjugate of the Thevenin's impedance. The maximum power consumed by the load is equal to the square of the Thevenin's voltage divide by the 8 times of the Thevenin's resistance.
8. For the resistive load, the load resistance is equal to the magnitude of the Thevenin's impedance.
9. The reciprocity theorem is applicable only when the circuit consists of only one source.
10. The reciprocity theorem states that the current flowing through any branch (first branch) of the circuit due to a voltage source or a current source is equal to the current flowing through the branch (source branch) when a voltage source or a current source is positioned at the first branch.
11. Duality concept will help to solve the problem very quickly.

EXERCISES

Multiple Choice Questions

1. The equivalent circuit of the given circuit is shown in Fig. 2.67

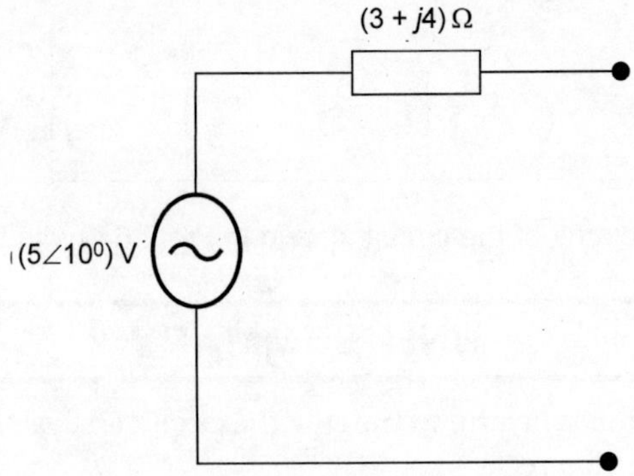

Fig. 2.67: Source transformation of question 1

(a) Current source of $1\angle -43.13^0\,A$ is in parallel with (3+j4) Ω

(b) Current source of $1\angle 43.13^0\,A$ is in parallel with (3+j4) Ω

(c) Current source of $1\angle -43.13^0\,A$ is in parallel with $(5\angle -53.13)\Omega$

(d) Current source of $1\angle 43.13^0\,A$ is in parallel with $(5\angle -53.13)\Omega$

2. The equivalent circuit of the given circuit is shown in Fig. 2.68

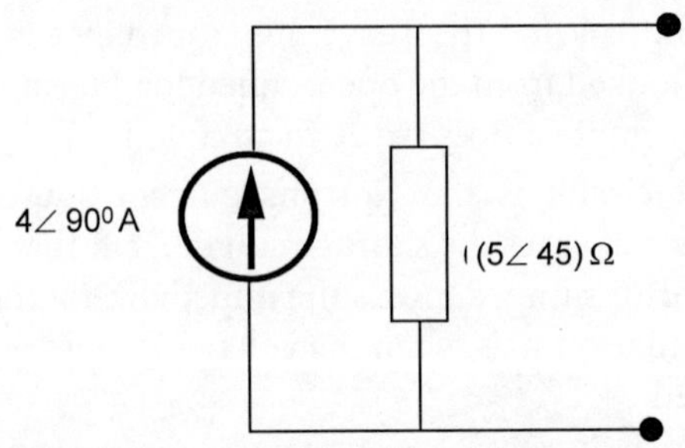

Fig. 2.68: Source transformation of question 2

(a) Voltage source of $20\angle 135^0\,V$ is in series with (3.5+j4) Ω

(b) Voltage source of $20\angle 135^0\,V$ is in series with (3.5+j4) Ω

(c) Voltage source of $20\angle -135^0\,V$ is in series with $5\angle 45^0\,\Omega$

(d) Voltage source of $20\angle 135^0\,V$ is in series with $5\angle 45^0\,\Omega$

3. Three impedances Z_{RY} = (20+j30) Ω, Z_{YB} = (10+j10) Ω, Z_{BR} = (15+j10) Ω are delta-connected impedances which is shown in Fig.2.69. Find Z_R, Z_Y and Z_B.

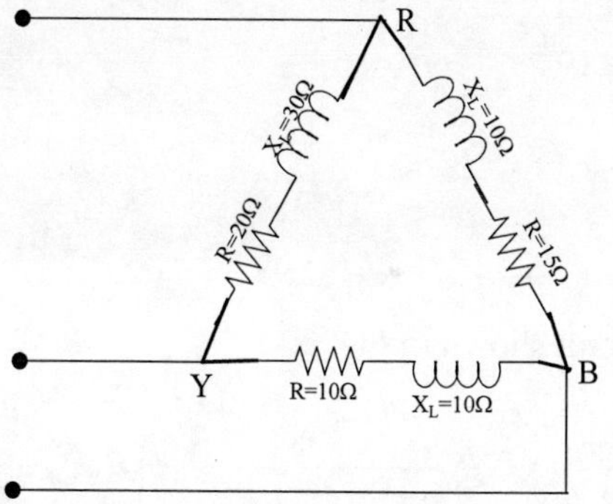

Fig. 2.69: Source transformation of question 3

(a) ZR = 9.67∠41.99 Ω; Z_Y = 7.57∠53.29 Ω; Z_B = 3.71∠41.99 Ω
(b) Z_R = 19.67∠90 Ω; Z_Y = 17.57∠3.29 Ω; Z_B = 3.71∠4.199 Ω
(c) Z_R = 9.67∠– 41.99 Ω; Z_Y = 7.57∠–53.29 Ω; Z_B = 3.71∠–41.99 Ω
(d) None of the above

4. Find the current flowing through the circuit shown in Fig.2.70.

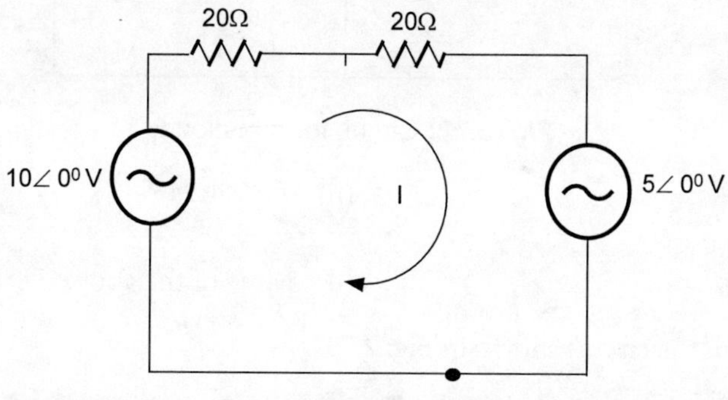

Fig. 2.70: Source transformation of question 4

(a) 125mA
(b) 25mA
(c) 125∠5° mA
(d) None of the above

5. Find the voltage and current in the circuit shown in Fig. 2.71.

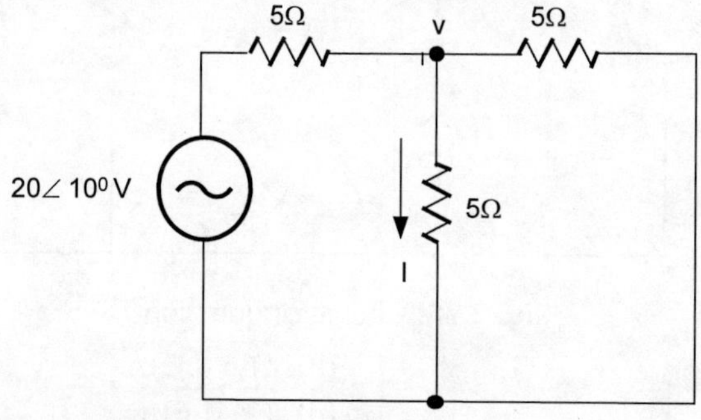

Fig. 2.71: Circuit for question 5

(a) $V = \frac{20}{3}\angle -30^0 V; I = \frac{4}{3}\angle -30^0$

(b) $V = \frac{20}{3}\angle 10^0 V; I = \frac{4}{3}\angle 10^0$

(c) $V = \frac{20}{3}\angle 0^0 V; I = \frac{4}{3}\angle 0^0$

(d) None of the above

6. Find the current in the circuit shown in Fig. 2.72.

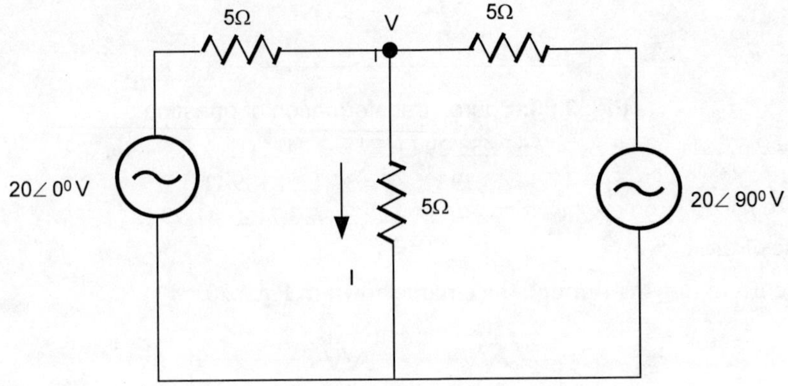

Fig. 2.72: Circuit for question 6

(a) $I = \frac{4}{3}\angle -45^0$

(b) $I = 9.4\angle 45^0$

(c) $I = 4.9\angle -45^0$

(d) None of the above

7. Find the current in the circuit shown in Fig. 2.73.

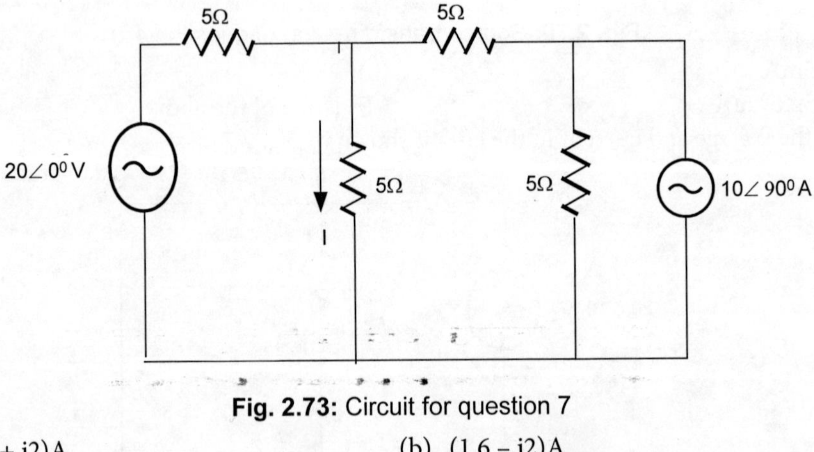

Fig. 2.73: Circuit for question 7

(a) (1.6 + j2)A

(b) (1.6 – j2)A

(c) (2 – j1.6) A

(d) (2+j1.6)A

8. Find the load current in the circuit shown in Fig. 2.74 using Thevenin's theorem.

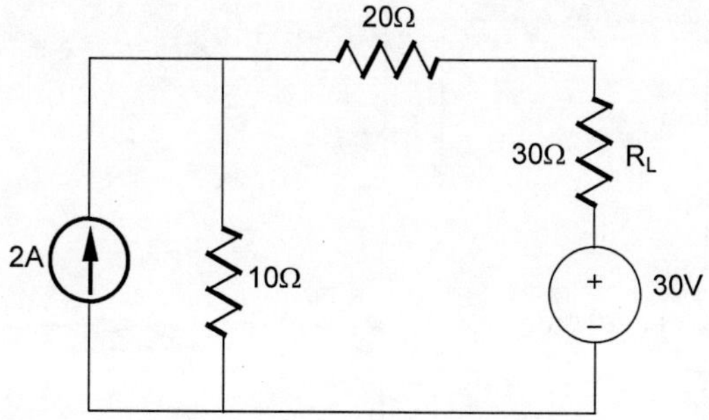

Fig. 2.74: Circuit for question 8, 9

(a) (0.166)A (b) (−0.166)A
(c) (−1.166)A (d) (1.166)A

9. Find the Thevenin's equivalent circuit of the circuit shown in Fig. 2.74.

(a)

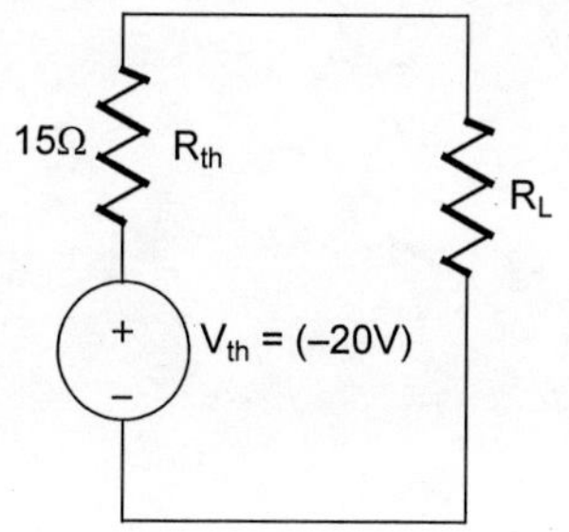

(b)

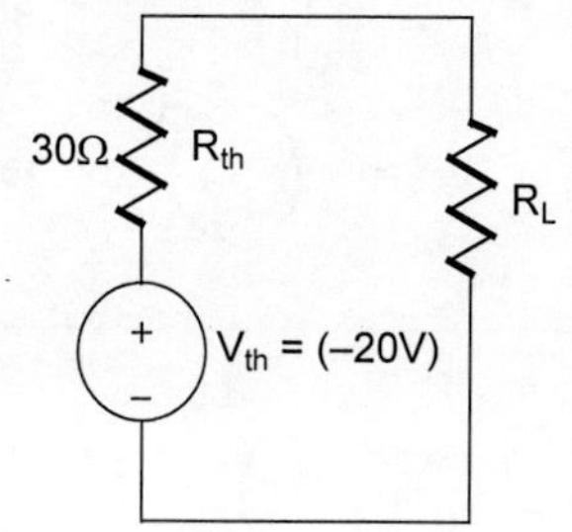

(c)

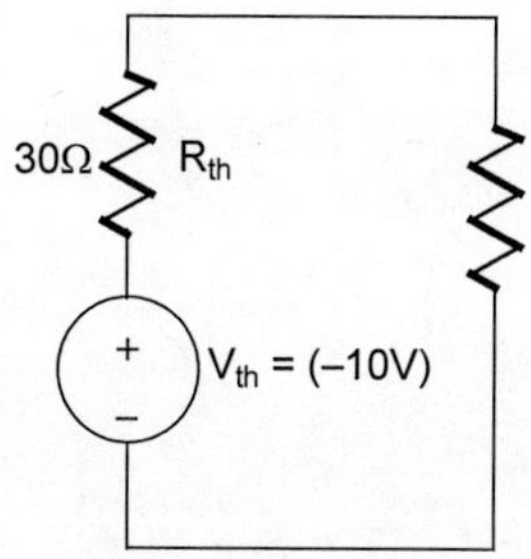

(d)

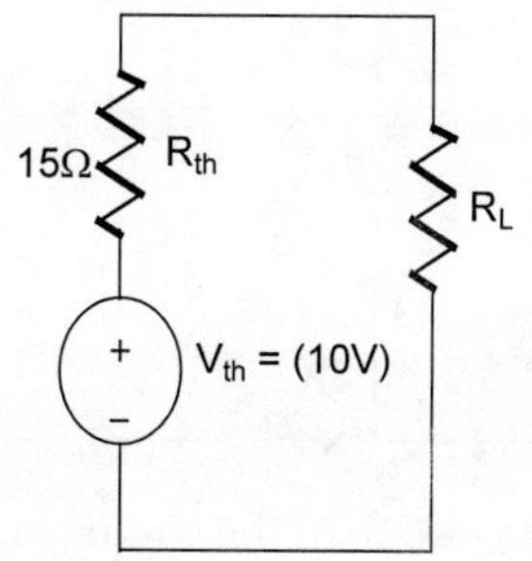

10. Which circuit shown below is applicable to the reciprocity theorem?

(a)

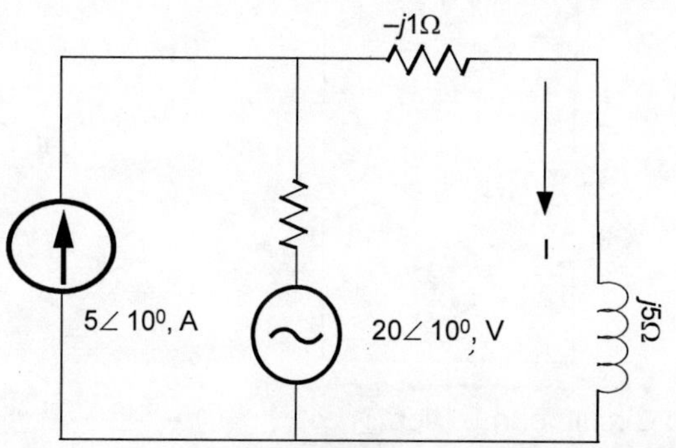

(b)

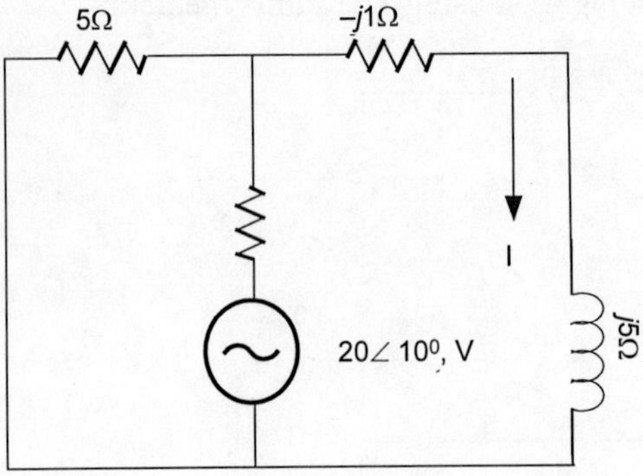

(c)

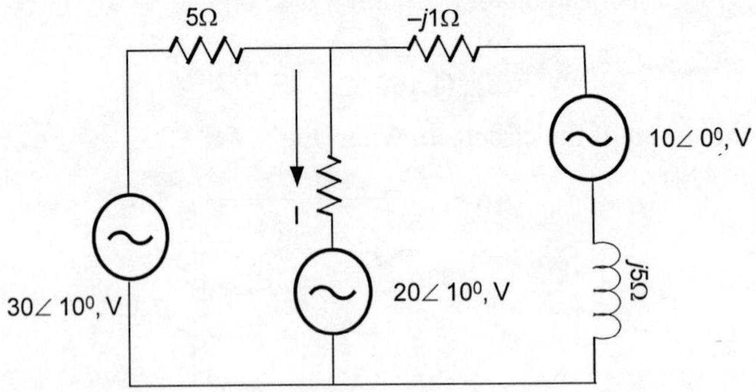

(d)

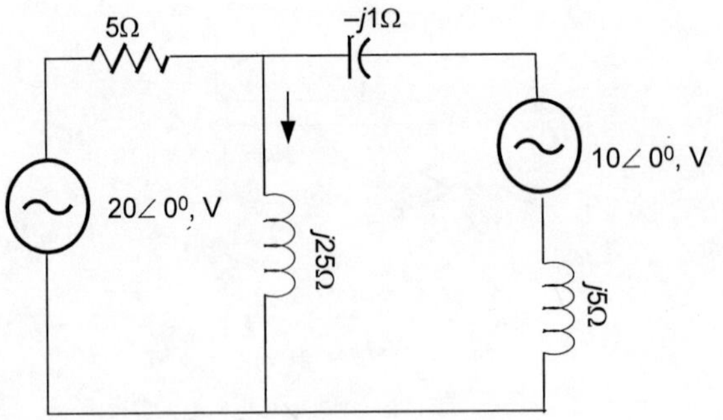

11. Obtain the duality circuit of the circuit shown in Fig,2.75.

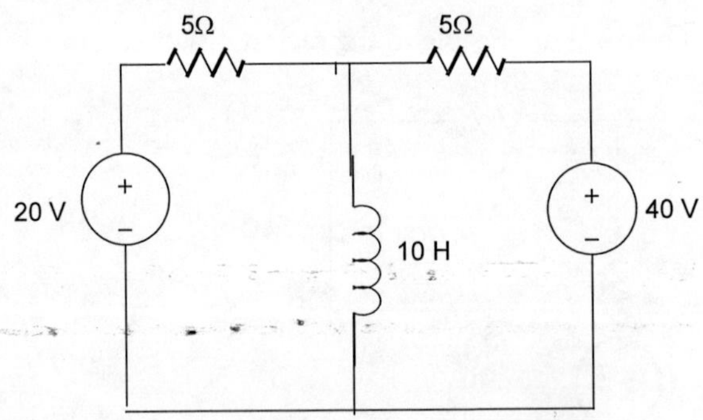

Fig. 2.75: Circuit for question 11.

(a)

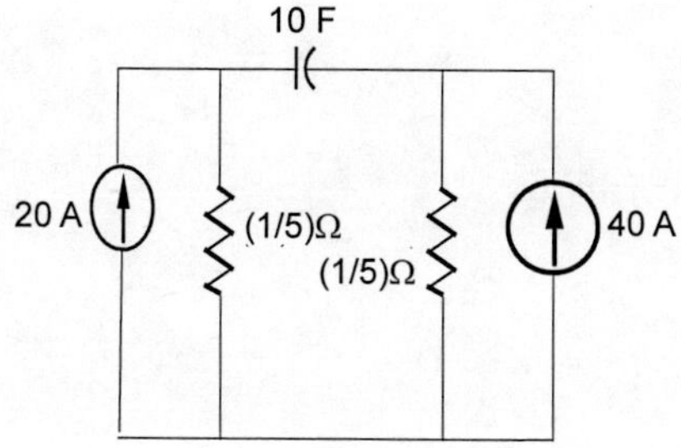

(b)

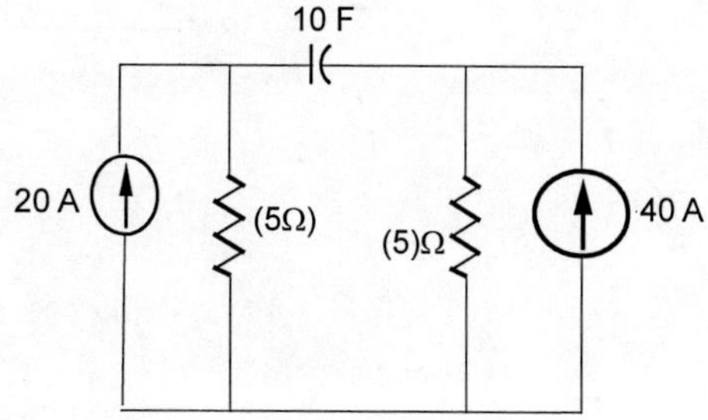

(c)

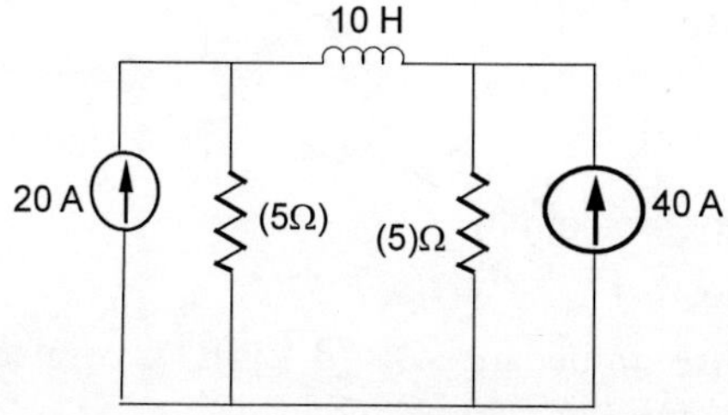

(d) None of the above

ANSWERS															
1.	(a)	2.	(d)	3.	(a)	4.	(a)	5.	(b)	6.	(b)	7.	(a)	8.	(b)
9.	(c)	10.	(b)	11.	(a)										

Short Answer Questions:

1. What is the need of source transformation in the electric circuits?
2. Find out the equivalent star connected impedances of the delta connected impedances presented in Fig.2.76.

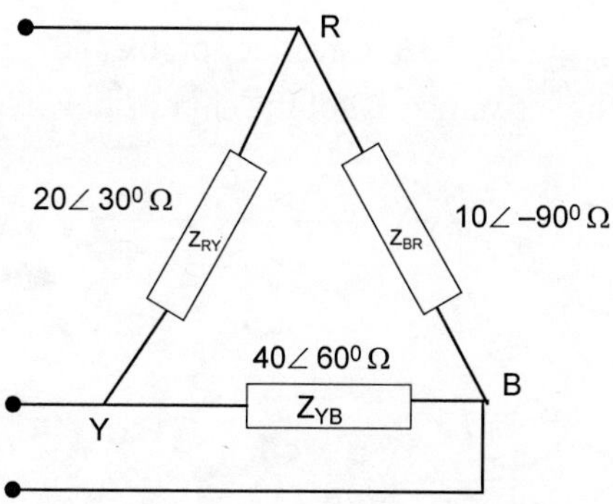

Fig. 2.76: Circuit for problem 2

3. Find out the equivalent delta connected impedances of the star connected impedances presented in Fig.2.77.

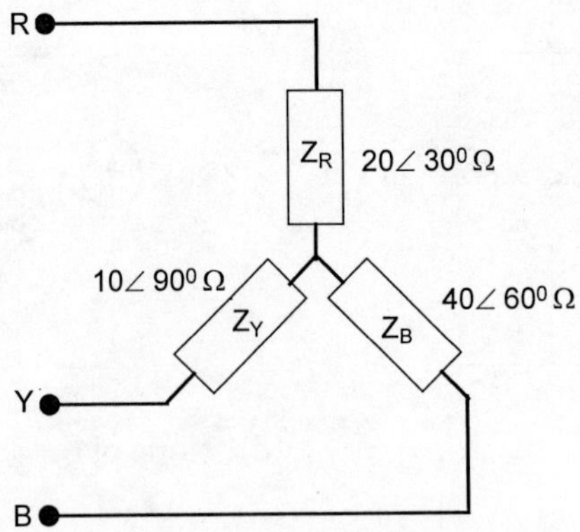

Fig. 2.77: Circuit for problem 3

4. Differentiate the balanced load and unbalanced load.
5. The balanced star connected load formed with three impedance $Z = (3 + j4)\Omega$ is connected to the supply. Find the equivalent delta connected load of the star connected load.
6. Find the current flowing the circuit shown in Fig.2.78 using mesh analysis.

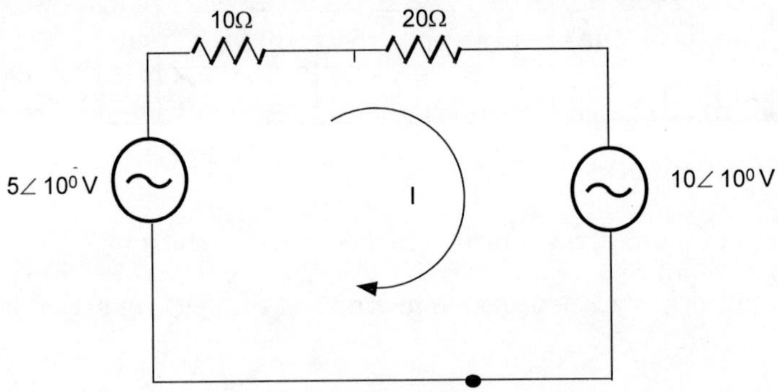

Fig. 2.78: Circuit for problem 6

7. Find the potential across the resistance 5Ω of the circuit shown in Fig.2.79 using node analysis.

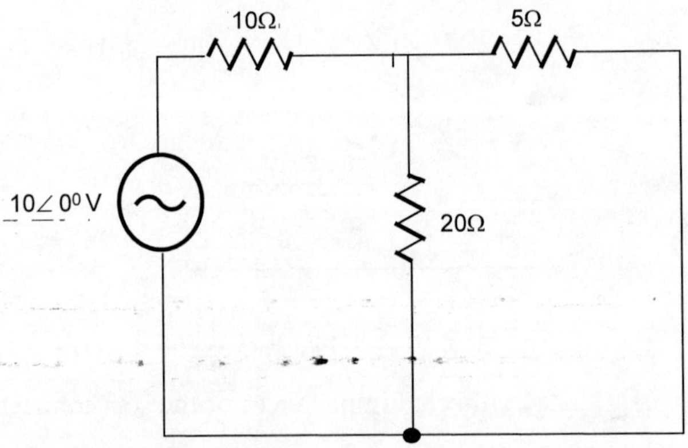

Fig. 2.79: Circuit for problem 7

8. State: Superposition theorem
9. Define the Thevenin's voltage, Thevenin's impedance.
10. Describe the Norton's current, Norton's Equivalent circuit.
11. What is the condition for transferring maximum power from the circuit to load?
12. What is the value of the load impedance when the load is resistive type during the maximum power transfer condition?
13. State the reciprocity theorem and explain with the simple circuits.
14. Prove the reciprocity theorem to the circuit shown in Fig.2.80.

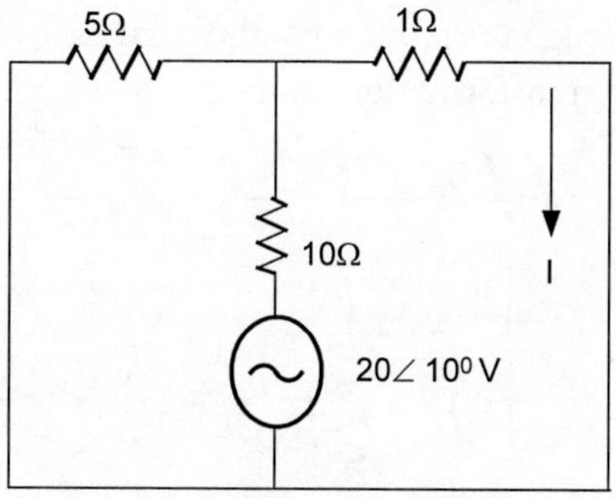

Fig. 2.80: Circuit for problem 14

15. What is principle of duality in electric circuits?
16. List out the dual pairs in an electric circuit.
17. Write the steps to formulate the duality circuit.

Long Answer Questions:

1. Describe the method of converting the star connected impedances into the delta connected impedances.
2. Discuss the method of converting the delta connected impedances into the star connected impedances.
3. Explain the steps to be performed in the source transformation method with suitable examples.
4. For the circuit shown in Fig.2.81 with the angular velocity of "ω", apply the mesh analysis procedure and find the loop currents.

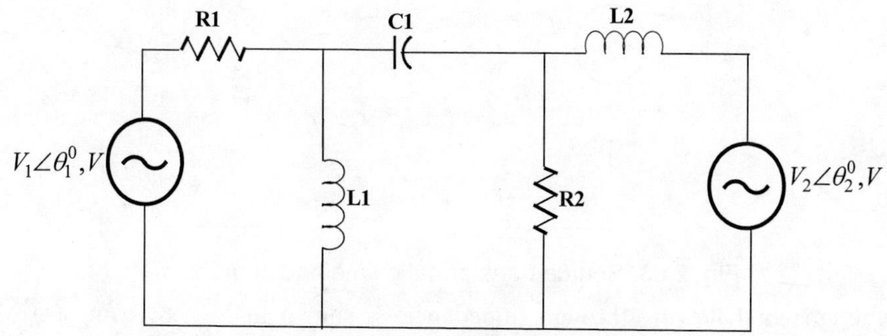

Fig.2.81: Circuit for problem 4

5. For the circuit shown in Fig.2.81 with the angular velocity of "ω", apply the node analysis procedure and find the node potentials.
6. Derive the condition for the maximum power transfer from the circuit to the load. Also calculate the maximum power consumed by the load.
7. Prove the reciprocity theorem with suitable example.
8. Explain the concept of duality with example circuit.

Numerical Problems

1. Find the current flowing through the inductor of the circuit shown in Fig. 2.82 if the circuit operates at 50Hz. Solve using source transformation method.

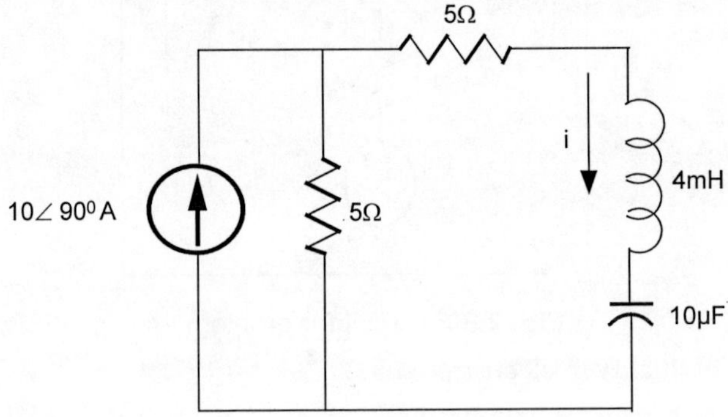

Fig. 2.82: Source transformation method of problem 1

2. Find the voltage across the resistance as the circuit shown in Fig. 2.83 using source transformation method.

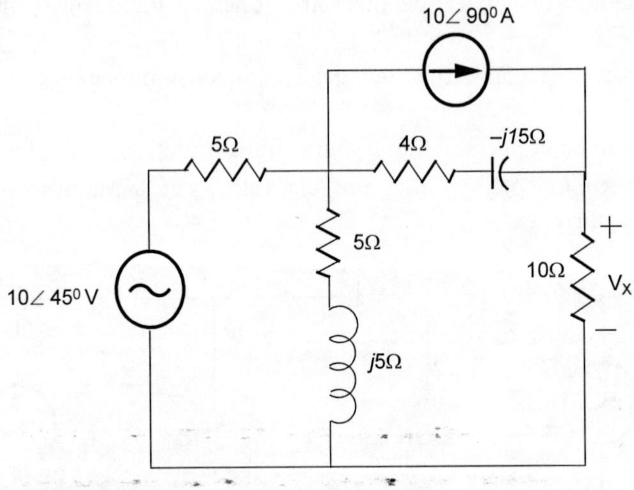

Fig. 2.83: Source transformation method of problem 2

3. Find the current I flowing through impedance Z_4 shown in Fig.2.84 using (a) Mesh analysis (b) Node analysis methods. Verify the results.

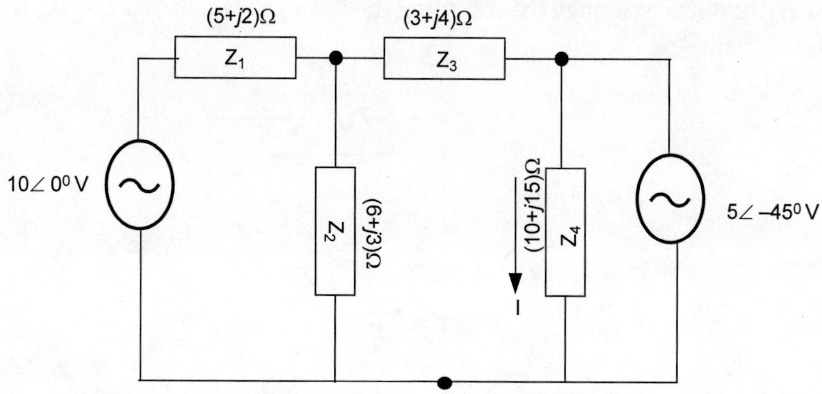

Fig. 2.84: AC network for node and mesh analysis of problem 3

4. Find the voltage across the impedance as the circuit shown in Fig. 2.85 using (a) source transformation method (b) Node analysis (c) Mesh analysis

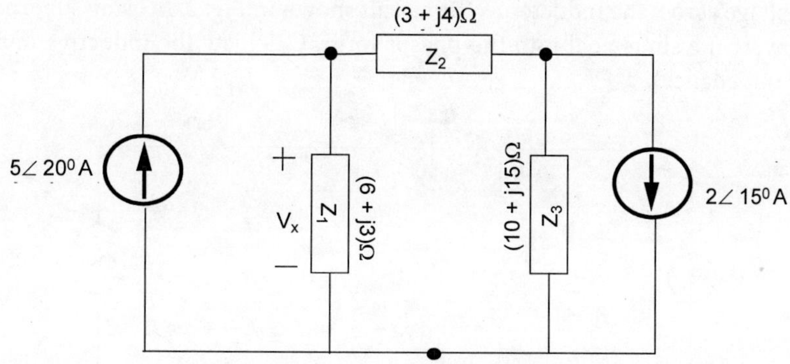

Fig. 2.85: AC network for problem 4

5. Find the current flowing through the inductance of the circuit shown in Fig. 2.86 using (a) Node analysis (b) Mesh analysis. (Hint: Convert the sinusoidal into the phasor form, Calculate the inductive impedance and the capacitive impedance).

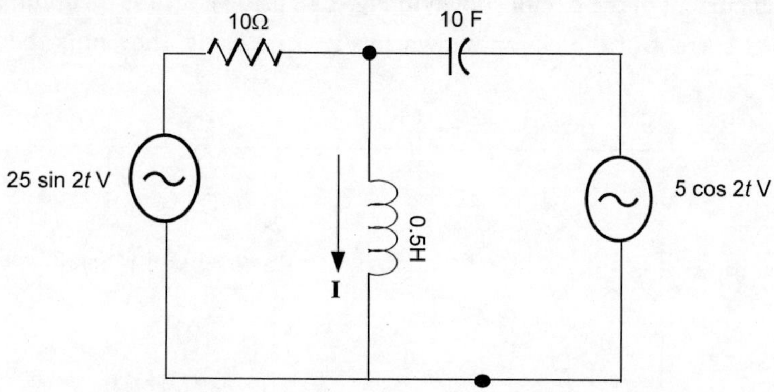

Fig. 2.86: AC network for node and mesh analysis of problem 5

6. Find the current flowing through the resistor of the circuit shown in Fig. 2.87 using superposition theorem

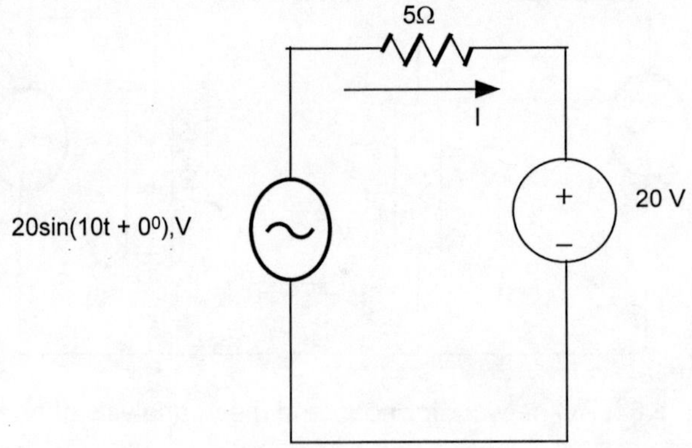

Fig.2.87: Circuit for problem 6

7. Find the voltage across the inductor as the circuit shown in Fig. 2.88 using superposition theorem. (Hint: Convert the sinusoidal into the phasor form, Calculate the inductive impedance and the capacitive impedance).

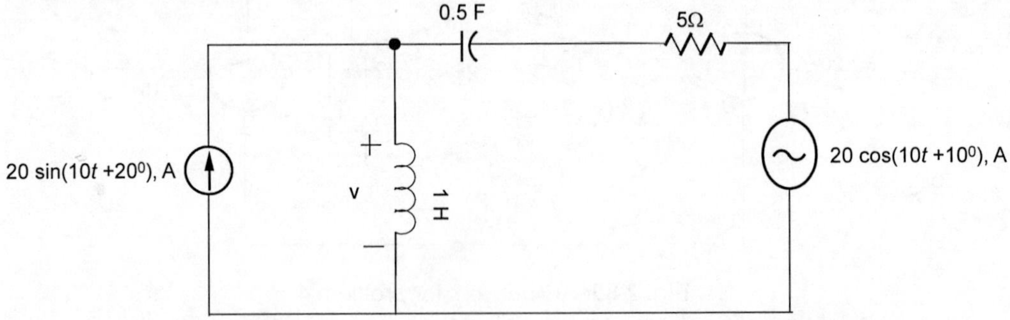

Fig.2.88: Find the voltage across the inductor using superposition theorem of Problem 7

8. Find the load current of the circuit shown in Fig. 2.88 using Thevenin's theorem.
9. Find the load current of the circuit shown in Fig. 2.88 using Norton's theorem.
10. Find the load current of the circuit shown in Fig. 2.89 using Thevenin's theorem and Norton's theorem.

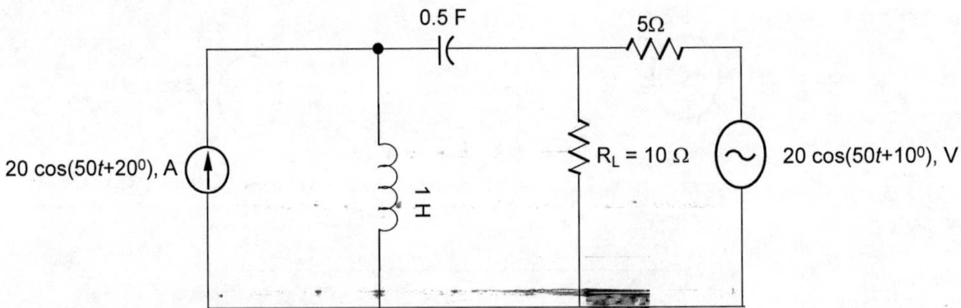

Fig.2.89: AC Circuit for problem 10

11. Find the load resistance of the circuit shown in Fig. 2.90 using maximum power transfer theorem.

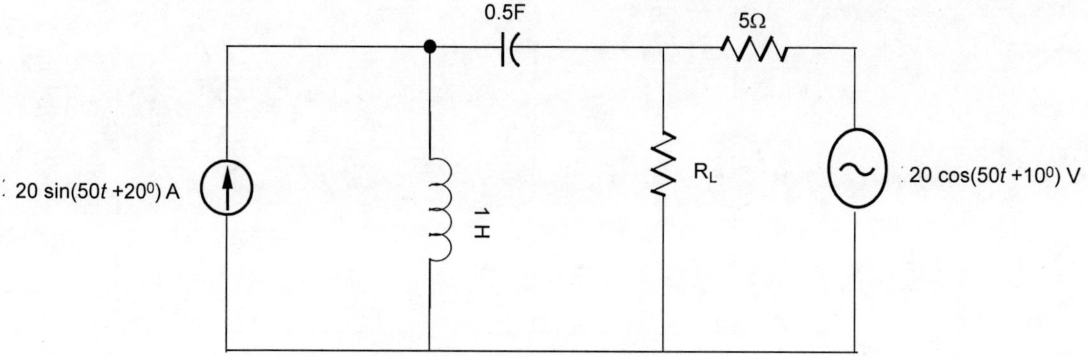

Fig.2.90: AC circuit for problem 11

12. Find the load resistance of the circuit shown in Fig.2.91 using maximum power transfer theorem. Also calculate the maximum power absorbed by the circuit.

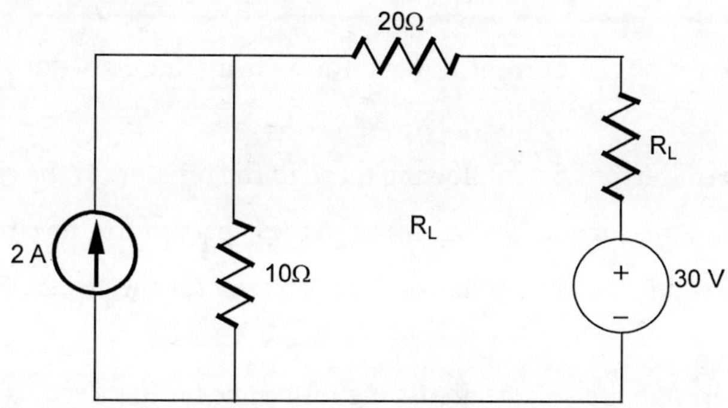

Fig.2.91: AC circuit for problem 12

13. Find the load impedance of the circuit shown in Fig.2.92 using maximum power transfer theorem. Also calculate the maximum power absorbed by the circuit. (Hint: Two frequency sources present in the circuit. Obtain two Thevenin's equivalent circuit and two load impedances and two maximum power will be obtained).

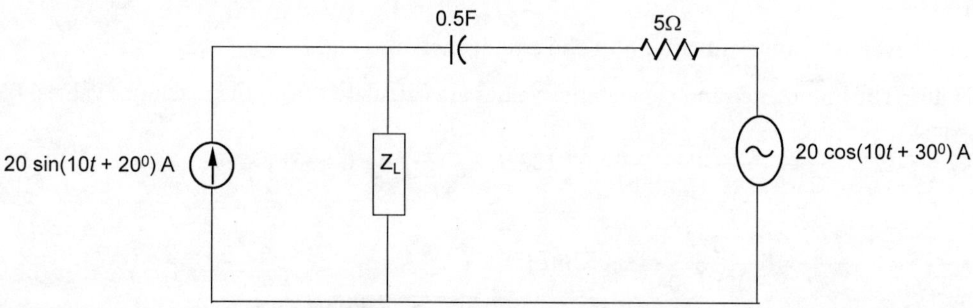

Fig.2.92: Find the load impedance using maximum power transfer theorem of Problem 13

14. Find the current flowing any element of the circuit shown in Fig. 2.93 using reciprocity theorem.

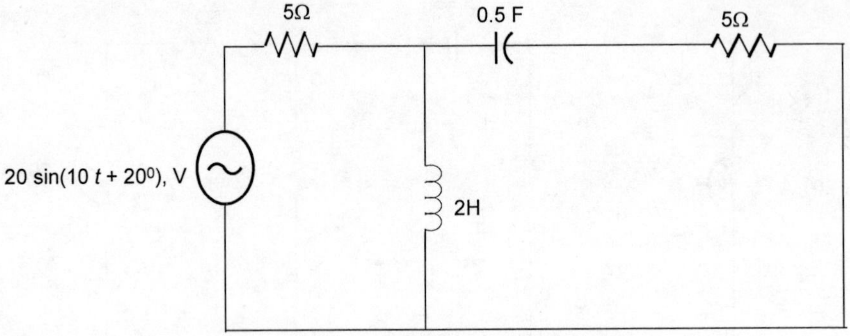

Fig. 2.93: AC circuit for Problem 14

15. Find the duality circuit for the circuit shown in Fig.2.93.

PRACTICAL

1. Use PSpice to determine the current flowing through the inductor for the circuit shown in Fig. 2.84.
2. Use PSpice to determine the current flowing through the resistor for the circuit shown in Fig. 2.85.
3. Apply PSpice to find the current flowing through the capacitor for the circuit shown in Fig. 2.88.
4. Use PSpice to determine the voltage across the resistance for the circuit shown in Fig. 2.89. Verify the results.
5. Use PSpice to determine the voltage across the resistance for the circuit shown in Fig. 2.90. Verify the results.
6. Apply PSpice for the reciprocity theorem circuits and verify the results.

KNOW MORE

Example 2.11:

Find the Thevenin's equivalent circuit for rhe circuit shown in Fig. 2.41

[Note1: The inductance and capacitance values are calculated from the reactance values.

$$X_L = j4\Omega; L = \frac{X_L}{\omega} = \frac{X_L}{2\pi f} = \frac{4}{2\pi(50)} = 0.0127H; X_C = -j6\Omega; C$$

$$= \frac{1}{X_C\omega} = \frac{1}{X_C 2\pi f} = \frac{1}{(6)2\pi(50)} = 0.053mF]$$

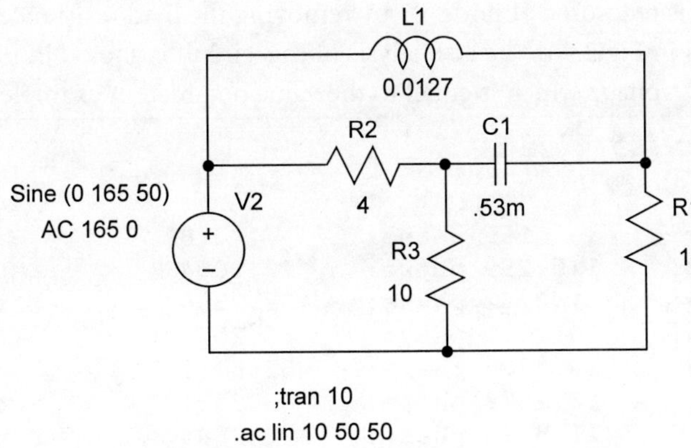

Fig. 2.94: LTSpice circuit model for the example 2.12

```
--- AC Analysis ---

frequency:    50           Hz
V(n001):      mag:       165 phase:          0°    voltage
V(n002):      mag:    131.87 phase:   -38.3661°    voltage
V(n003):      mag:   148.393 phase:   -8.02477°    voltage
I(C1):        mag:   12.4974 phase:   -35.4623°    device_current
I(L1):        mag:   25.6762 phase:   -36.9532°    device_current
I(R3):        mag:   14.8393 phase:   -8.02477°    device_current
I(R2):        mag:    6.8707 phase:   -131.081°    device_current
I(R1):        mag:    13.187 phase:   -38.3661°    device_current
I(V2):        mag:   27.0531 phase:     157.72°    device_current
```

Fig. 2.95: LTSpice AC analysis of the example 2.12

When the load resistance is 10Ω, the load current (current through R3 resistance) is that is shown in Fig.2.95. The value is equal to the theorical value obtained in example 2.11.

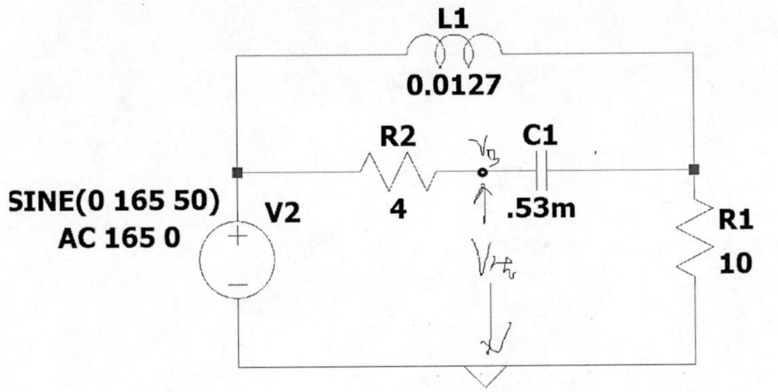

Fig. 2.96: LTSpice circuit for Thevenin's voltage calculation of the example 2.12

Thevenin's voltage is measured at node V3 by removing the load resistance and the corresponding LTspice circuit is shown in Fig.2.96. The ac analysis of this circuit is shown in Fig.2.97. The highlighted voltage is the Thevenin's voltage which is equal to the value obtained in example 2.11.

```
        --- AC Analysis ---

frequency:    50              Hz
V(n001):      mag:      165 phase:         0°        voltage
V(n002):      mag:  115.299 phase:  -22.9804°        voltage
V(n003):      mag:   172.62 phase:  -13.7344°        voltage
I(C1):        mag:  10.2679 phase:  -86.2521°        device_current
I(L1):        mag:  18.5705 phase:  -52.5734°        device_current
I(R2):        mag:  10.2679 phase:  -86.2521°        device_current
I(R1):        mag:  11.5299 phase:  -22.9804°        device_current
I(V2):        mag:  11.5299 phase:    157.02°        device_current
```

Fig. 2.97: AC analysis of the LTSpice circuit presented in Figure 2.96 of the example 2.12

REFERENCES AND SUGGESTED READINGS

1. Charles K. Alexander and Matthew N. O. Sadiku, "Fundamentals of Electric Circuits", McGraw Hill Publishers, 7th Edition, 2021.
2. D P Kothari and I J Nagrath, "Basic Electrical Engineering", McGraw Hill Publishers, 4th Edition, 2019.
3. C L Wadhwa, "Basic Electrical Engineering", New Age International Publishers, 4th Edition, 2007.
4. S P Ghosh and A K Chakraborty, "Network Analysis and Synthesis", McGraw Hill Education, 2009.
5. M E Van Valkenburg and T S Rathore, "Network Analysis", Prentice Hall India Ltd, 3rd edition, 2019.

□□

3 Three Phase Circuits

UNIT SPECIFICS

Through this unit we have discussed the following aspects:

- Phasors and complex representation of three phase supply
- Phase sequence and polarity
- Types of three phase connections: star, delta
- Line and phase quantities
- Loads: balanced and unbalanced load
- Power: active, reactive and apparent

RATIONALE

The single-phase systems are discussed in the previous chapters. For increasing the power generation and transmission, there is a need for the new system model. Two phase system model was developed and the angle between those windings is 90^0. Whereas, the angle is 120^0 in the three-phase windings of the three-phase systems. Three-phase power supply is generated using a three-phase generator. Even the higher order supply systems are available which could be chosen based on the applications. But the three-phase systems are the basic for the higher order systems and it is more efficient, requires less material for the given power capacity and is low cost than the single-phase systems. Hence it is necessary to study the three-phase systems.

PRE-REQUISITES

Vector calculus

Single-phase systems

UNIT OUTCOMES

List of outcomes of this unit is as follows:

U3-O1: Understand the three-phase system representation and terminology

U3-O2: Describe the types of connection: star and delta and their concepts

U3-O3: Describe the balanced and unbalanced loads

U3-O4: Realize the power calculation in three-phase systems

U3-O5: Apply vector calculus to solve complex problems

Unit-3 Outcome	EXPECTED MAPPING WITH COURSE OUTCOMES (1 - Weak Correlation; 2 - Medium Correlation; 3 - Strong Correlation)				
	CO-1	CO-2	CO-3	CO-4	CO-5
U3-O1	-	-	2	-	-
U3-O2	1	1	2	2	-
U3-O3	2	1	3	1	-
U3-O4	-	-	3	1	-
U3-O5	1	1	3	-	-

3.1 FUNDAMENTALS OF THREE PHASE SYSTEMS

Three-phase electrical power is generated by alternators in large power plants. This three-phase power is utilized for powering the network through the transmission and distribution systems. Three-phase power supply is obtained by the formation of the three coils 1200 apart from each other and rotated in a uniform magnetic field. These coils are arranged in star or delta fashion (Refer. Sec. 3.2.1 and 3.2.2). In three-phase systems, voltage/current phasors are 1200 apart fromeach other. Voltage in each phase coil is represented as a phasor. For three-phase coils' phase is shown by a letter with arrow on it such as $\overrightarrow{V_R}, \overrightarrow{V_Y}$ & $\overrightarrow{V_B}$, and its magnitude is denoted by V_R, V_Y, V_B respectively (Fig. 3.1).

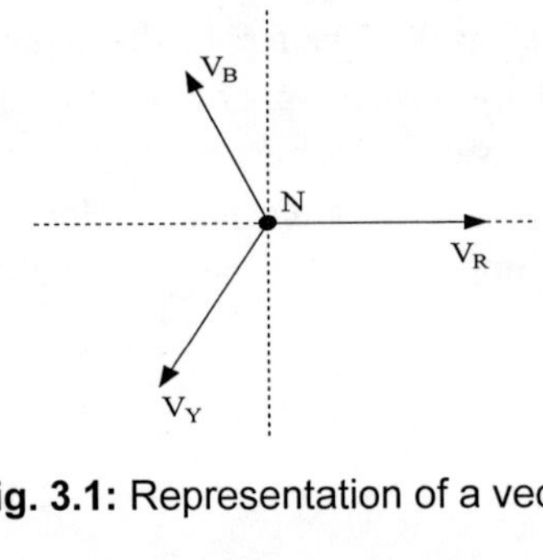

Fig. 3.1: Representation of a vector

3.1.1 Generation of the three-phase Supply

In the single phase system, both line voltage and phase voltage have been introduced in Chapter 1. The single-phase alternator is used to generate a single phase power supply which has only one rotor windings. The number of rotor windings are increased and independent voltage waveforms are produced according to the number of windings. When three windings (with dedicated colours red (R), yellow (Y) and blue(B)) are present, three voltage waveforms are obtained. Each phase of the three-phase supply is represented as phasor $\overrightarrow{V_R}$, $\overrightarrow{V_Y}$ and $\overrightarrow{V_B}$. These phasors are rotating with angular frequency in anticlockwise direction with 120^0 angular displacement, based on the waveform generated and is shown in Fig. 3.2. The three-phase waveforms are obtained and represented in Fig.3.3. These phase voltages are presented by equations (3.1) to (3.3) as follows.

$$v_R(t) = V_m \sin \omega t \tag{3.1}$$

$$v_Y(t) = V_m \sin(\omega t - 120^0) \tag{3.2}$$

$$v_B(t) = V_m \sin(\omega t - 240^0) \tag{3.3}$$

For the balanced systems, the addition of all three voltages produces zero. The balanced systems are discussed in the later part of this chapter.

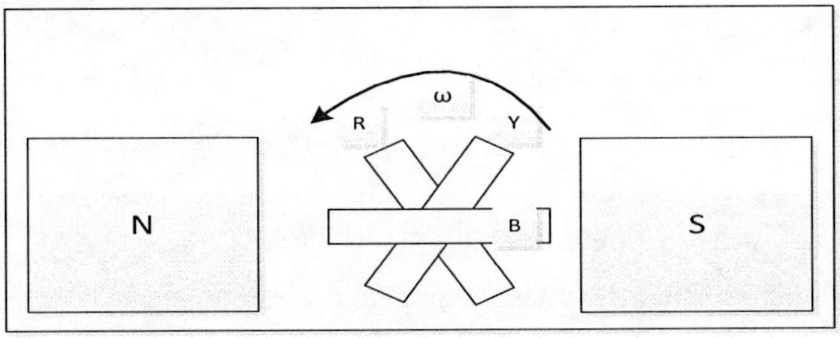

Fig.3.2: Three windings (red, yellow and blue) are rotating between magnetic bars (N and S) with angular velocity of ω.

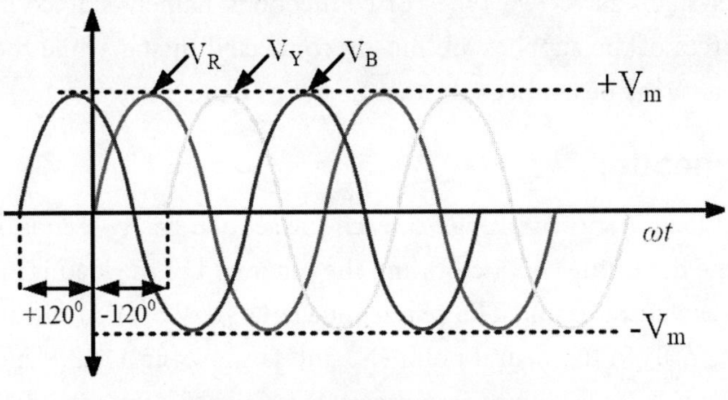

Fig.3.3: Three-phase waveforms

3.1.2 Phase Sequence

The phase sequence is the order in which each of the three phases reached either zero crossing or the maximum value of the quantity. The standard phase sequence is RYB, it means that R reaches the maximum value and then Y, finally B that is obtained only when phasors rotate in anticlockwise direction. It is also denoted as a positive phase sequence. When phasors rotate in clockwise direction, the obtained phase sequence is RBY that is known as negative phase sequence. The phase sequence could be reversed by interchanging any two among the three phases.

Example 3.1:

Determine the phase sequence of the given three-phase voltages $v_{RN}(t) = 310 \sin(\omega t + 30^0)$, $v_{YN}(t) = 310 \sin(\omega t - 90^0)$ and $v_{BN}(t) = 310 \sin(\omega t - 210^0)$.

Solutions:

The voltages are represented in phasor form as follows. The phasor diagram is shown in Fig.3.4.

$V_{RN} = 310\angle 30^0; V_{YN} = 310\angle -90^0; V_{BN} = 310\angle -210^0$

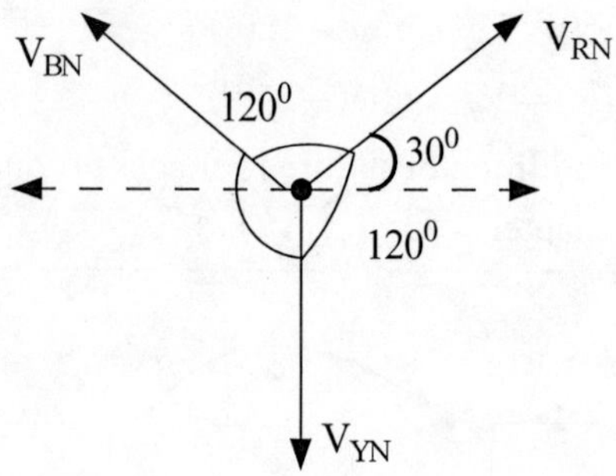

Fig.3.4: Phasor diagram of the three-phase voltages

From this, it is understood that the phase sequence is RYB.

3.2 TYPES OF THREE-PHASE CONNECTIONS

Three phase supply systems have two types of connections namely star connections (Y) and delta connections (Δ). Sources of the supplies are mostly connected in star while the transformers and the motor or any other loads may be connected either in star or delta.

3.2.1 Star Connection

A star connected load is shown in Fig.3.5, each phase-load has two ends. One end of the load is connected to one of the three-line conductors and the other end of the load is joined together with two loads a point named as neutral point. The phase voltage (V_{RN}, V_{YN}, V_{BN}) measured between any one of line conductors (R, Y, B) to the neural point (N) and is shown in Fig.3.3 and hereafter these phase voltages are denoted as V_R, V_Y, V_B. The line voltage (V_{RY}, V_{YB}, V_{BR}) measured between any of the two-line conductors (R-Y, Y-B, B-R), which is shown in Fig. 3.5. From the pictorial representation of the star systems, it is understood that phase voltage and line voltage carry different values. The instantaneous voltages are explained in the next paragraphs.

For a balanced system, the loads are equal Z_R=Z_Y=Z_B and the current in the neutral wire is zero. Since the neutral current is zero, the neutral conductor is not needed and it is often evaded. The phase voltages/currents and line voltages are equal in magnitude which is presented in equation (3.4) - (3.6). The phasor addition of the line currents is zero which is presented as equation in (3.7).

$$|I_R| = |I_Y| = |I_B| \tag{3.4}$$

$$|V_R| = |V_Y| = |V_B| \tag{3.5}$$

$$|V_{RY}| = |V_{YB}| = |V_{BR}| \tag{3.6}$$

$$I_R + I_Y + I_B = 0 \tag{3.7}$$

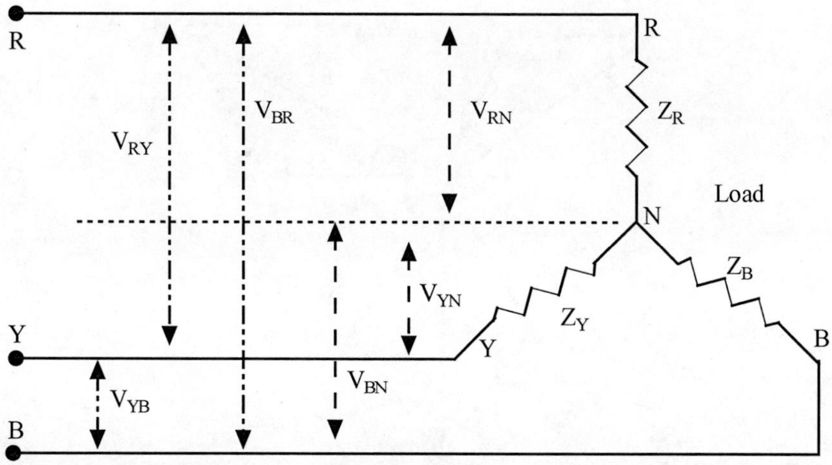

Fig.3.5: Star connected load with the information of the phase voltages and line voltages

The instantaneous value or time domain representation of the phase voltage is shown in equation (3.8) - (3.10). The phasor diagram of the phase voltages is presented Fig.3.6 and also the phasor diagram for obtaining the line voltages is shown in Fig.3.7. The phase voltage is represented in phasor form as shown in equation (3.11). The line voltage (V_{RY}) is the instantaneous line voltage ($v_{RY}(t)$, $v_{YB}(t)$, $v_{BR}(t)$) of the three-phase star connected loads are calculated by the vector subtraction of the phase voltage $v_R(t)$, $v_Y(t)$ and $v_B(t)$ which isstated in equations (3.12) to (3.14). The line voltage is expressed in phasor form as shown in equation (3.15) using the equations (3.12a) to (3.14a). For the balanced system, the line voltages are also 1200 apart from each other.

$$v_R(t) = V_m \sin \omega t \tag{3.8}$$

$$v_Y(t) = V_m \sin(\omega t - 120^0) \tag{3.9}$$

$$v_B(t) = V_m \sin(\omega t - 240^0) \tag{3.10}$$

$$V_R = |V_m|\angle 0^0; \quad V_Y = |V_m|\angle -120^0; \quad V_B = |V_m|\angle -240^0 \tag{3.11}$$

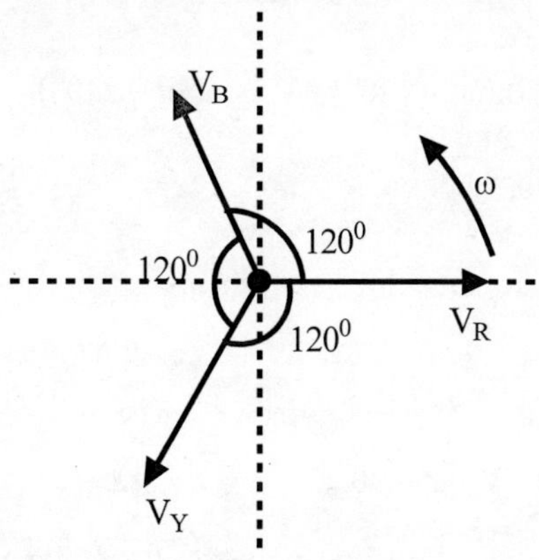

Fig.3.6: Phasor diagram of the phase voltage in Y connected 3Φ systems

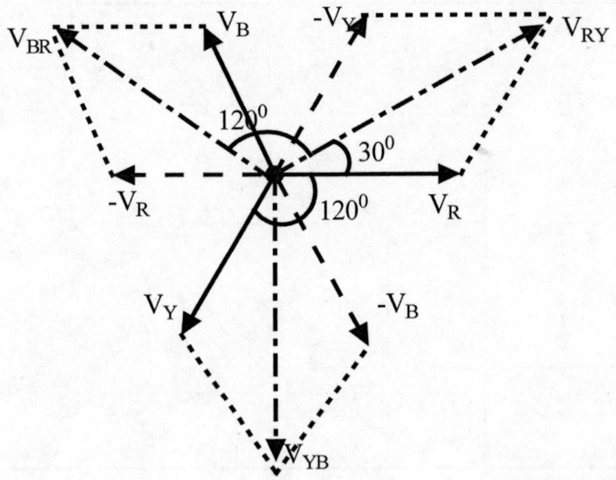

Fig.3.7: Phasor diagram of the line voltage in Y connected 3Φ systems

$$v_{RY}(t) = v_R(t) - v_Y(t) = V_m sin\omega t - V_m \sin(\omega t - 120^0) \tag{3.12}$$

$$v_{RY}(t) = V_m sin\omega t - V_m[sin\omega t * cos120 - cos\omega t * sin120]$$

$$= V_m sin\omega t + V_m sin\omega t * \frac{1}{2} + V_m cos\omega t * \frac{\sqrt{3}}{2}$$

$$= \frac{3}{2} V_m sin\omega t + V_m cos\omega t * \frac{\sqrt{3}}{2}$$

$$= \sqrt{3} V_m [\frac{\sqrt{3}}{2} sin\omega t + \frac{1}{2} cos\omega t]$$

$$= \sqrt{3} V_m \left[\cos 30 sin\omega t + \sin 30 cos\omega t\right]$$

$$= \sqrt{3} V_m \left[\sin(\omega t + 30^0)\right] \tag{3.12a}$$

$$v_{YB}(t) = v_Y(t) - v_B(t) = V_m \sin(\omega t - 120^0) - V_m \sin(\omega t - 240^0) \tag{3.13}$$

$$v_{YB}(t) = \sqrt{3} V_m \left[\sin \sin\left(\omega t - 90^0\right)\right] \tag{3.13a}$$

$$v_{BR}(t) = v_B(t) - v_R(t) = V_m \sin\left(\omega t - 240\right) - V_m \sin \omega t \tag{3.14}$$

$$v_{BR}(t) = \sqrt{3} V_m \left[\sin \sin\left(\omega t - 210^0\right)\right] \tag{3.14a}$$

$$V_{RY} = \sqrt{3}|V_m|\angle 30^0; V_{YB} = \sqrt{3}|V_m|\angle -90^0; V_B = \sqrt{3}|V_m|\angle -210^0 \tag{3.15}$$

By Comparing the phase voltage and line voltage of the star connected systems with equations (3.11) and (3.15), the magnitude of the line voltages and phase voltages could be written as in equation

(3.16) and (3.17) respectively. It is understood that the magnitude of the line voltage is $\sqrt{3}$ times of the magnitude of the phase voltage. Hence it could be simply represented by the equation (3.18).

$$|V_{RY}| = |V_{YB}| = |V_{BR}| = \sqrt{3}|V_m| = |V_L| \quad (3.16)$$

$$|VR| = |VY| = |VB| = |Vm| = |VP| \quad (3.17)$$

$$|V_L| = \sqrt{3}|V_m| = \sqrt{3}|V_P| \quad (3.18)$$

The current flows in any of the line conductors (R, Y, B) is known as line current(I_{L1}, I_{L2}, I_{L3}), which is denoted as (I_L), for balanced systems. The current flowing in the phase circuit is known as phase current which is represented as I_{RN}, I_{YN}, I_{BN} and these are represented as(I_P) for a balanced system. In the star connection, theline current is equal to the phase current which is presented in equation (3.19). The instantaneous value or time domain representation of the phase current is shown in equation (3.20)-(3.22).

$$I_L = I_P \quad (3.19)$$

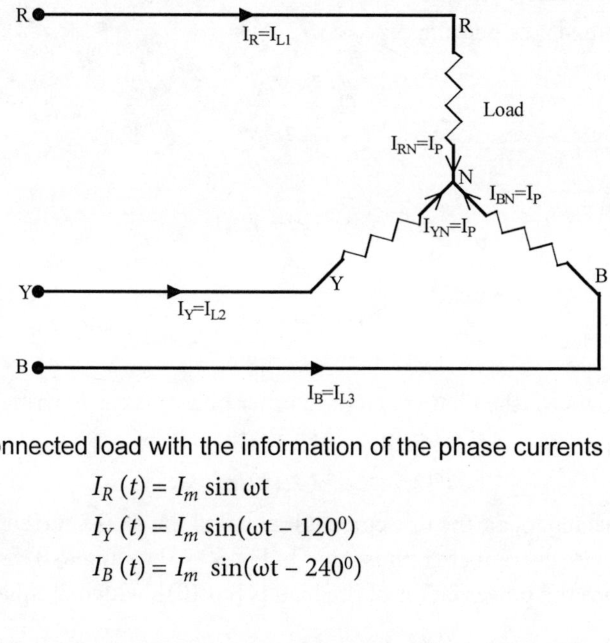

Fig.3.8: Star connected load with the information of the phase currents and line currents

$$I_R(t) = I_m \sin \omega t \quad (3.20)$$

$$I_Y(t) = I_m \sin(\omega t - 120^0) \quad (3.21)$$

$$I_B(t) = I_m \sin(\omega t - 240^0) \quad (3.22)$$

Example 3.2:

A balanced three-phase star connected load is supplied from 310V, 50Hz source. Find the line voltage of the load, the line current and phase current when the load per phase is resistance of 10Ω and inductance of 20mH. Find the real power, reactive power, apparent power and power factor of the load.

Solutions:

The supply voltage is 310V, which is the line voltage. Let us assume the phase sequence is RYB. Hence the line voltage of the star connected load is mentioned below and its phasor form is represented in Fig.3.9.

$$V_{RY} = 310\angle 0^0 V; V_{YB} = 310\angle -120^0 V; V_{BR} = 310\angle -240^0 V$$

The phase voltage of the load is V_{RN}, V_{YN}, V_{BN}. Phase voltage = Line voltage/ $\sqrt{3}$

$$V_{RN} = \frac{310}{\sqrt{3}} = 178.9V;$$

The magnitude of all three phases is equal and it is 178.9V. The R phase voltage is lagging behind the line voltage V_{RY} by 300 for the Y connected load. Hence the phase voltages are found as follows and the relation between the phase voltage and line voltage phasor are shown in Fig. 3.10.

$$V_{RN} = 178.9\angle -30^0 V; V_{YN} = 178.9\angle -150^0 V; V_{BN} = 178.9\angle -270^0 V$$

The load per phase is resistance of 10Ω and inductance of 20mH.

R/phase = 10Ω; L = 20mH

Inductive reactance $\frac{X_L}{phase}$ =2πfL=2π (50) (20*10-3) =(2π)

Impedance/phase (Z/Φ) = (10 + j2π)Ω = 11.81∠32.14Ω

$$\text{Phase current} = \frac{\text{Phase Voltage}}{\text{Impedance per phase}}$$

$$I_R = \frac{V_{RN}}{Z/\phi}; I_Y = \frac{V_{YN}}{Z/\phi}; I_B = \frac{V_{BN}}{Z/\phi}$$

$$I_R = \frac{V_{RN}}{Z/\phi} = \frac{178.9\angle -30}{(10 + j2\pi)}$$

$$= \frac{178.9\angle -30}{11.81\angle 32.14}$$

$$= 15.14\angle -62.14^0 A$$

Since the load is balanced, the phase current in other phases is easily written as

$$I_Y = 15.14\angle -182.14^0 A$$

$$I_B = 15.14\angle -302.14^0 A$$

Whereas in Y connected loads, the line current is equal to the phase current. The phasor diagram of the phase voltage and phase current are represented in Fig.3.11. The angle (θ) between these parameters is 32.140 (lagging). Hence the power factor of the load is [cos(θ)], which is equal to 0.8467(lagging).

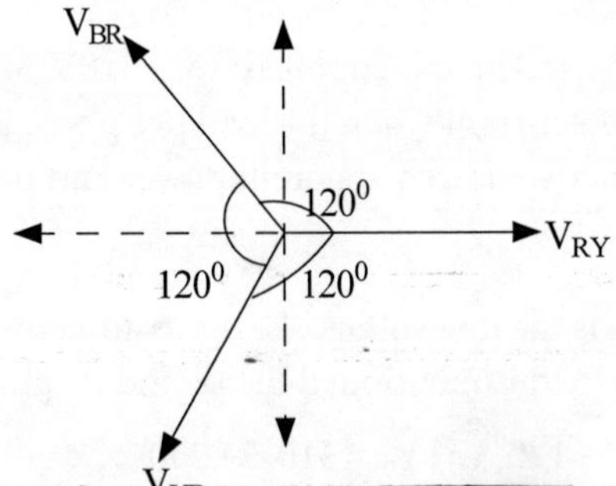

Fig.3.9: Phasor diagram of the three-phase line voltages

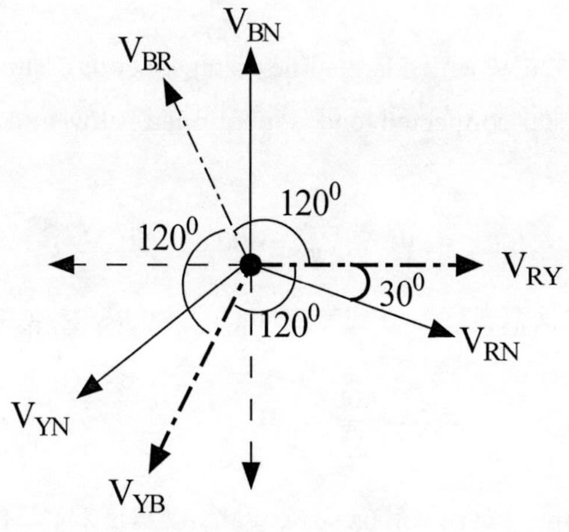

Fig.3.10: Relation between three-phase line voltages and phase voltages

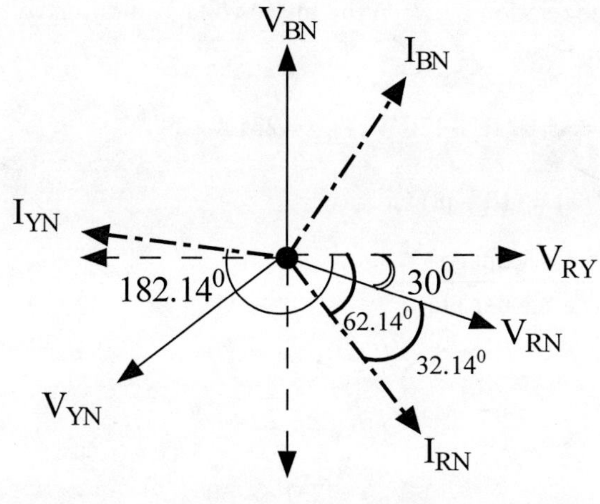

Fig. 3.11: Relation between the phase voltages and phase currents

Power:

Active power of the load $P_{3\varphi} = \sqrt{3}V_L I_L cos\varphi = \sqrt{3} * 310 * 15.14 * \cos(32.14) = 6883.41\text{W}$

Reactive power of the load $Q_{3\varphi} = \sqrt{3}V_L I_L sin\varphi = \sqrt{3} * 310 * 15.14 * \sin(32.14) = 4324.65\text{VAR}$

Apparent power of the load $S_{3\varphi} = \sqrt{P_{3\varphi}^2 + Q_{3\varphi}^2} = 8129.2\text{VA}.$

Example 3.3:

A balanced three-phase star connected load of (10 + j8) Ω/phase is supplied from a balanced three-phase supply of $400\angle -20^0$ V, 50Hz. Find the line current, the real power, reactive power, apparent power and power factor the load.

Solutions:

The supply voltage is 400∠ – 20⁰V, which is the line voltage. Let us assume the phase sequence is RYB. Hence the line voltage of the star connected load is mentioned below and its phasor form is represented in Fig.3.12.

$$V_{RY} = 400\angle -20^0 \text{V}; V_{YB} = 400\angle -140^0 \text{V}; V_{BR} = 400\angle -260^0 \text{V}$$

The phase voltage of the load is V_{RN}, V_{YN}, V_{BN}. Phase voltage =Line voltage/ $\sqrt{3}$

$$V_{RN} = \frac{400}{\sqrt{3}} = 231V;$$

The magnitude of all three phase voltages is equal and it is 231V. The R phase voltage is lagging behind the line voltage V_{RY} by 300 according to the star (Y) connected load. Hence the phase voltages are found as follows and the relation between the phase voltage and line voltage phasor form are shown in Fig.3.13.

$$V_{RN} = 231\angle -50^0 V; V_{YN} = 231\angle -170^0 V; V_{BN} = 231\angle -290^0 V$$

The load per phase (Z/φ) = (10 + j8) Ω

$$\text{Phase current} = \frac{\text{Phase Voltage}}{\text{Impedance per phase}}$$

$$I_R = \frac{V_{RN}}{Z/\phi}; I_Y = \frac{V_{YN}}{Z/\phi}; I_B = \frac{V_{BN}}{Z/\phi}$$

$$I_R = \frac{V_{RN}}{Z/\phi} = \frac{231\angle -50}{(10 + j8)}$$

$$= \frac{231\angle -50}{12.81\angle 38.65}$$

$$= 18.03\angle - 88.65^0 \text{A}$$

Since the load is balanced, the phase current in other phases is easily written as

$$I_Y = 18.03\angle - 208.65^0 \text{ A}$$

$$I_B = 18.03\angle - 328.65^0 \text{ A}$$

Whereas in Y connected loads, the line current is equal to the phase current. The phasor diagram of the phase voltage and phase current is represented in Fig.3.13. The angle (θ) between these parameters is 38.650 (lagging). Hence the power factor of the load is [cos(θ)] lagging, which is equal to 0.781.

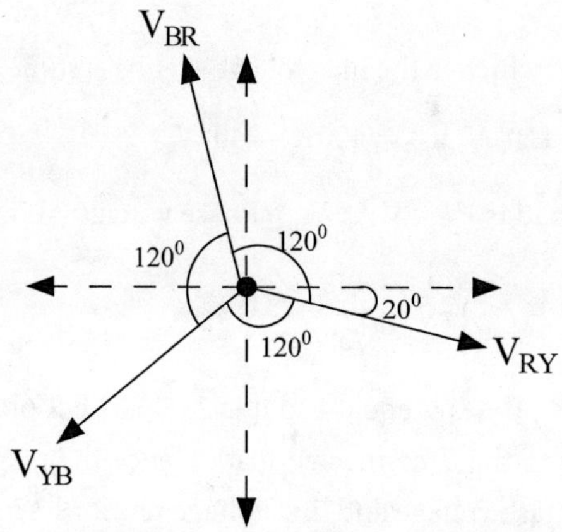

Fig. 3.12: Phasor diagram of the three-phase line voltages

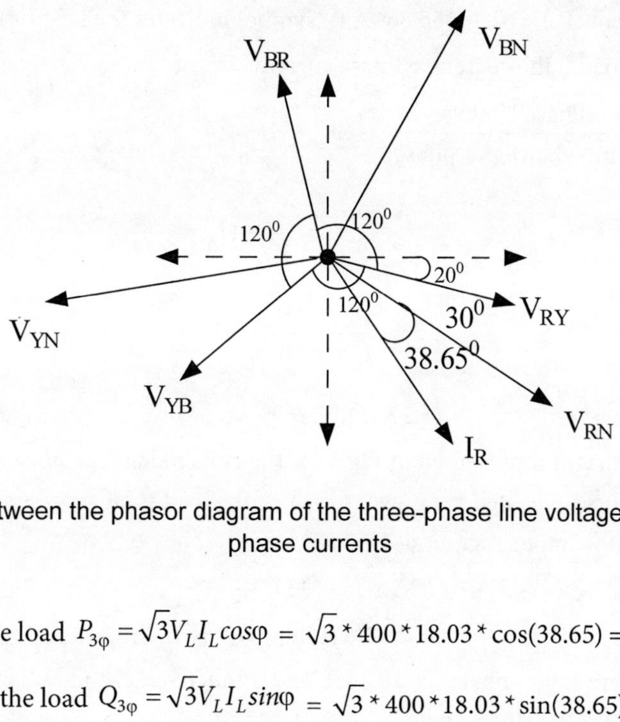

Fig. 3.13: Relation between the phasor diagram of the three-phase line voltages, phase voltages and phase currents

Power:

Active power of the load $P_{3\varphi} = \sqrt{3}V_L I_L cos\varphi = \sqrt{3} * 400 * 18.03 * \cos(38.65) = 9755.59W$

Reactive power of the load $Q_{3\varphi} = \sqrt{3}V_L I_L sin\varphi = \sqrt{3} * 400 * 18.03 * \sin(38.65) = 7801.74VAR$

Apparent power of the load $S_{3\varphi} = \sqrt{P_{3\varphi}^2 + Q_{3\varphi}^2} = 12491.54VA.$

Example 3.4:

A balanced Y connected three-phase load is supplied from a 3 phase 3 wire 400V, 50Hz system which takes a line current of 10A at 66.86^{0}lagging with respect to the line voltage. Calculate (a) load components and its parameters (b) power factor of the load and (c) the active power consumed by the load. Draw the phasor diagram.

Solutions:

The supply voltage is $400\angle 0^0$ V, which is the line voltage. Let us assume the phase sequence is RYB.

$$V_{RY} = 400\angle 0^0\,\text{V}; V_{YB} = 400\angle -120^0\,\text{V}; V_{BR} = 400\angle -240^0\,\text{V}$$

The phase voltage of the load is V_{RN}, V_{YN}, V_{BN}. Phase voltage =Line voltage/ $\sqrt{3}$

$$V_{RN} = \frac{400}{\sqrt{3}} = 231V;$$

The magnitude of all three phases is equal and it is 231V. The R phase voltage is lagged by 30^0 with the line voltage V_{RY} according to the Y connected load. Hence the phase voltages are found as follows and the relation between the phase voltage and line voltage phasor form are shown in Fig.3.14.

$$V_{RN} = 231\angle -30^0\,V; V_{YN} = 231\angle -150^0\,V; V_{BN} = 231\angle -270^0\,V$$

Line current current (IL) $=10\angle -66.86^0$ A (– symbol indicates the lagging)

For Y connected load, Line current = Phase current

$$\text{Phase current} = \frac{\text{Phase Voltage}}{\text{Impedance per phase}}$$

$$I_R = \frac{V_{RN}}{Z/\phi}; I_Y = \frac{V_{YN}}{Z/\phi}; I_B = \frac{V_{BN}}{Z/\phi}$$

$$Z/\phi = \frac{V_{RN}}{I_R} = \frac{231}{10}$$

$$= 23.1\Omega$$

From the phasor diagram presented in Fig.3.14, the current lags the line voltage by 66.86^0 and lags the phase voltage by 36.86^0. Hence the power factor of the load is lagging one and the load will be the inductive load. Hence the impedance angle is 36.86^0.

$$Z/\Phi = 23.1\angle 36.86^0\ \Omega$$

$$= (18.48 + j13.85)\ \Omega = R + jX_L$$

Hence the resistance per phase is 18.48Ω and inductive reactance is 13.85Ω; Therefore, the inductance per phase is calculated as follows.

$$X_L = 2\pi f L$$

$$L = 44.08\ \text{mH}$$

Inductance per phase = 44.08mH

The power factor of the load is cos (36.86) = 0.8 (lagging).

Active power consumed by the load $P_{3\varphi} = \sqrt{3}V_L I_L \cos\cos\varphi = \sqrt{3} * 400 * 10 * 0.8 = 5542.56W$

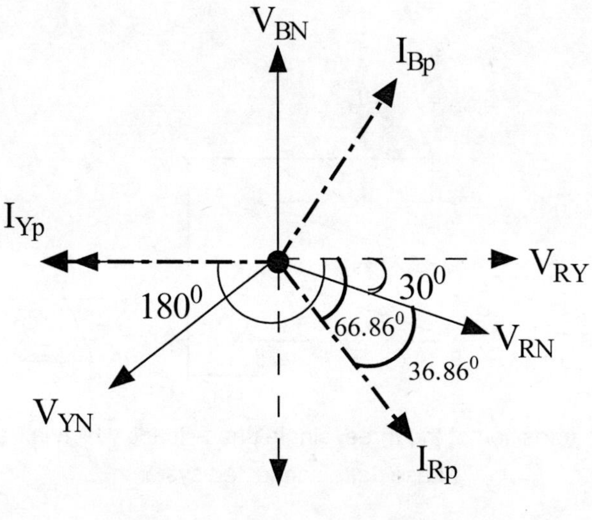

Fig.3.14: Phasor diagram of Example 3.4

3.2.2 Delta Connection

A three-phase delta connected system is shown in Fig.3.15, each single-phase load has two ends. One end of the load is connected to another load's one end. Similarly, all three loads are connected which forms the closed loop and looks like a delta symbol. Hence it is named as delta connection. In delta connection, the neutral of the one phase is joined with another phase of the three phase system which is indicated clearly in Fig.3.15. In delta connection, there is no physical neutral point. Hence, the line voltage (V_{RY}, V_{YB}, V_{BR}) measured between any of the two-line conductors (R-Y, Y-B, B-R), which is shown in Fig.3.16 that is equal to the phase voltage. From the pictorial representation, it is understood that the line voltage is equal to the phase voltage of the delta connected systems and presented in equations (3.23).

$$V_L = V_P \tag{3.23}$$

The current flows in any of the line conductors (R, Y, B) is known as line current($I_{R\ line}$, $I_{Y\ line}$, $I_{B\ line}$). The current flows through the phase coil are known as phase current and represented as(I_{Rp}, I_{Yp}, I_{Bp}) which are presented in Fig. 3.17. In order to obtain the relation between the line currents and phase currents, apply the KCL at the nodes of the delta connected loads. The KCL is applied at node R, the incoming currents is equal to the outgoing currents which is stated in equation (3.24). Similarly, KCL is applied to other nodes Y and B, the corresponding equations (3.25) and (3.26) are obtained respectively. The line currents are obtained by the phasor of the two phase currents which is shown in equation (3.24a) to (3.26a) and the corresponding phasor diagram is presented in Fig. 3.18.

$$I_{Rline} + I_{Bp} = I_{Rp} \tag{3.24}$$

$$I_{Yline} + I_{Rp} = I_{Yp} \tag{3.25}$$

$$I_{Bline} + I_{Yp} = I_{Bp} \tag{3.26}$$

$$I_{Rline} = I_{Rp} - I_{Bp} \tag{3.24a}$$

$$I_{Yline} = I_{Yp} - I_{Rp} \tag{3.21a}$$

$$I_{Bline} = I_{Bp} - I_{Yp} \tag{3.22a}$$

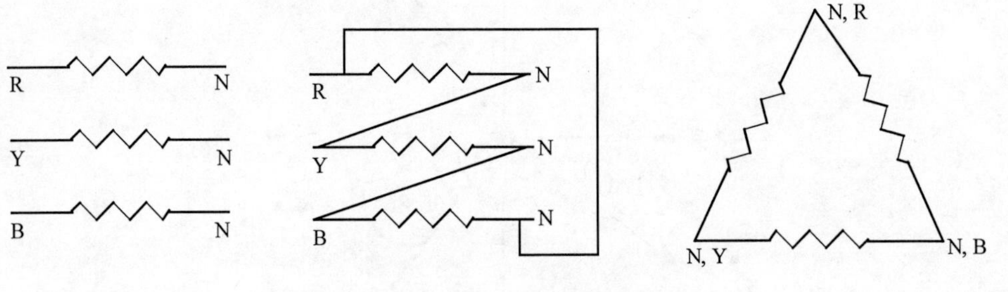

Fig. 3.15: A connection transition of the three, single phase loads which is transformed into the three-phase delta connected systems

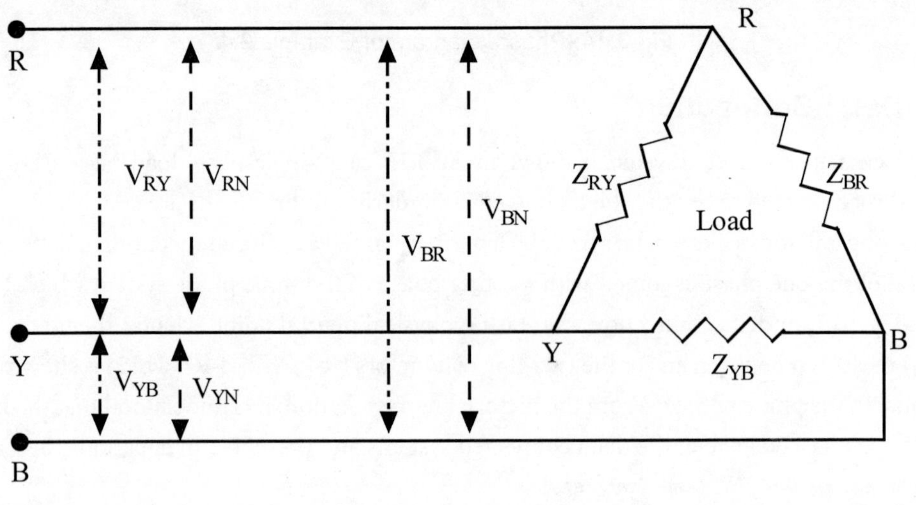

Fig. 3.16: Delta connected load with the information of the phase voltages and line voltages

The instantaneous value or time domain representation of the phase current is shown in equation (3.27)- (3.29). The phasor diagram and phasor form of these currents are represented in Fig. 3.18, and in equation (3.30) and also the phasor diagram for obtaining the line current is shown in Fig.3.19. The magnitude of the phase current is $|I_P|$ which is equal to I_m and explained in equation (3.20a).

$$i_{Rp}(t) = I_m \sin\omega t \tag{3.27}$$

$$i_{Yp}(t) = I_m \sin(\omega t - 120^0) \tag{3.28}$$

$$i_{Bp}(t) = I_m \sin(\omega t - 240^0) \tag{3.29}$$

$$I_{Rp} = I_m\angle 0;\ I_{Yp} = I_m \angle -120^0;;\ I_{Bp} = I_m \angle -240^0 \tag{3.30}$$

$$|I_{Rp}| = |I_{Yp}| = |I_{Bp}| = |I_P| = I_m \tag{3.30a}$$

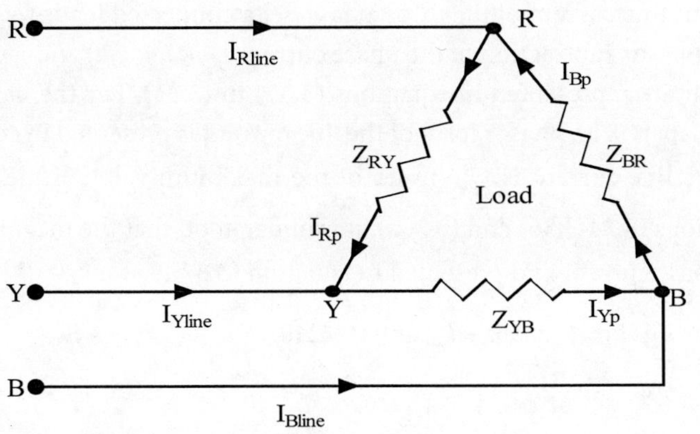

Fig. 3.17: Delta connected load with the information of the phase currents and line currents

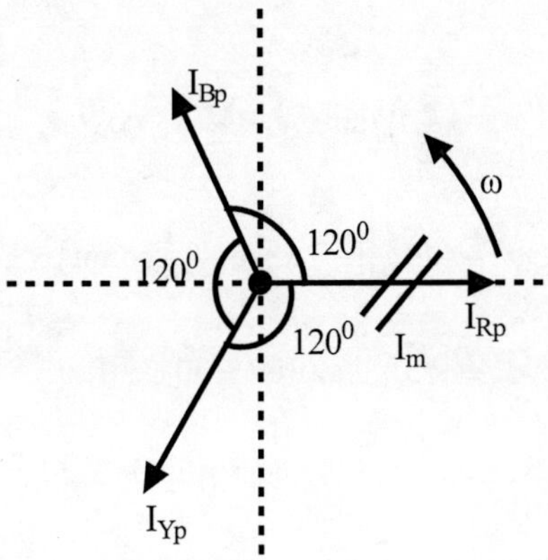

Fig. 3.18: Phasor diagram of the phase current in Δ connected 3Φ systems

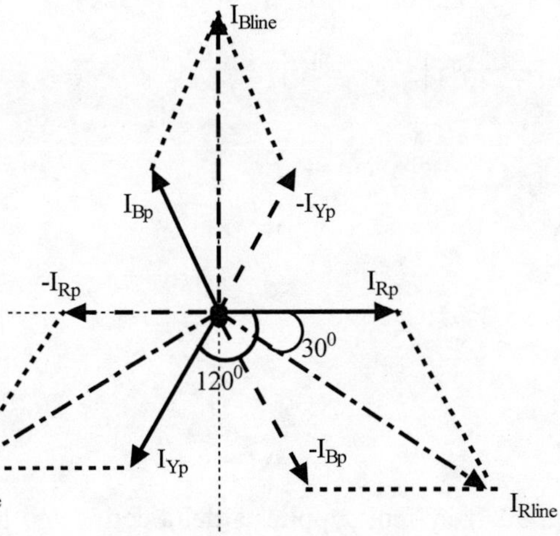

Fig. 3.19: Phasor diagram of the line current inΔ connected 3Φ systems

The instantaneous line current of the three-phase delta connected loads ($i_{Rline}(t)$, $i_{Yline}(t)$, $i_{Bline}(t)$) are calculated by the phasor subtraction of the phase current $i_{Rp}(t)$, $i_{Yp}(t)$ and $i_{Bp}(t)$ using the equations (3.27) to (3.29) which are represented in equations (3.31) to (3.33). For the balanced system, the line currents are also 1200apart. The phasor form of the line current is presented in equation (3.34) to (3.36). The magnitude of the line current is $\sqrt{3}$ times of the maximum value of the phase current (I_m). By comparing the equations (3.34-3.36) and (3.30a), it is understood that the magnitude of the line current is $\sqrt{3}$ times of the phase current and presented in equation (3.37).

$$i_{Rline}(t) = i_{Rp}(t) - i_{Bp}(t) = I_m sin\omega t - I_m \sin(\omega t - 240) \tag{3.31}$$

$$i_{Rline}(t) = I_m sin\omega t - I_m \sin\omega t\ cos240 + I_m \cos\omega t \sin 240$$

$$= I_m sin\omega t - I_m \sin\omega t * \left(-\frac{1}{2}\right) + I_m \cos\omega t * (-\frac{\sqrt{3}}{2})$$

$$= \frac{3}{2} I_m \sin\omega t - \frac{\sqrt{3}}{2} I_m \cos\omega t$$

$$= \sqrt{3} I_m [\frac{\sqrt{3}}{2} sin\omega t - \frac{1}{2} cos\omega t] = \sqrt{3} I_m [sin\omega t\ cos30^0 - cos\omega t\ sin30^0]$$

$$= \sqrt{3} I_m \left[\sin(\omega t - 30^0)\right]$$

$$i_{Yline}(t) = i_{Yp}(t) - i_{Rp}(t) = I_m \sin(\omega t - 120^0) - I_m \sin(\omega t) \tag{3.32}$$

$$i_{Yline}(t) = \sqrt{3} I_m \left[\sin(\omega t - 150^0)\right]$$

$$i_{Bline}(t) = i_{Bp}(t) - i_{Yp}(t) = I_m \sin(\omega t - 240) - I_m \sin(\omega t - 120) \tag{3.33}$$

$$i_{Bline}(t) = \sqrt{3} I_m \left[\sin(\omega t - 270^0)\right]$$

$$I_{Rline} = \sqrt{3} |I_m| \angle -30^0 \tag{3.34}$$

$$I_{Yline} = \sqrt{3} |I_m| \angle -150^0 \tag{3.35}$$

$$I_{Bline} = \sqrt{3} |I_m| \angle -270^0 \tag{3.36}$$

$$|I_{line}| = \sqrt{3} |I_P| \tag{3.37}$$

Example 3.5:

A three-phase 3 wire 415V, 50Hz system supplies a delta connected load which consists of resistance 20Ω and inductance 10mH in series. Determine the phase current, line current and also draw the phasor diagram.

Solutions:

The supply voltage is $415\angle 0^0$ V and the system is described by Fig. 3.20. Let us assume the phase sequence is RYB. Hence the line voltage of the delta connected load is mentioned below and its phasor diagram is shown in Fig.3.21.

$$V_{RY} = 415\angle 0^0\,\text{V}; V_{YB} = 415\angle -120^0\,\text{V}; V_{BR} = 415\angle -240^0\,\text{V}$$

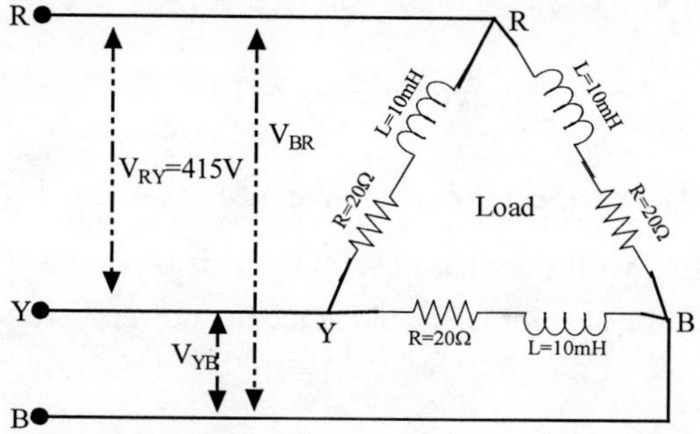

Fig. 3.20: Delta connected load (Example 3.5)

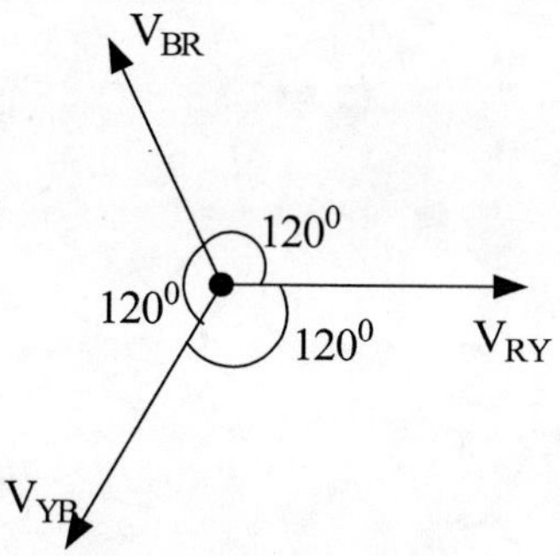

Fig. 3.21: The phasor diagram of the three-phase line voltages of Example 3.5

The inductive load per phase $\left(\frac{L}{\varphi}\right)$ = 10mH;

The inductive reactance per phase $X_L / \varphi) = (2\pi fL)\Omega \;\left(2\pi * 50 * 10 * 10^{-3}\right) = 3.14\Omega$

The load per phase (Z/φ) = (20 + j3.14)Ω

In delta connected load, the line voltage is equal to the phase voltage. Hence,

$$\text{Phase current} = \frac{\text{Phase Voltage}}{\text{Impedance per phase}}$$

$$I_{Rp} = \frac{V_{RY}}{Z/\phi}; I_{Yp} = \frac{V_{YB}}{Z/\phi}; I_{Bp} = \frac{V_{BR}}{Z/\phi}$$

$$I_{Rp} = \frac{V_{RY}}{Z/\phi} = \frac{415\angle 0}{(20 + j3.14)}$$

$$= \frac{415\angle 0}{20.24\angle 8.92}$$

$$= 20.5\angle -8.92^0 \text{ A}$$

Since the load is balanced, the phase current in other phases is easily written as

$$I_{Yp} = 20.5\angle -128.92^0 \text{ A}$$

$$I_{Bp} = 20.5\angle -248.92^0 \text{ A}$$

Whereas in Δ connected loads, the line current magnitude is equal to $\sqrt{3}$ times of the phase current. The line current is 300 lagging with the phase current which is represented in Fig.3.22. The angle (Φ) between the line voltage and line current is 8.920. Hence the power factor of the load is (cos(Φ)), which is equal to 0.987 (lagging).

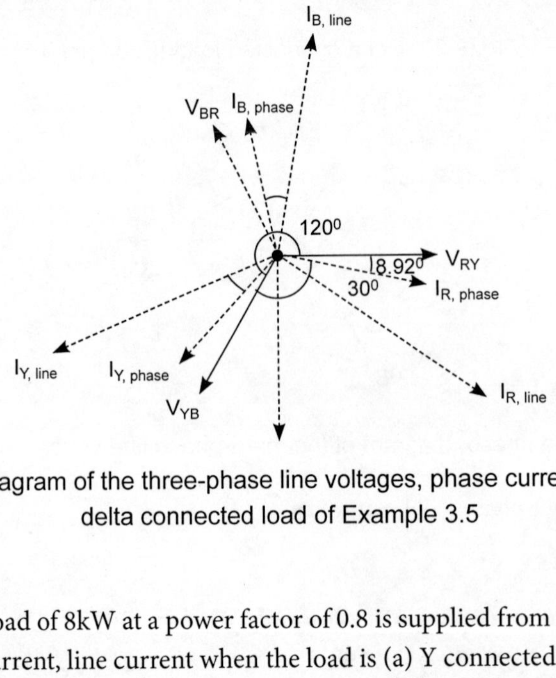

Fig. 3.22: The phasor diagram of the three-phase line voltages, phase current and line current of the delta connected load of Example 3.5

Example 3.6:

A balanced three-phase load of 8kW at a power factor of 0.8 is supplied from 3 phase 3 wire 415V, 50Hz system. Find the phase current, line current when the load is (a) Y connected (b) Δ connected.

Solutions:

$$\text{Real power } (P_{3\Phi}) = \sqrt{3}V_L I_L \cos\varphi = \sqrt{3}(415)I_L(0.8)$$

$$(P_{3\Phi}) = 8000 = \sqrt{3}(415)I_L(0.8)$$

$$I_L = \frac{8000}{\sqrt{3}(415)(0.8)} = 13.91A$$

(a) Y connected load:

Line current = Phase current

$I_L = 13.91\text{A}$

(b) Δ connected load:

Line current = $\sqrt{3}$ * Phase current

$I_L = \sqrt{3} * I_{Ph} = 13.91\text{A}$

$I_{Ph} = \frac{I_L}{\sqrt{3}} = 8.03A$

3.3 BALANCED AND UNBALANCED LOAD

In general, unbalanced system occurs due to unbalanced voltage sources or unbalanced loads. In this section, the assumption is made that the unbalanced system occurs due to the unbalanced loads only. The balanced load indicates that the load on each phase of the three-phase load has an equal value. The three-phase balanced load could be connected in the form of Y or Δ, the calculation of the voltage and current are similar to the Y connected or Δ connected systems discussed in the earlier sections.

The unbalanced load indicates that the unequal value of the load is connected in each phase of the three-phase loads. When the unbalanced load is powered by the three-phase balanced supply, the behaviour is different for Y and Δ connected loads.

3.3.1 Star connected unbalanced load

The star or Y connected unbalanced load (shown in Fig.3.23) is connected to the three phase balanced supply systems (not shown in Fig.3.23). The loads R_R, R_Y and R_B have unequal values; hence it is named as unbalanced load. The line currents are calculated according to Ohm's law and presented in equation (3.38). These values are not equal, since the load values are different. Hence, in order to find the neutral current, apply KCL at the neutral junction and the corresponding equation is in equation (3.39) and find the neutral current in equation (3.40). The neutral current is not zero as in balanced three phase systems. The neutral wire should have the low impedance value in order to carry the neutral current and this value should be equal to or less than the line current of the system. The neutral of the load should be connected to the neutral of the supply, in order to maintain the same potential. Three-phase connected load shown in Fig.3.23 is a three-phase four wire model. Whereas the three-phase three wire model has the unbalanced load, the neutral point is known as the floating neural point which is rarely found in practice. The analysis of such a system is different from the three-phase 4 wire load model.

$$I_R = \frac{V_{RN}}{R_R}; I_Y = \frac{V_{YN}}{R_Y}; I_B = \frac{V_{BN}}{R_B} \tag{3.38}$$

$$I_R + I_Y + I_B + I_N = 0 \tag{3.39}$$

$$I_N = -(I_R + I_Y + I_B) \tag{3.40}$$

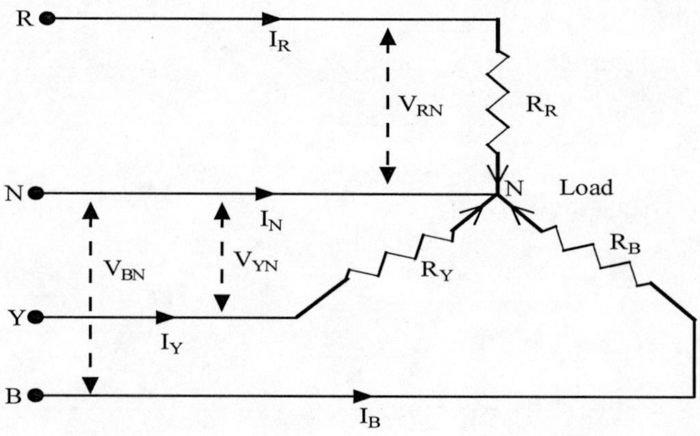

Fig. 3.23: Star or Y connected unbalanced three-phase load

3.3.2 Delta Connected Unbalanced Load

The unbalanced Δ connected load does not present the new problems like the unbalanced Y connected loads which is presented in Fig. 3.24. The loads are not equal ($Z_{RY} \neq Z_{YB} \neq Z_{RR}$). The current flowing in the load depends on the load which are different for each phase. The voltages across each phase or load remains the same but having the phase angle difference of 1200 like the balanced system. Hence the load current of each phase (phase current) is calculated according to the equation (3.41). The line current of the unbalanced Δ connected load is found using equations (3.42) to (3.44).

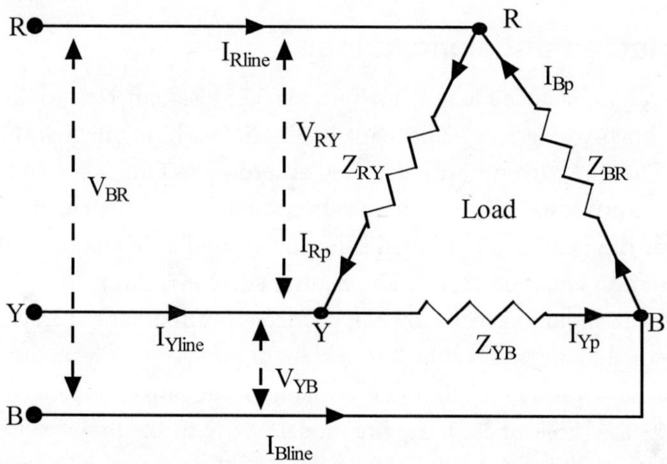

Fig. 3.24: Delta or Δ connected unbalanced three-phase load

$$I_{Rp} = \frac{V_{RY}}{Z_{RY}}; I_{Yp} = \frac{V_{YB}}{Z_{YB}}; I_{Bp} = \frac{V_{BR}}{Z_{BR}} \tag{3.41}$$

$$I_{R\,line} = I_{RP} - I_{BP} \tag{3.42}$$

$$I_{Y\,line} = I_{YP} - I_{RP} \tag{3.43}$$

$$I_{B\,line} = I_{BP} - I_{Yp} \tag{3.44}$$

3.4 POWER IN THREE-PHASE SYSTEMS (BALANACED SYSTEM)

A power is consumed by three-phase load is the sum of the power consumed by each phase. For the balanced system, the three-phase real power ($P_{3\varphi}$) is equal to three times of the single-phase real power ($P_{1\varphi}$).

$$P_{3\varphi} = 3 \times P_{1\varphi} \tag{3.45}$$

$$= 3 \times V_{ph} \times I_{ph} \times \text{coss}(\theta)$$

Where V_{ph}, I_{ph} are the single-phase voltage and phase current respectively. θ is the angle between the single-phase voltage and phase current.

For star connected systems, $V_L = \sqrt{3}V_{ph}; I_L = I_{ph}$ according to equation (3.15) and (3.19). Hence, the three-phase power could be calculated as,

$$P_{3\varphi} = 3 \times V_{ph} \times I_{ph} \times \text{coss}(\theta)$$

$$= 3 \times \frac{V_L}{\sqrt{3}} \times I_L \times \cos(\theta)$$

$$P_{3\varphi} = \sqrt{3} V_L I_L \cos(\theta) \tag{3.46}$$

For delta connected systems, $V_L = V_{ph}; I_L = \sqrt{3}I_{ph}$ according to equation (3.23) and (3.37). Hence, the three-phase power could be calculated as,

$$P_{3\varphi} = 3 \times V_{ph} \times I_{ph} \times \text{coss}(\theta)$$

$$= 3 \times XV_L \times \frac{I_L}{\sqrt{3}} \times \cos(\theta)$$

$$P_{3\varphi} = \sqrt{3} V_L I_L \cos(\theta) \tag{3.47}$$

Hence, the reactive power for the three-phase systems could be written as shown in equation (3.48) for both star and delta connection.

$$Q_{3\varphi} = \sqrt{3} V_L I_L \sin(\theta) \tag{3.48}$$

The total apparent power of three-phase system ($S_{3\varphi}$) is calculated using the equation (3.49) or (3.50).

$$S_{3\varphi} = \sqrt{3} V_L I_L \tag{3.49}$$

$$S_{3\varphi} = \sqrt{P_{3\varphi}^2 + Q_{3\varphi}^2} \tag{3.50}$$

3.5 MEASUREMENT OF POWER IN THREE-PHASE SYSTEMS

A wattmeter is the device which is used for measuring the average or real power in AC circuits. The measurement of single-phase power is discussed in Chapter 1. Three-phase power measurement is presented in this section. The active power (real power) measurement of three-phase system is performed with three methods namely,

- Three wattmeter method
- Two wattmeter method
- One wattmeter method

Both three wattmeter and two wattmeter methods are used for balanced as well as for unbalanced systems. One wattmeter method is used for the balanced systems only.

3.5.1 Three Wattmeter Method

In order to measure the three-phase real power, three wattmeters are used and those have been connected as per the circuit shown in Fig.3.25. The load shown in Fig.3.25 is the star connected load. But this circuit is applicable to the delta or Δ connected load also. When the load is balanced, the real power of each phase remains the same which could be easily measured with single wattmeter from any one of the phases. P_1, $P_2$and P_3 are the power measured in phase R, phase Y and Phase B respectively. Hence the three-phase real power for the balanced load is the sum of powers in each phase, $P_{3\Phi} = P_1+P_2+P_3$. The measured power from each wattmeter is namely W_1, $W_2$and W_3 from the three-wattmeter method for the three-phase balanced load. Hence the three-phase real power for the balanced load is the sum of wattmeter values $W_1+W_2+ W_3$ which is equivalent to $P_{3\Phi}$. The three-phase power could be written as per the circuit in equation (3.51).

$$P_{3\varphi} = real part of \left(e_R i_R + e_Y i_Y + e_B i_B\right) = \sqrt{3} V_L I_L \cos(\varphi) \tag{3.51}$$

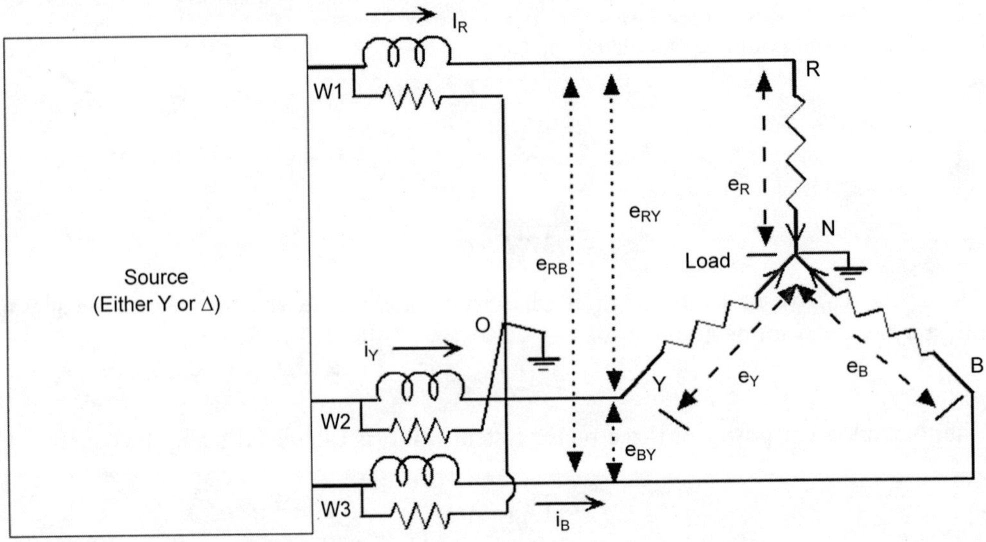

Fig. 3.25: Three wattmeter method for measuring three-phase power

3.5.2 Two Wattmeter Method

In this method, the current coils of the two wattmeters are connected to any two-phase lines among the three phase lines and the potential coil is connected between any two of the phases. As shown in Fig. 3.26, the current coil of the first and second wattmeter is connected to R and B phase respectively. The potential coil of the first and second wattmeter is connected between R and Y, B and Y phase respectively.

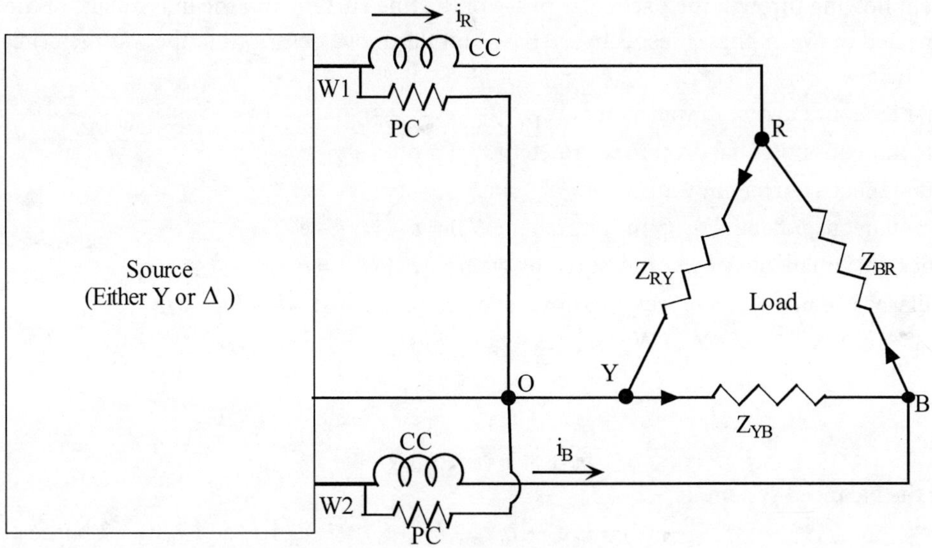

(a) Y connected load

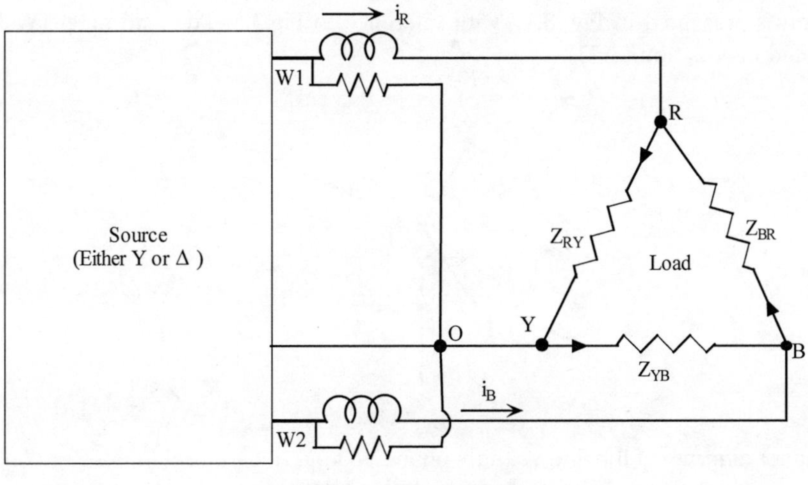

(b) Δ connected load

Fig. 3.26: Two wattmeter method for measuring three-phase power

The potential coil of the first and second wattmeter is connected between R and Y, B and Y phase respectively.

The current coils of the two phases are connected in series with the phase and the potential coils are connected between two phases as presented in Fig. 3.26.

Here, a star-connected load is considered that is explained for the two-wattmeter method. The same concept could be applied to Δ-connected loads. Since, a Δ-connected load could be replaced by an equivalent Y-connected load. The proof is true whether the load is balanced or unbalanced. The two-wattmeter method measures the average power consumed by the three-phase loads, due to the inertia of the moving system. The currents through and potential difference across each wattmeter measures the current flowing through the particular phase of the line current and the line voltage of the potential coil connected between phases respectively. Based on the above description, the wattmeter readings are indicated below.

Instantaneous current in wattmeter (W_1) = i_R

Instantaneous potential drop in wattmeter (W_1) = $e_{RY} = e_R - e_Y$

Instantaneous current in wattmeter (W_2) = i_B

Instantaneous potential drop in wattmeter (W_2) = $e_{BY} = e_B - e_Y$

Hence, instantaneous power measured by wattmeter (W_1) = $i_R * (e_R - e_Y)$

Similarly, instantaneous power measured by wattmeter (W_2) = $i_B * (e_B - e_Y)$

$$W_1 + W_2 = i_R^*(e_R - e_Y) + i_B^*(e_B - e_Y)$$
$$= i_R e_R - i_R e_Y + i_B e_B - i_B e_Y$$
$$= i_R e_R + i_B e_B - i_R e_Y - i_B e_Y$$
$$= i_R e_R + i_B e_B - e_Y (i_R + i_B)$$

For the balanced system, $i_R + i_Y + i_B = 0$;

$$W_1 + W_2 = i_R e_R + i_B e_B - e_Y(-i_Y) = i_R e_R + i_Y e_Y + i_B e_B = 3 \text{ phase power}$$

Let us assume that the phase sequence is RYB and load impedance is $Z_R = Z_R \angle\theta$. The phase angle of that phase voltage iszero and hence, the current in that phase is lagging by an angle θ and the concerned phasor diagram is presented in Fig. 3.27 with reference to Fig.3.6. The wattmeter (W_1) reads the real power (P_1) stated in equation (3.52).

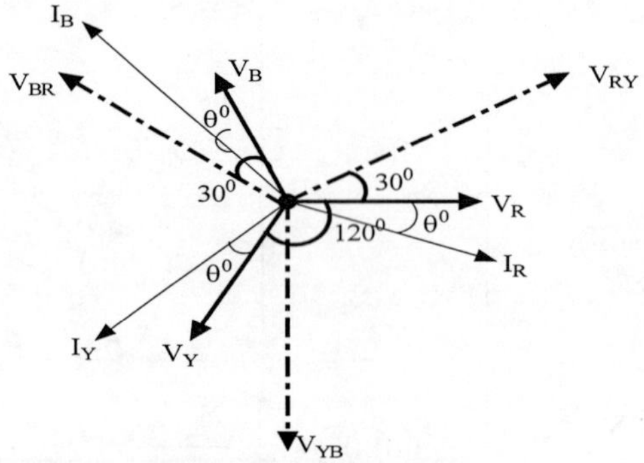

Fig. 3.27: Phasor diagram of the line voltage, phase voltage and current in the star connected load for two-wattmeter method

$$P_1 = Real\left[V_{RY} I_R^*\right] = V_{RY} I_R \cos(\theta + 30) = V_L I_L \cos(\theta + 30) \tag{3.52}$$

Similarly, the second wattmeter (W_2) reads the real power (P_2) which could be represented in equation (3.53).

$$P_2 = Real\left[V_{BR} I_B^*\right] = V_{BR} I_B \cos(30 - \theta) = V_{BR} I_B \cos(\theta - 30) = V_L I_L \cos(\theta - 30) \tag{3.53}$$

By the addition of these two wattmeter readings, the equation (3.54) is obtained. After the simplification (using cos(A+B) and cos(A-B) trigonometry formulae), the obtained value is equal to the three-phase average power of the system.

$$P_1 + P_2 = V_L I_L \cos\left(\theta + 30^0\right) + V_L I_L \cos\left(\theta - 30^0\right) = \sqrt{3} V_L I_L \cos\theta \tag{3.54}$$

Using the measured values, perform P_1-P_2. Three-phase reactive power $Q_{3\Phi}$ is obtained using equation (3.55). Whereas the total apparent power for the three-phase systems could be obtained using equation (3.56). The tangent angle and power factor is also calculated using (3.54) and (3.55) and those are stated in equations (3.57) and (3.58) respectively. The two-wattmeter method is capable to provide three-phase power and is also able to obtain the power factor.

$$P_1 - P_2 = V_L I_L \cos\left(\theta + 30^0\right) - V_L I_L \cos\left(\theta - 30^0\right) = -V_L I_L \sin\theta$$

OR

$$P_2 - P_1 = V_L I_L \sin\theta$$

$$Q_{3\varphi} = \sqrt{3} V_L I_L \sin\theta = \sqrt{3}(P_2 - P_1) \tag{3.55}$$

$$S_{3\varphi} = \sqrt{P_{3\varphi}^2 + Q_{3\varphi}^2} \tag{3.56}$$

$$tan\theta = \frac{Q_{3\varphi}}{P_{3\varphi}} = \frac{\sqrt{3}(P_2 - P_1)}{(P_1 + P_2)} \tag{3.57}$$

$$cos\theta = \cos[\tan^{-1}\left(\frac{\sqrt{3}\left(P_2 - P_1\right)}{\left(P_1 + P_2\right)}\right)] \tag{3.58}$$

Using (3.58), the types of loads connected system could be identified as follows.

- When $P_1 = P_2$, it is resistive load. Since, the power factor becomes unity.
- When $P_2 > P_1$, it is inductive load. Since, the power factor becomes positive.
- When $P_2 < P_1$, it is capacitive load. Since, the power factor becomes negative.

3.5.3 One Wattmeter Method

One wattmeter method is used only for the balanced load systems and the connection diagram is shown in Fig. 3.28. The current coil is connected to one of the phase lines. As per the connection diagram, current in R phase is measured. Whereas one end of the potential coil is connected to the R phase and the other end of the potential coil is connected to node 1. Node 1 is connected to either node 2 or node 3. When the node 1 is connected to node 2 and node 3, the corresponding measured potential is and respectively. Hence two wattmeter readings are obtained, which could be denoted as W_1 and W_2. The equation (3.59) is similar to the equations used in the two-wattmeter method; hence the measured power is the three-phase real power or active power. This is proved by a phasor diagram which is shown in Fig. 3.29. The power measured by the wattmeter when connected to node 2 and node 3 is active power and denoted as P_1 and P_2which is represented in equation (3.60) and (3.61) respectively.

$$W_1 + W_2 = i_R{}^*(e_{RY}) + i_R{}^*(e_{RB})$$

$$W_1 + W_2 = i_R{*}(e_R - e_Y) + i_R{*}(e_R - e_B) \quad (3.59)$$

$$= i_R\, e_R - i_R\, e_Y + i_R\, e_R - i_R\, e_B)$$

$$= 2i_R\, e_R - i_R\,(e_Y + e_B)$$

[For balanced system, $e_R + e_Y + e_B = 0$]

$$= 2i_R\, e_R + i_R\, e_R$$

$$= 3i_R\, e_R = 3{*}\text{single phase power}$$

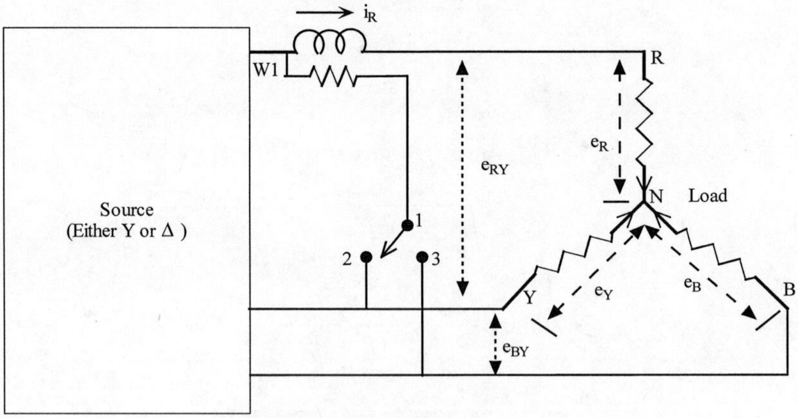

Fig. 3.28: One wattmeter method for measuring three-phase power

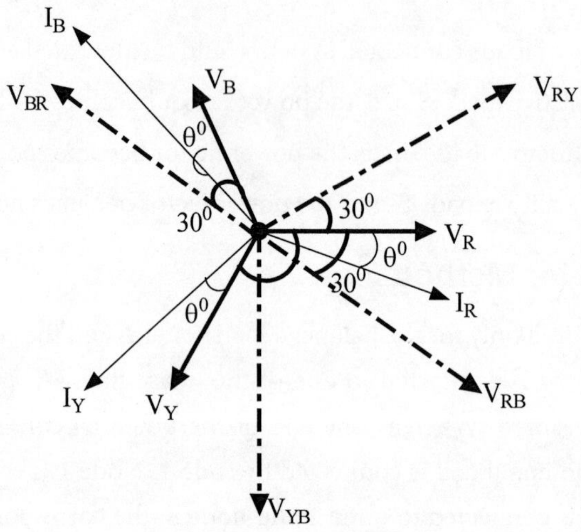

Fig. 3.29: Phasor diagram of the line voltage, phase voltage and current in the star connected load for one wattmeter method

$$P_1 = Real\left[V_{RY} I_R^*\right] = V_{RY} I_R \cos\left(\theta + 30^0\right) = V_L I_L \cos\left(\theta + 30^0\right) \quad (3.60)$$

$$P_2 = Real\left[V_{RB} I_R^*\right] = V_{RB} I_R \cos\left(30 - \theta\right) = V_{RB} I_R \cos\left(\theta - 30^0\right) = V_L I_L \cos\left(\theta - 30^0\right) \quad (3.61)$$

By the addition of these powers, the equation (3.62) is obtained. After the simplification (using cos(A+B) and cos(A-B) trigonometry formulae), the obtained value is equal to the three-phase average power of the system.

$$P_1 + P_2 = V_L I_L \cos\left(\theta + 30^0\right) + V_L I_L \cos\left(\theta - 30^0\right) = \sqrt{3} V_L I_L cos\theta \qquad (3.62)$$

Example 3.7:

A balanced Y connected three-phase load of $20\angle 30\Omega$ per phase is supplied from a 3 phase 3 wire 400V, 50Hz system. Calculate (a) the phase voltage (b) the line current. Draw the phasor diagram. Find the wattmeter readings if two wattmeters are used for the power measurement. Also calculate the reactive power and the power factor using the wattmeter readings.

Solutions:

The supply voltage is $400\angle 0^0$ V. Let us assume the phase sequence is RYB. Hence the line voltage of the delta connected load is mentioned below and its phasor form is represented in Fig.3.30.

$$V_{RY} = 400\angle 0^0 \text{V}; V_{YB} = 400\angle -120^0 \text{V}; V_{BR} = 400\angle -240^0 \text{V}$$

The load per phase $(Z/\varphi) = 20\angle 30\Omega$

In star connected load, the line voltage is equal to $\sqrt{3}$ times of the phase voltage by magnitude. Hence,

$$\text{Phase voltage} = \frac{line\ Voltage}{\sqrt{}} = 231\text{V}$$

The angle between the line voltage and phase voltage is 30^0. For Example, the line voltage leads the phase voltage by 30^0.

$$V_{RN} = 231\angle -30^0 \text{V}; V_{YN} = 231\angle -150^0 \text{V}; V_{BN} = 231\angle -270^0 \text{V}$$

$$\text{Phase current} = \frac{\text{Phase Voltage}}{\text{Impedance per phase}}$$

$$I_{Rp} = \frac{V_{RN}}{Z/\phi}; I_{Yp} = \frac{V_{YN}}{Z/\phi}; I_{Bp} = \frac{V_{BN}}{Z/\phi}$$

$$I_{Rp} = \frac{V_{RN}}{Z/\phi} = \frac{231\angle -30^0}{20\angle 30^0}$$

$$= 11.55\angle -60^0 \text{ A}$$

Since the load is balanced, the phase current in other phases is easily written as

$$I_{YP} = 11.55\angle -180^0\text{A}$$

$$I_{BP} = 20.5\angle -300^0\text{A}$$

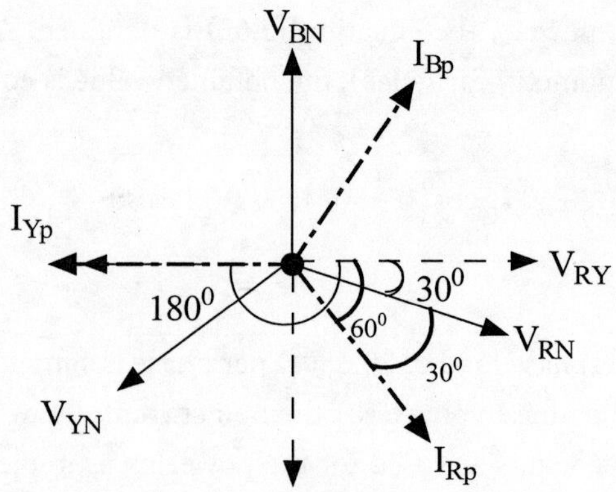

Fig. 3.30: Phasor diagram of Example 3.7

The angle (Φ) between the phase voltage and phase current is 30^0 (lagging). Hence the power factor of the load is ($\cos(\Phi)$), which is equal to 0.866. W_1 and W_2 are the wattmeter readings and they represented as P_1 and P_2 respectively.

Real power $P = \sqrt{3}V_L I_L \cos\varphi = \sqrt{3}(400)(11.55)\cos(30^0) = 6930W$

Real power $P = P_1 + P_2 = 6930W$

As we know, $\tan\varphi = \dfrac{\sqrt{3}(P_2 - P_1)}{P_1 + P_2}$

$$P_2 - P_1 = \frac{\tan\varphi * (P_1 + P_2)}{(\sqrt{3})} = 2310\text{W}$$

Hence the wattmeter readings are P_1 = 2310W; P_2 = 4620W.

Reactive power $Q = \sqrt{3}V_L I_L \sin\varphi = \sqrt{3}(400)(11.55)\sin(30^0) = 4001.04$ VAR

$$\text{Power factor} = cos\theta = \cos[\tan^{-1}\left(\frac{\sqrt{3}(P_2 - P_1)}{(P_1 + P_2)}\right)]$$

$$cos\theta = \cos[\tan^{-1}\left(\frac{\sqrt{3} \times 2310}{6930}\right) = \cos\left(29.99^0\right) = 0.866 \text{ (lagging)}$$

Example 3.8:

A balanced Δ connected three-phase load of $10\angle 60\Omega$ per phase is supplied from a 3 phase 3 wire 400V, 50Hz system. Calculate (a) the phase voltage (b) the line current. Draw the phasor diagram.

Find the wattmeter readings if the two wattmeter is used for the power measurement. Also calculate reactive power and the power factor using the wattmeter readings.

Solutions:

The supply voltage is $400\angle 0^0$ V. Let us assume the phase sequence is RYB. Hence the line voltage of the delta connected load is mentioned below and its phasor form is represented in Fig.3.31.

$V_{RY} = 400\angle 0^0\,V; V_{YB} = 400\angle -120^0\,V; V_{BR} = 400\angle -240^0\,V$

The load per phase (Z/φ)= $10\angle 60^0\ \Omega$

In delta connected load, the line current is equal to $\sqrt{3}$ times of the phase current by magnitude. The phase voltage is equal to the line voltage.

$V_{RN} = 400\angle 0^0\,V; V_{YN} = 400\angle -120^0\,V; V_{BN} = 400\angle -240^0\,V$

$$\text{Phase current} = \frac{\text{Phase Voltage}}{\text{Impedance per phase}}$$

$$I_{Rp} = \frac{V_{RN}}{Z/\phi}; I_{Yp} = \frac{V_{YN}}{Z/\phi}; I_{Bp} = \frac{V_{BN}}{Z/\phi}$$

$$I_{RPhase} = \frac{V_{RN}}{Z/\phi} = \frac{400\angle 0^0}{10\angle 60^0}$$

$$= 40\angle -60^0\,A$$

Since the load is balanced, the phase current in other phases is easily written as

$$I_{YPhase} = 40\angle -180^0 A$$
$$I_{BPhase} = 40\angle -300^0 A$$

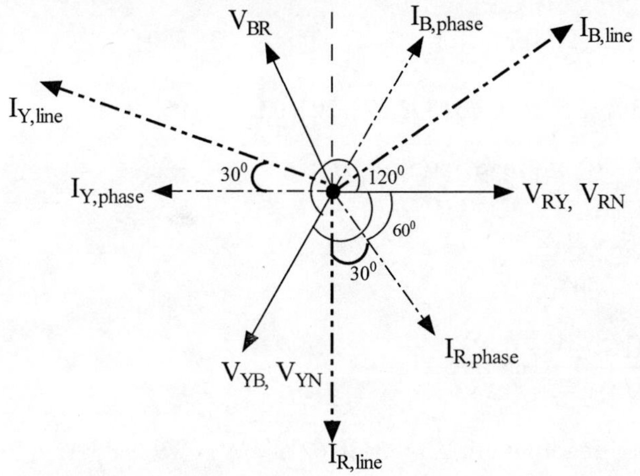

Fig. 3.31: Phasor diagram of Example 3.8

In delta connected load, the line current lags 300 with the phase current. Therefore, the line currents are mentioned as follows.

$I_{R,line} = 40\sqrt{3}\angle -90^0\,A; I_{Y,line} = 40\sqrt{3}\angle -210^0\,A; I_{R,line} = 40\sqrt{3}\angle -330^0\,A$

The angle (Φ) between the phase voltage and phase current is 600. Hence the power factor of the load is (cos(Φ)), which is equal to 0.5 (lagging).

Real power $P = \sqrt{3}V_L I_L \cos\varphi = \sqrt{3}(400)(40\sqrt{3})\cos(60) = 24000W$ =24kW

Real power $P = W_1 + W_2 = 24kW$

As we know, $\tan\varphi = \dfrac{\sqrt{3}(W_2 - W_1)}{W_1 + W_2}$

$W_2 - W_1 = \dfrac{\tan\varphi * (W_1 + W_2)}{(\sqrt{3})}$ =24000W

Hence the wattmeter readings are W_1 = 0W; W_2 = 24000W.

Reactive power $Q = \sqrt{3}V_L I_L \sin\varphi = \sqrt{3}(400)(40\sqrt{3})\sin(60) = 41569.21VAr$

Example 3.9:

A phase voltage and current of the balanced Y connected three-phase load are 230V, 20A and the power factor is 0.8(lag). If the power of the load is measured by two wattmeters, find the wattmeter readings.

Solutions:

The phase voltage (V_{ph}) is 230V and the phase current (I_{ph}) is 20A. For the Y connected load, $I_L = I_{Ph}$ means that the line current is equal to the phase current. $V_L = \sqrt{3}V_{Ph}$, The line voltage is $\sqrt{3}$ times of the phase voltage.

$V_L = \sqrt{3}V_{Ph}$ = 398.37V; Power factor (cos φ) = 0.8; φ = 36.860

Real power $P = \sqrt{3}V_L I_L \cos\varphi = \sqrt{3}(398.37)(20)(0.8) = 11039.95$ W

Real power $P = W_1 + W_2 = 11039.95W$

As we know, $\tan\varphi = \dfrac{\sqrt{3}(W_2 - W_1)}{W_1 + W_2}$

$W_2 - W_1 = \dfrac{\tan\varphi * (W_1 + W_2)}{(\sqrt{3})}$ =4778.71W

Hence the wattmeter readings are W_1 = 3130.62W; W_2 = 7909.33W.

Example 3.10:

A phase voltage and current of the balanced Δ connected three-phase load is 230V, 20A and the power factor is 0.8(lag). If the power is measured by two wattmeters, find the wattmeter readings.

Solutions:

The phase voltage (V_{ph}) is 230V and the phase current (I_{ph}) is 20A. For the Δ connected load, $I_L = \sqrt{3}I_{Ph}$ means that the line current is equal to $\sqrt{3}$ times of the phase current. $V_L = V_{Ph}$, The line voltage is equal to the phase voltage.

$V_L = V_{Ph}; I_L = \sqrt{3} I_{Ph} = 20\sqrt{3} A$; Power factor (cos φ) = 0.8; φ = 36.860

$$\text{Real power } P = \sqrt{3} V_L I_L \cos\varphi = \sqrt{3}(230)(20\sqrt{3})(0.8) = 11040 \text{ W}$$

$$\text{Real power } P = W_1 + W_2 = 11040W$$

$$\text{As we know, } \tan\varphi = \frac{\sqrt{3}(W_2 - W_1)}{W_1 + W_2}$$

$$W_2 - W_1 = \frac{\tan\varphi * (W_1 + W_2)}{(\sqrt{3})} = 4778.71\text{W}$$

Hence the wattmeter readings are W_1 =3130.62W; W_2 =7909.33W.

Example 3.11:

The power consumed by the load is measured by two wattmeter methods and its readings are 4.5kW and 10kW and the line voltage of the system is 410V. Find (a) the power factor and (b) the line current.

Solutions:

The supply voltage is $410\angle 0^0$ V and the wattmeter readings are W_1 = 4.5kW; W_2 = 10kW.

$$W_1 + W_2 = 14.5kW; W_2 - W_1 = 5.5kW;$$

$$\tan\varphi = \frac{\sqrt{3}(W_2 - W_1)}{W_1 + W_2}$$

$$\tan\varphi = \frac{\sqrt{3}(5.5k)}{14.5k}$$

φ = 33.26; Power factor, cosφ = 0.836

$$\text{Real power } P = W_1 + W_2 = \sqrt{3} V_L I_L \cos\varphi$$

$$14.5k = \sqrt{3}(410)(I_L)(0.836)$$

$$I_L = 24.42\text{A}$$

Hence line current value is 24.42A.

Example 3.12:

A balanced star connected load, each phase having a resistance of 10 Ω and inductive reactance of 30 Ω is connected to 400V, 50Hz supply. The phase sequence is red, yellow and blue. Wattmeters connected to read total power have their current coils in the red and blue lines respectively. Calculate the reading on each wattmeter and draw a phasor diagram in explanation.

Solutions:

The supply voltage is $400\angle 0^0$ V, 50Hz and the phase sequence is RYB. Hence the line voltage of the star connected load is mentioned below and its phasor form is represented in Fig.3.32.

$V_{RY} = 400\angle 0^0\,V; V_{YB} = 400\angle -120^0\,V; V_{BR} = 400\angle -240^0\,V$

$Z_{ph}\text{per phase} = (10 + j30)\Omega = 31.62\angle 71.56\Omega$

In star connected load, the line voltage is equal to $\sqrt{3}$ times of the phase voltage by magnitude. The phase current is equal to the line current.

$V_{RN} = 231\angle -30^0\,V; V_{YN} = 231\angle -150^0\,V; V_{BN} = 231\angle -270^0\,V$

$$\text{Phase current} = \frac{\text{Phase Voltage}}{\text{Impedance per phase}}$$

$$I_{Rp} = \frac{V_{RN}}{Z/\phi}; I_{Yp} = \frac{V_{YN}}{Z/\phi}; I_{Bp} = \frac{V_{BN}}{Z/\phi}$$

$$I_{RPhase} = \frac{V_{RN}}{Z/\phi} = \frac{231\angle -30}{31.62\angle 71.56}$$

$$= 7.31\angle - 101.56^0\ A$$

Since the load is balanced, the phase current in other phases is easily written as

$$I_{YPhase} = 7.31\angle - 221.56^0\ A$$

$$I_{BPhase} = 7.31\angle - 341.56^0\ A$$

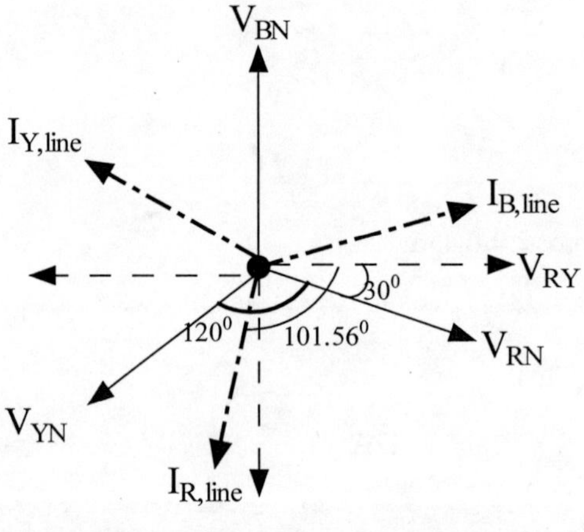

Fig. 3.32: Phasor diagram of Example 3.12

The angle (Φ) between the phase voltage and phase current is 71.560 (lagging). Hence the power factor of the load is (cos(Φ)), which is equal to 0.316.

Real power $P = \sqrt{3}V_L I_L \cos\varphi = \sqrt{3}(400)(7.31)\cos(71.56) = 1601.96W$

Real power $P = W_1 + W_2 = 1601.96W$

As we know, $\tan\varphi = \dfrac{\sqrt{3}(W_2 - W_1)}{W_1 + W_2}$

$$W_2 - W_1 = \frac{\tan\varphi * (W_1 + W_2)}{(\sqrt{3})} = 2773.86\text{W}$$

Hence the wattmeter readings are W_1 = –585.95W; W_2 = 2187.91W.

3.6 TYPES OF CONNECTIONS IN THREE-PHASE SYSTEMS

The three-phase power supply is generated from the three coils which is mentioned at the beginning of this chapter. The generated power is utilized by the three-phase loads. The different connections could be achieved between three-phase source and load as follows.

- star (Y) – star (Y)
- star (Y) - delta(Δ)
- delta (Δ)- star (Y)
- delta (Δ)- delta (Δ)

The pictorial representation of different connections are depicted in Fig.3.33 to Fig.3.36.

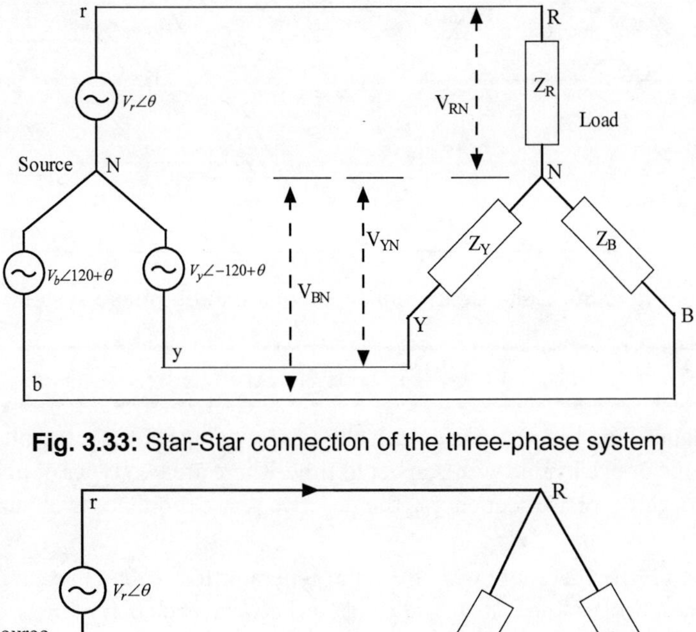

Fig. 3.33: Star-Star connection of the three-phase system

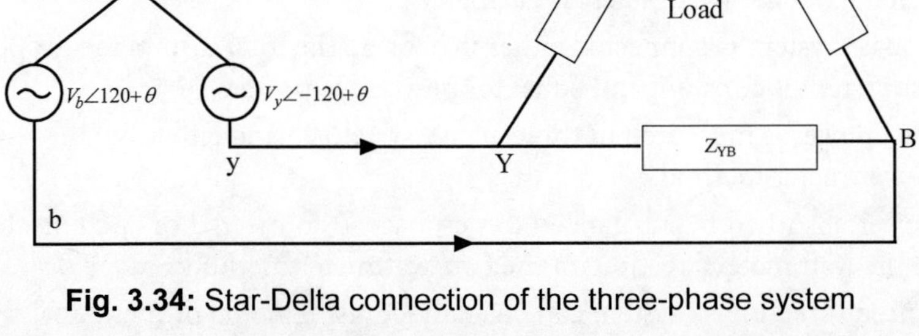

Fig. 3.34: Star-Delta connection of the three-phase system

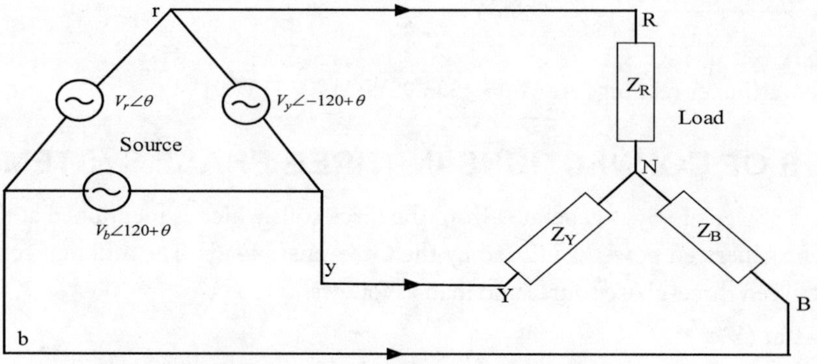

Fig. 3.35: Delta-Star connection of the three-phase system

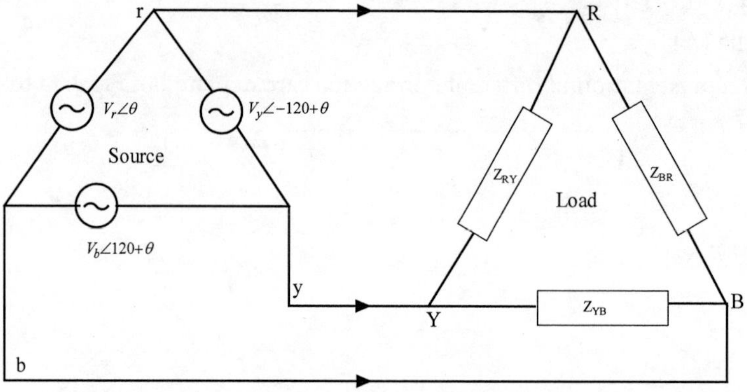

Fig. 3.36: Delta-Delta connection of the three-phase system

UNIT SUMMARY

1. The phase sequence is the order in which the phase voltages of a three-phase supply reach the maximum value or peak value with respect to time. There are two types of phase sequence namely positive and negative phase sequences. The negative phase sequence is obtained by interchanging any two phases.
2. The line current is the current flowing from the generator to the load in each transmission line in a three-phase system. The line voltage is the voltage between each pair of lines, excluding the neutral line if it exists. The phase current is the current flowing through each phase in a three-phase load. The phase voltage is the voltage across each phase.
3. The three-phase system is connected either in Y or Δ. The total instantaneous power in a balanced three-phase system is constant and equal to the average power.
4. The total real power is measured in three-phase systems using either the three-wattmeter method or the two-wattmeter method.
5. The real power absorbed by a balanced three-phase Y connected or Δ connected load is equal to the sum of the watt meters' reading in the two-wattmeter method.
6. An unbalanced three-phase system can be analysed using nodal or mesh analysis.

EXERCISES

Multiple Choice Questions

1. A balanced Y connected load phase sequence is RYB and its R phase to neutral voltage is $5\angle 10^0$ V. Find the line voltage between the phase Y to phase B.

   (a) $\sqrt{3} * 5\angle -110^0$ V (b) $\sqrt{3} * 5\angle -80^0$ V
   (c) $5\angle -110^0$ V (d) $5\angle -80^0$ V

2. A balanced delta connected load of resistance 10Ω, inductance 20mH, capacitance 10μH per phase is connected to a phase supply of 400V, 50Hz. Find the phase current of the load.

   (a) $1.28\angle -88.16^0$ A (b) $1.28\angle -188.16^0$A
   (c) $1.28\angle 88.16^0$ A (d) $1.28\angle -10^0$ A

3. The three-phase delta connected load is supplied by the balanced three-phase supply. With the same load, the load is reconnected into Y and supplied by the same source.

   (a) Phase currents will be 1/3 of the previous value
   (b.) Line currents will be 1/3 of the previous value
   (c) Power consumed will be $1/\sqrt{3}$ of the previous value
   (d) Power consumed will be $\sqrt{3}$ times of the previous value
   (e) Power consumed will be same as the previous value

4. In a balanced three phase AC circuit, the sum of all three generated voltages is _______?

   (a) Zero (b) Infinite
   (c) 1 (d) None of the above

5. For a star connection network, consuming power of 1.8kW and power factor 0.5, the inductance and resistance of each coil at a supply voltage of 230 Volts, 60 Hz is ——-

   (a) 0.01H, 8 Ohms (b) 0.05H, 10 Ohms
   (c) 0.03H, 7.4 Ohms (d) 1H, 7 Ohms

6. A three-phase star connected balanced load of (4 + j3) Ω per phase is connected across three-phase supply of 400V, 50Hz supply. Find the current drawn from the supply.

   (a) 20.23A (b) 46.19A
   (c) 50.54A (d) 50.52A

7. In a three-phase, 3 wire system, the sum of the line currents is equal to.........

   (a) 3 times of phase current (b) times of phase current
   (c) Zero (d) None of the above

8. What is the minimum number of wattmeters required for measuring power of a balanced three-phase load?

   (a) 1 (b) 2
   (c) 3 (d) None

9. In the two-wattmeter method with the load balanced, the readings of the wattmeters will be identical if the power factor is
(a) 0.5 (b) 0.707
(c) 0.866 (d) 1

10. In a balanced star-connected system, line voltages are …….ahead of their respective phase voltages.
(a) 30^0 (b) 60^0
(c) 45^0 (d) 120^0

11. Three identical capacitances, each of 45μF, are connected in star. The value of capacitance in each phase of the equivalent delta-connected load would be ....
(a) $45\sqrt{3}$ μF (b) $15\sqrt{3}$ μF
(c) 135μF (d) $15\sqrt{3}$ μF

12. A three-phase Δ connected generator is driving a balanced load such that each phase current is 12 A in magnitude. When $I_R\angle\theta = 12\angle 30^0$ A, what are the polar expressions for the other phase currents?
(a) $I_Y\angle\theta = 12\angle 150^0$A, $I_B\angle\theta = 12\angle -90^0$A
(b) $I_Y\angle\theta = 12\angle 120^0$A, $I_B\angle\theta = 12\angle 30^0$A
(c) $I_Y\angle\theta = 12\angle 30^0$A, $I_B\angle\theta = 12\angle 120^0$A
(d) $I_Y\angle\theta = 12\angle -90^0$A, $I_B\angle\theta = 12\angle 150^0$A

13. The line R to neutral voltage is $10\angle 15^0$ V, for a balanced three-phase starconnected load with phase sequence RYB. The voltage of line Y with respect to line B is given by
(a) $\sqrt{3}*10\angle 105^0 V$ (b) $10\angle 105^0 V$
(c) $\sqrt{3}*10\angle -75^0 V$ (d) $\sqrt{3}*10\angle -90^0 V$

14. The R phase current is $2\angle -30^0$ A, for a balanced three-phase delta connected load with phase sequence RYB. The current of line B with respect to line R is given by
(a) $\sqrt{3}*2\angle 0^0$ A (b) $2\angle 0^0$ A
(c) $\sqrt{3}*2\angle -120^0$ A (d) $\sqrt{3}*2\angle 120^0$ A

15. Two wattmeters are used for measuring power of the three-phase star connected load. The meter readings are 10W and 20W. Find the real power consumed by the load and identify the type of load.
(a) 30W, Inductive load (b) 10W, Capacitive load
(c) 30W, Inductive load (d) None of the above

16. Two wattmeters are used for measuring power of the three-phase delta connected load. The meter readings are 50W and 20W. Find the real power consumed by the load and power factor of the load.
(a) 30W, 0.8Lag (b) 70W, 0.8Lead
(c) –30W, 0.8Lead (d) None of the above

17. A star connected three-phase load is supplied by the star connected three-phase four wire system. The line current of the load is $5\angle 10^0$A, $5\angle -10^0$A and $4\angle 130^0$A. Identify that whether load is a balanced or unbalanced. Calculate the current carrying in the neutral line.
(a) Balanced Load, Zero (b) Unbalanced Load, Zero
(c) Unbalanced Load, A (d) Unbalanced Load, A

18. The three impedances $Z_1 = 20\angle 30^\circ\Omega$, $Z_2 = 40\angle 60^\circ\Omega$, $Z_3 = 10\angle -90^\circ\Omega$ are delta-connected to a 400V, 3 Φ system as shown in Fig.3. 37. Find the line current IY.

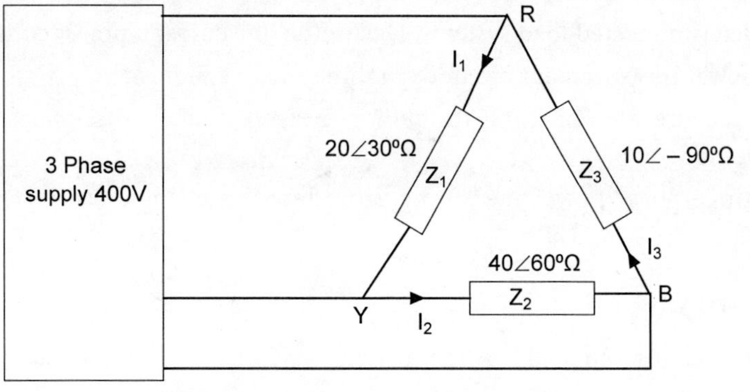

Fig. 3.37: Circuit for Problem 18

(a) $11.55\angle -30^0$A
(b) $5.775\angle -60^0$A
(c) $7.15\angle -6.2^0$A
d. None of the above

19. The three impedances $Z_1 = 20\angle 30^\circ\Omega$, $Z_2 = 40\angle 60^\circ\Omega$, $Z_3 = 10\angle -90^\circ\Omega$ are delta-connected to a 400V, 3 Φ system as shown in Fig.3.37. Find the current flowing through the impedance Z_2.

(a) $11.55\angle -30^0$A
(b) $5.775\angle -60^0$A
(c) $7.15\angle -6.2^0$A
(d) None of the above

20. A three-phase supply supplies a load consisting of three equal star connected resistors. If one of the resistors is removed, the load power is

(a) reduced by 25%
(b) reduced by 33.3%
(c) reduced by 50%
(d) reduced by 66.6%

ANSWERS															
1.	(b)	2.	(c)	3.	(c)	4.	(a)	5.	(c)	6.	(b)	7.	(c)	8.	(a)
9.	(d)	10.	(a)	11.	(c)	12.	(d)	13.	(c)	14.	(d)	15.	(a)	16.	(b)
17.	(d)	18.	(c)	19.	(b)	20.	(c)								

Short Answer Questions:

1. What is the difference between star and delta connections?
2. Define phase sequence.
3. What is the positive phase sequence and negative phase sequence?
4. What is the meaning of balanced load in three-phase systems?
5. Differentiate the balanced load and unbalanced load.
6. What are the different circuit connections possible in three-phase systems?
7. State: the phase voltage and line voltage of three-phase systems.
8. State: the phase current and line current of three-phase systems.
9. How does the real power measured in three-phase systems?
10. What are the methods available to measure the unbalanced load power in three-phase systems?

Long Answer Questions:

1. Explain the star connected load systems. Derive the line current, power consumed by the load.
2. Explain the delta connected load systems. Derive the line current, power consumed by the load.
3. Discuss the power measurement methods in three-phase systems.
4. Write the procedure for obtaining the power measurement in three-phase systems with two-wattmeter method.
5. Explain the three-phase power measuring procedure using single-wattmeter method under the balanced load.

Numerical Problems

1. A balanced deltaconnected load has line current $I_R = 5\angle 25^0$A. Find the phase currents I_{RY}, I_{YB} and I_{BR} in phasor form.
2. A balanced star connected load has phase voltage $V_R = 215\angle -20^0$V. Find the line voltages V_{RY}, V_{YB} and V_{BR} in phasor form.
3. A three-phase star connected load of (6 + j8)Ω per phase is supplied by three-wire star connected supply with a phase voltage of 240V, 50Hz. Find the line current of the load, power factor of the load. Also draw the phasor diagram.
4. A three-phase star connected load of per phase is supplied by three-wire star connected supply with a phase voltage of 240V, 50Hz. Find the line current of the load, power factor of the load. Also draw the phasor diagram.
5. If $V_{RN} = 220\angle 60$V in the given network of Fig.3.38, Find the phase current of the load.

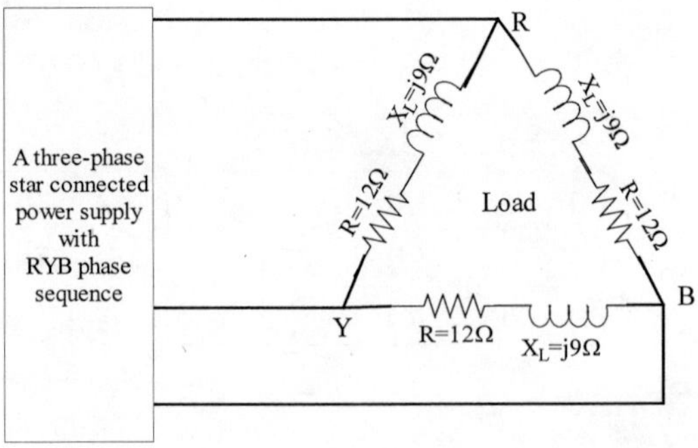

Fig.3.38: Delta connected load of Problem 5

6. A three-phase delta connected load of (6+j8)Ω per phase which is supplied by three-wire star connected supply with a phase voltage of 240V, 50Hz. Find the line current of the load and power factor of the load. Also draw the phasor diagram.
7. A three-phase delta connected load is supplied by three-phase star connected generator with a phase voltage of 240V, 50Hz. It draws a line current of 5A with a lagging power factor of 0.966. Draw the phasor diagram.
8. A three-phase delta connected load with a resistance of 10Ω per phase and an inductance of 10mH is supplied by balanced three-phase supply of 415V, 50Hz. Calculate the load phase and line voltage, phase current, power factor at the load end. Draw the phasor diagram.

9. A three-phase motor rating 35kW, 0.85PF, with an efficiency of 90% is supplied by three-phase supply of 500V, 50Hz, Find the reading of the two-wattmeters connected to measure the motor's input power.
10. A load of impedances each (9+j21)Ω is supplied through a line to which a voltage of 415V is applied. The impedance of each line is (2 + j4) Ω. Determine the power input and output when the load is (a) star connected (b) mesh connected.
11. A three-phase balanced system with a line voltage of 202V supplies a delta connected load with $Z_P = 25\angle 60^0\Omega$.
    (i) Find the line current.
    (ii) Determine the total power supplied to the load using two wattmeters connected to the R and B lines.
12. Two wattmeters are used for measuring the power input and the power factor of an over-excited synchronous motor which is powered by three-phase supply. If the readings of the meters are (– 2.0 kW) and (+ 7.0 kW) respectively, calculate the input power and power factor of the motor.
13. A balanced star connected inductive load across 415-V, 50-Hz three-phase supply takes a line current of 25A. The phase sequence is RYB. A single-phase wattmeter has its current coil connected in the R line and its voltage coil across the line YB. With these connections, the reading is 8 kW. Draw the vector diagram and find (i) the kW (ii) the kVAR (iii) the kVA and (iv) the power factor of the load.
14. Two wattmeters are connected to measure the input to a 400 V; 3-phase, star connected motor outputting 24.4 kW at a power factor of 0.4 (lag) and 80% efficiency. Calculate the (i) resistance and reactance of motor per phase (ii) reading of each wattmeter.
15. Two wattmeters connected to measure the input to a balanced three-phase circuit indicate 2500 W and 500 W respectively. Find the power factor of the circuit (a) when both readings are positive and (b) when the later reading is obtained after reversing the connections to the current coil of one instrument.

PRACTICAL

1. Use LTspice to determine the line current of Fig. 3.39. Consider the load per phase is (a) 10Ω (b) (5+j10) Ω

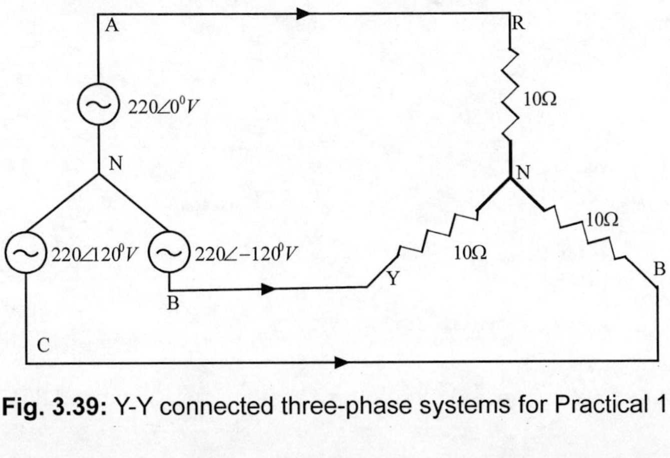

Fig. 3.39: Y-Y connected three-phase systems for Practical 1

2. Use LTspice to determine the line current, phase current, phase voltage of Fig. 3.40. Consider the load per phase is (a) 10Ω (b) (5+j10) Ω

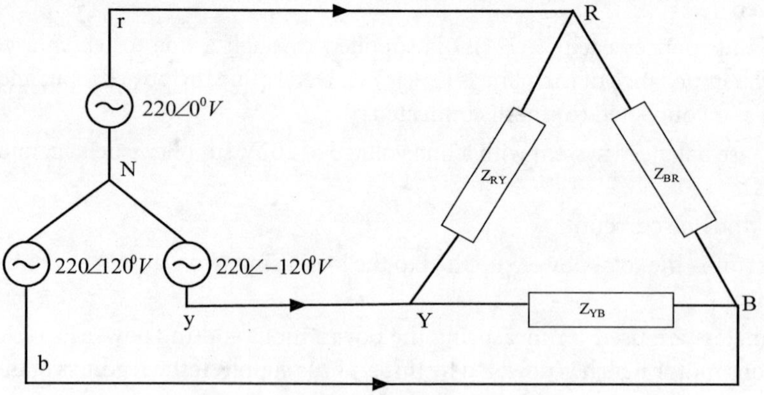

Fig. 3.40: Y-Δ connected three-phase systems for Practical 2

KNOW MORE

1. Example 3.3: A balanced three-phase star connected load of (10+j8) Ω/phase is supplied from a balanced three-phase supply of 400∠-20⁰V, 50Hz. Find the line current, the real power, reactive power, apparent power and power factor the load. [Note1: The load components are found to do it in LTSpice.

[**Note2:** The obtained current values are in magnitude (LTspice).

$R_L = 10\Omega; X_L = 8\Omega; L = \frac{X_L}{\omega} = \frac{X_L}{2\pi f} = \frac{8}{2\pi(50)} = 0.0255H]$ convert into the rms values].

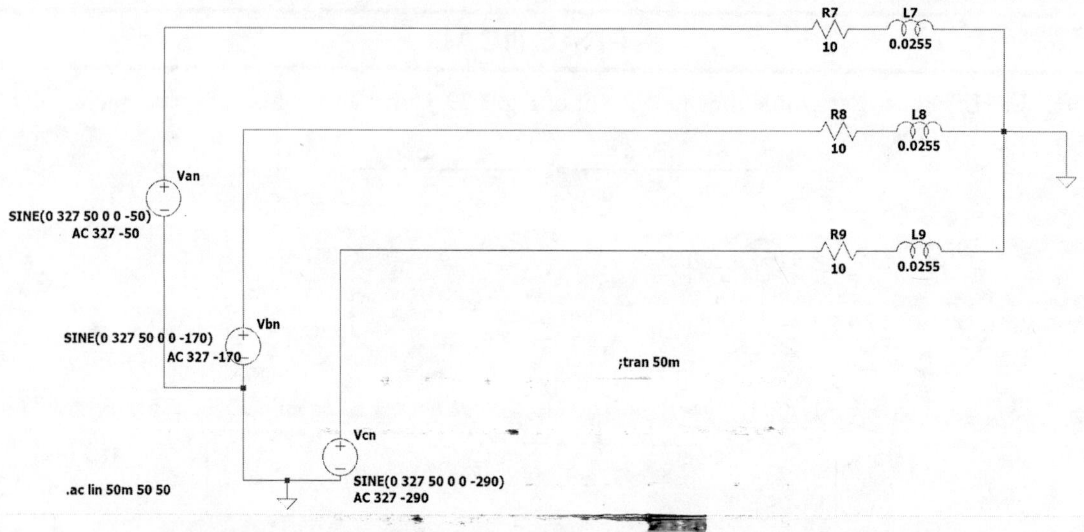

Fig. 3.41: Y-Y connected three-phase systems for Example 3.3 (LTSpice)

```
--- AC Analysis ---

frequency:     50            Hz
V(n001):       mag:       327 phase:        -50°        voltage
V(n003):       mag:       327 phase:       -170°        voltage
V(n005):       mag:       327 phase:         70°        voltage
V(n002):       mag:   204.435 phase:    1.29721°        voltage
V(n004):       mag:   204.435 phase:   -118.703°        voltage
V(n006):       mag:   204.435 phase:    121.297°        voltage
I(L9):         mag:   25.5191 phase:    31.3044°        device_current
I(L8):         mag:   25.5191 phase:    151.304°        device_current
I(L7):         mag:   25.5191 phase:   -88.6956°        device_current
I(R9):         mag:   25.5191 phase:    31.3044°        device_current
I(R8):         mag:   25.5191 phase:    151.304°        device_current
I(R7):         mag:   25.5191 phase:   -88.6956°        device_current
I(Vcn):        mag:   25.5191 phase:   -148.696°        device_current
I(Vbn):        mag:   25.5191 phase:   -28.6956°        device_current
I(Van):        mag:   25.5191 phase:    91.3044°        device_current
```

Fig. 3.42: Y-Y connected three-phase systems for Example 3.3 solution (LTSpice)

2. Example 3.5: A three-phase 3 wire 415V, 50Hz system supplies a delta connected load which consists of resistance 20Ω and inductance 10mH in series. Determine the phase current, line current and also draw the phasor diagram.

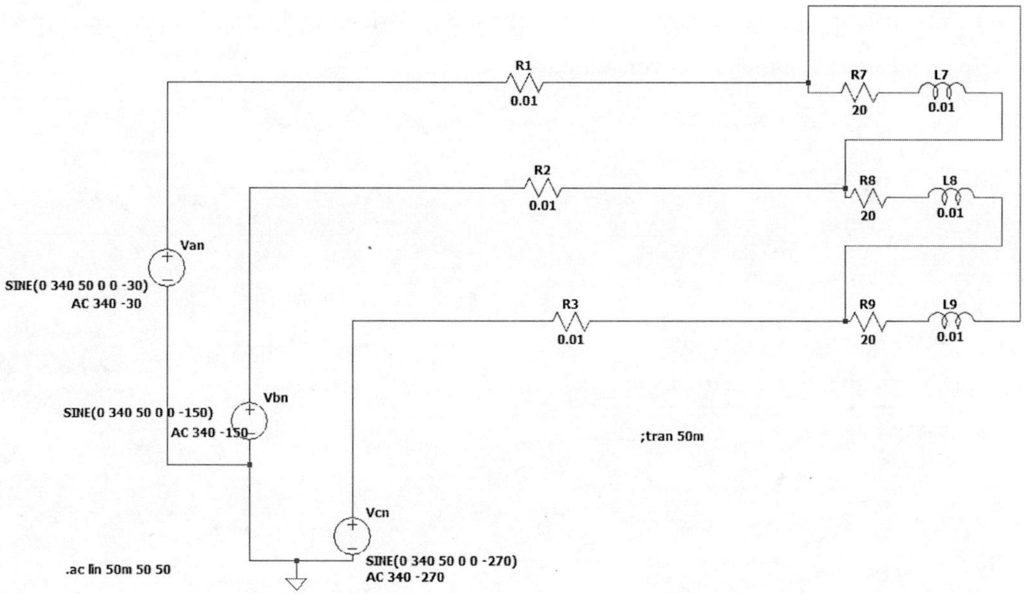

Fig. 3.43: Y-Δ connected three-phase systems for Example 3.5 (LTSpice)

```
--- AC Analysis ---

frequency:    50              Hz
V(n002):      mag:      340 phase:       -30°     voltage
V(n005):      mag:      340 phase:      -150°     voltage
V(n008):      mag:      340 phase:        90°     voltage
V(n001):      mag:  339.503 phase:  -29.9868°     voltage
V(n003):      mag:  290.937 phase:  -164.105°     voltage
V(n004):      mag:  339.503 phase:  -149.987°     voltage
V(n006):      mag:  290.937 phase:    75.895°     voltage
V(n007):      mag:  339.503 phase:   90.0132°     voltage
V(n009):      mag:  290.937 phase:   -44.105°     voltage
I(L9):        mag:  29.0443 phase:   111.087°     device_current
I(L8):        mag:  29.0443 phase:  -128.913°     device_current
I(L7):        mag:  29.0443 phase:  -8.91346°     device_current
I(R3):        mag:  50.3061 phase:   81.0865°     device_current
I(R2):        mag:  50.3061 phase:  -158.913°     device_current
I(R1):        mag:  50.3061 phase:  -38.9135°     device_current
I(R9):        mag:  29.0443 phase:   111.087°     device_current
I(R8):        mag:  29.0443 phase:  -128.913°     device_current
I(R7):        mag:  29.0443 phase:  -8.91346°     device_current
I(Vcn):       mag:  50.3061 phase:  -98.9135°     device_current
I(Vbn):       mag:  50.3061 phase:   21.0865°     device_current
I(Van):       mag:  50.3061 phase:   141.087°     device_current
```

Fig. 3.44: Y-Δ connected three-phase systems for Example 3.5 solution (LTSpice)

3. A three-phase 3 wire system of maximum voltage of 120V/phase, 50Hz having self-impedance of (0.2+j0.3) Ω/phase supplies a star connected load consists of resistance 24Ω and inductance 60.4mH in series that is connected through the transmission line having impedance of (0.6+j0.7) Ω/phase. Determine the power consumed by the load.

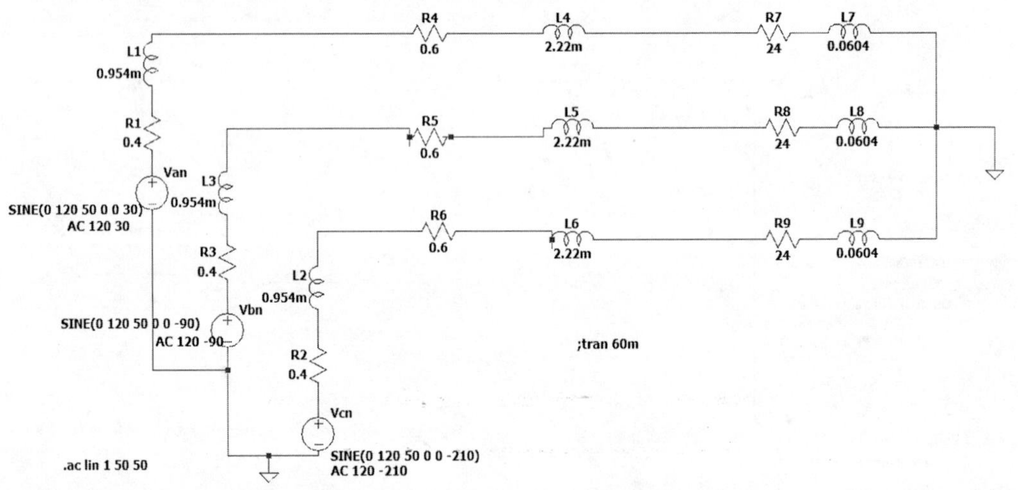

Fig. 3.45: Three phase power measurement for Y-Y connected three-phase systems (LTSpice)

```
;Power Measurement for three phase circuits
.meas start param 20ms
.meas end param 40ms
.meas Vph RMS V(N003) from start to end
.meas Vline RMS V(N003, N008) from start to end
.meas Iline RMS I(R7) from start to end
.meas RT param sqrt(3)
.meas S param RT*Vline*Iline
.meas P1P avg (V(N003)*I(R7)) from start to end
.meas P3P param P1P*3
.meas PF param P3P/S
.meas S2 param S*S
.meas P2 param P3P*P3P
.meas Q2 param S2-P2
.meas Q3Q param sqrt(Q2)
```

Fig. 3.46: Three phase power measurement Formulas for Y-Y connected three-phase systems (LTSpice)

```
       --- AC Analysis ---

frequency:      50            Hz
V(n010):        mag:        120 phase:        30°          voltage
V(n005):        mag:    118.832 phase:   30.4514°          voltage
V(n001):        mag:    118.124 phase:   30.0293°          voltage
V(n017):        mag:    118.832 phase:   150.451°          voltage
V(n011):        mag:    118.124 phase:   150.029°          voltage
V(n013):        mag:    118.832 phase:  -89.5486°          voltage
V(n006):        mag:    118.124 phase:  -89.9707°          voltage
V(n018):        mag:        120 phase:       150°          voltage
V(n016):        mag:        120 phase:       -90°          voltage
V(n002):        mag:    116.375 phase:   30.7211°          voltage
V(n007):        mag:    116.375 phase:  -89.2789°          voltage
V(n012):        mag:    116.375 phase:   150.721°          voltage
V(n003):        mag:    114.732 phase:   29.7121°          voltage
V(n008):        mag:    114.732 phase:  -90.2879°          voltage
V(n014):        mag:    114.732 phase:   149.712°          voltage
V(n004):        mag:    71.1555 phase:   81.3792°          voltage
V(n009):        mag:    71.1555 phase:  -38.6208°          voltage
V(n015):        mag:    71.1555 phase:  -158.621°          voltage
I(L9):          mag:    3.74992 phase:   111.382°          device_current
I(L8):          mag:    3.74992 phase:  -128.618°          device_current
I(L7):          mag:    3.74992 phase:  -8.61781°          device_current
I(L6):          mag:    3.74992 phase:   111.382°          device_current
I(L5):          mag:    3.74992 phase:  -128.618°          device_current
I(L4):          mag:    3.74992 phase:  -8.61781°          device_current
I(L3):          mag:    3.74992 phase:  -128.618°          device_current
I(L2):          mag:    3.74992 phase:   111.382°          device_current
I(L1):          mag:    3.74992 phase:  -8.61781°          device_current
I(R9):          mag:    3.74992 phase:   111.382°          device_current
I(R8):          mag:    3.74992 phase:  -128.618°          device_current
```

Fig. 3.47: Voltages and Currents for the Y-Y connected three-phase systems' simulated output (LTSpice)

```
start: 20ms=(-33.9794dB,0°)
end: 40ms=(-27.9588dB,0°)
vph: RMS(v(n003))=114.732 FROM 0.02 TO 0.04
vline: RMS(v(n003, n008))=198.722 FROM 0.02 TO 0.04
iline: RMS(i(r7))=3.74992 FROM 0.02 TO 0.04
rt: sqrt(3)=(4.77121dB,0°)
s: rt*vline*iline=(62.2165dB,0°)
p1p: AVG(v(n003)*i(r7))=430.235 FROM 0.02 TO 0.04
p3p: p1p*3=(62.2165dB,0°)
pf: p3p/s=(-2.03473e-013dB,0°)
s2: s*s=(124.433dB,0°)
p2: p3p*p3p=(124.433dB,0°)
q2: s2-p2=(-142.158dB,0°)
q3q: sqrt(q2)=(-71.0792dB,0°)


Date: Tue Feb 28 19:11:10 2023
Total elapsed time: 0.032 seconds.
```

Fig. 3.48: Active power, Reactive power and Apparent power for the Y-Y connected three-phase systems' simulated output according to the formula used (LTSpice)

REFERENCES AND SUGGESTED READINGS

1. Charles K. Alexander and Matthew N. O. Sadiku, "Fundamentals of Electric Circuits", McGraw Hill Publishers, 7th Edition, 2021.
2. D P Kothari and I J Nagrath, "Basic Electrical Engineering", McGraw Hill Publishers, 4th Edition, 2019.
3. C L Wadhwa, "Basic Electrical Engineering", New Age International Publishers, 4th Edition, 2007.

❑❑

4 Transient Response of First and Second Order Electric Circuits

UNIT SPECIFICS

Through this unit the following topics have been discussed:

- Transient response of first order source free circuits.
- Concept of first order electric circuits.
- Response of first order circuits for step dc voltage or current input.
- Concept of second order electric circuits.
- Response of second order circuits for step dc voltage or current input.
- Application of Laplace Transform to find the responses of first and second order electric circuits.

RATIONALE

This unit deals with transient responses of first order and second order electric circuits. These concepts help the students to find out the responses of practical circuits during transient conditions. It also provides a knowledge of time constant of electric circuits. Time constant is important to understand how fast the response of the circuits reaches to it's steady state value. Transient analysis also provides an idea about the natural response and forced response of the electric circuits. Different combinations of resistance, capacitance and inductance are considered in this unit to understand this concept. The transient response also includes the behaviour of resistance, capacitance and inductance for different circuit configurations. The related problems are solved and included for insight knowledge of this unit. This unit is one of the foundations to understand and pursue the knowledge in the domain of electrical and electronics engineering.

PRE-REQUISITES

Calculus

Laplace Transform

UNIT OUTCOMES

List of outcomes of this unit is as follows:

U4-O1: Identify the first and second order electric circuits

U4-O2: Apply mathematical formula to calculate the response of a source free first order electric circuits

U4-O3: Apply mathematical formula to calculate the step response of first and second order electric circuits.

U4-O4: Apply the concepts of Laplace Transform to solve the first and second order electric circuits.

Unit-4 Outcome	EXPECTED MAPPING WITH COURSE OUTCOMES (1 - Weak Correlation; 2 - Medium Correlation; 3 - Strong Correlation)				
	CO-1	CO-2	CO-3	CO-4	CO-5
U4-O1	2	2	-	2	-
U4-O2	1	1	-	3	-
U4-O3	2	2	-	3	-
U4-O4	-	-	-	3	-

4.1 INTRODUCTION

The response of electric circuits alters for a short span of time before settling to a steady state value due to switching actions or any other change in source or circuit elements. This alteration exists for a short span of time if any other switching action or circuit alternation does not occur again. This short span of time is known as transient time and response of the circuit during that time period is known as transient response. This unit deals with transient response of different first order and second order electric circuits.

4.2 First order differential equations

First order differential equation is represented by,

$$a\frac{dy}{dx}+by=cx \tag{4.1}$$

Where y is the output, x is input and a, b, c are constant. When the external input is absent the response of the first order differential equation is known as natural response. In that case, the first order differential equation is represented as,

$$a\frac{dy}{dx}+by=0$$

$$\frac{dy}{dx}+\frac{b}{a}y=0 \tag{4.2}$$

In case of electric circuits this condition occurs when an energy storage element (inductor or capacitor) is not connected to any external independent source rather it dissipates the stored energy to a resistor. For an example if the *R-L* circuit is not connected to any external independent source and the stored energy in inductor decays by dissipating it to the resistance and the response of the circuit is natural response. The stored energy in inductor results the flow of current through the circuit. These circuits are known as source free circuits. Source free circuits may include dependent source. The natural response corresponding to equation (4.2) is given by,

$$y_n = Ke^{-\alpha t} \tag{4.3}$$

K is a constant and α is the root of the characteristic equation and, $\alpha = \frac{b}{a}$. If the value of K is evaluated by using the initial condition of the system then it becomes particular solution. The particular solution of equation (4.2) is given by,

$$y = y_0\, e^{-\alpha t} \quad \text{for } t > 0 \tag{4.4}$$

Here $y(0) = y_0$ at $t = 0$. If the input (x) is present and is a constant or independent variable then the solution of y is known as forced solution (y_f). The complete solution of equation (4.1) is then given by,

$$y = Ke^{-\alpha t} + e^{\alpha t}\int \frac{c}{a} e^{-\alpha t}\, dt$$

$$y = y_n + y_f \tag{4.5}$$

Where the second term in equation (4.5) is forced response.

4.3 First order electric circuits

The response of the series and parallel R-L and R-C circuits are governed by first order differential equations. Therefore, these circuits are together known as first order electric circuits. If the external independent source is not connected to these circuits, then they are known as source free electric circuits and the response of these circuits are known as natural response.

4.3.1 Source free Series R-L Circuit

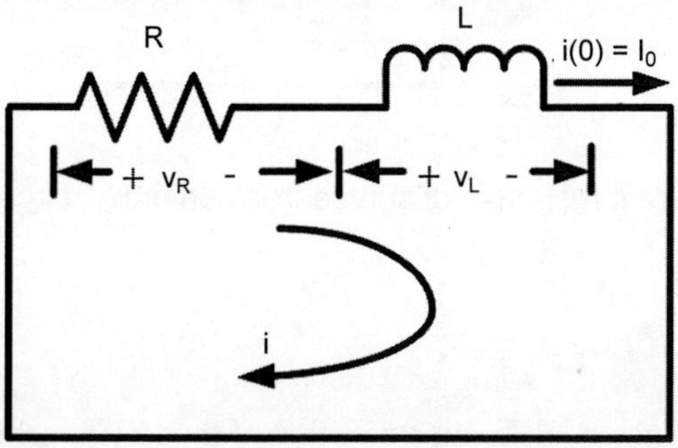

Fig. 4.1: Source free Series *R-L* circuit with $i(0) = I_0$

A circuit with a series combination of pure resistor (R) and pure charged inductor (L) is considered as shown in Fig. 4.1. This circuit is not connected to any independent external source. The charged inductor results a current flow through this source free circuit and decays the stored energy of it by dissipating the energy across the resistor. The response of this circuit is considered as current through the circuit. The initial current through the inductor due to the charged inductor at $t = 0$ is given by,

$$i(0) = I_0 \tag{4.6}$$

Applying KVL in the circuit shown in Fig. 4.1, the relation between voltage across the inductor and resistor is given by,

$$i(t)R + L\frac{di(t)}{dt} = 0 \tag{4.7}$$

Rearranging and integrating the equation (4.7), it is obtained as,

$$\int_{I_0}^{i(t)} \frac{di(t)}{i(t)} = -\int_0^t \frac{R}{L} dt \quad (4.8)$$

The solution of equation (4.8) is obtained as,

$$\ln \frac{i(t)}{I_0} = -\frac{Rt}{L} \quad (4.9)$$

Expressing the equation (4.9) in exponential form as,

$$i(t) = I_0 \, e^{-(Rt/L)} \quad (4.10)$$

The natural response of the series *R-L* circuit is given by equation (4.10) and it depicts that the current through a charged inductor in a source free series *R-L* circuit decays exponentially from the initial current value (I_0). This current is also known as transient current and it's value becomes zero as $t \to \infty$. The same response is illustrated by Fig. 4.2.

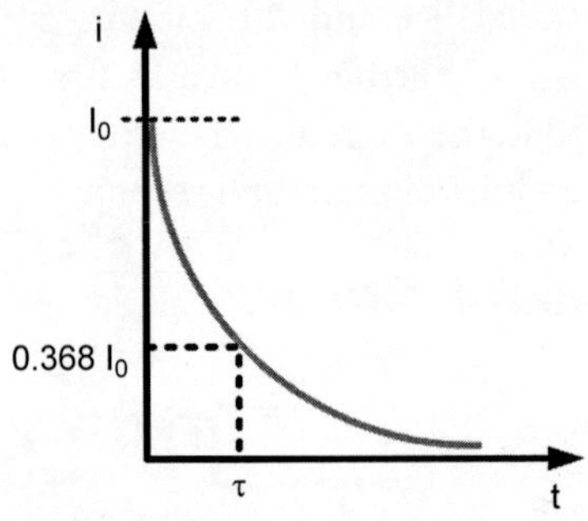

Fig. 4.2: Current response of source free series *R-L* circuit with $i(0) = I_0$

If *t* is considered as $\frac{L}{R}$ in equation (4.10), then the value of *i(t)* becomes,

$$i(t) = I_0 \, e^{-1} = 0.368 I_0 \quad (4.11)$$

The ratio of *L* and *R* is known as the time constant of the circuit, where time constant is defined as the time required for the circuit to reach a value of 36.8% of the initial value. Inverse of time constant is known as damping ratio. Smaller the value of time constant, faster the rate of decay in response. Generally, time constant is denoted by 'τ' and the time constant for a series *R-L* circuit is given by,

$$\tau = \frac{L}{R} \quad (4.12)$$

Voltage across the resistor is given by,

$$v_R(t) = i(t)R = I_0 \, R \, e^{-(Rt/L)} \quad (4.13)$$

Using equation (4.7), voltage across the inductor is found out as,

$$v_L(t) = L\frac{di(t)}{dt} = -i(t)R = -I_0 R e^{-(Rt/L)} \quad (4.14)$$

Power dissipated in the resistor is given by,

$$p(t)=v_R(t)i_R(t)=I_0^2Re^{-(2Rt/L)}=I_0^2Re^{-(2t/\tau)} \tag{4.15}$$

The energy absorbed by the resistor is given by,

$$w(t)=\int_0^t I_0^2Re^{-(2t/\tau)}dt=\frac{1}{2}LI_0^2\left(1-e^{-\left(\frac{2t}{\tau}\right)}\right) \tag{4.16}$$

Equation (4.16) clearly shows that as $t \to \infty$, the energy absorbed by the resistor becomes the initial energy available in the charged inductor and it is given by,

$$\lim_{t\to\infty} w(t)=\frac{1}{2}LI_0^2 \tag{4.17}$$

Example 4.1:

Find the current flowing through the circuit mentioned in Fig.4.1 if R = 10 ohm, L =0.5 H and initial current is 10 A. Also find the time constant of the circuit.

Solution:

$$R = 10 \text{ ohm}$$
$$L = 0.5 \text{ H}$$
$$I_0 = 10 \text{ A}$$

The current through the circuit for a source free R-L circuit is given by,

$$i(t) = I_0\, e^{-(Rt/L)}$$
$$i(t) = 10e^{-(10t/0.5)} = 10e^{-20t} \text{ A}$$

Time constant is given by,

$$\tau=\frac{L}{R}=\frac{0.5}{10}=0.05\,\text{sec}$$

Example 4.2:

Find the power dissipated and energy absorbed by the resistor for a source free R-L circuit with R = 2 ohm, L = 5 H and initial current is 2 A.

Solution:

$$R = 2 \text{ ohm}$$
$$L = 5 \text{ H}$$
$$I_0 = 2 \text{ A}$$

Power dissipated in R is calculated as,

$$p(t)=I_0^2Re^{-(2Rt/L)}=(2)^2\times 2e^{-(2\times 2t/5)}=8e^{-0.8t}\,\text{W}$$

The energy absorbed by R is calculated as,

$$w(t)=\frac{1}{2}LI_0^2\left(1-e^{-\left(\frac{2Rt}{L}\right)}\right)=\frac{1}{2}5\times(2)^2\times\left(1-e^{-\left(\frac{2\times 2t}{5}\right)}\right)=10\left(1-e^{-0.8t}\right)\text{J}$$

4.3.2 Source free series *R-C* circuit

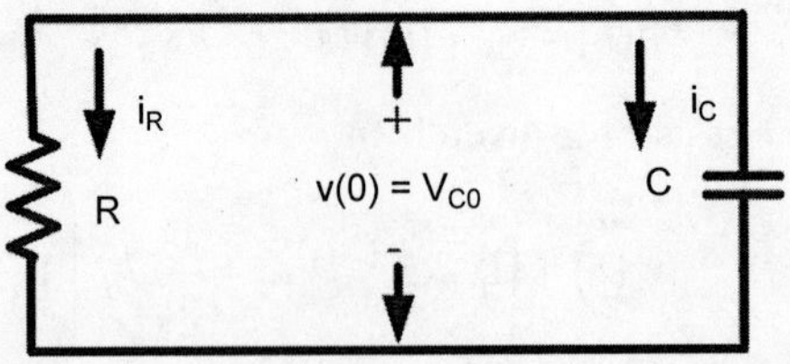

Fig. 4.3: Source free series *R-C* circuit with $v(0) = V_{CO}$

A circuit with a series combination of pure resistor (R) and pure charged capacitor (C) is considered as shown in Fig. 4.3. This circuit is not connected to any independent external source. The charged capacitor causes the flow of current through this source free circuit and decays the stored energy of it by dissipating the energy across the resistor. Voltage across the capacitor is considered to be the response of the circuit. Capacitor is fully charged initially and the voltage across the capacitor (at $t = 0$) is expressed as,

$$v(0) = V_{CO} \tag{4.18}$$

Applying KCL in the circuit shown in Fig. 4.3, the relation between current through the capacitor and resistor is given by,

$$i_C(t) + i_R(t) = 0$$

$$C\frac{dv(t)}{dt} + \frac{v(t)}{R} = 0 \tag{4.19}$$

Rearranging and integrating the equation (4.19), the voltage across the capacitor is obtained as,

$$\int_{V_{CO}}^{v(t)} \frac{dv(t)}{v(t)} = -\int_{0}^{t} \frac{1}{RC}dt \tag{4.20}$$

The solution of equation (4.20) is obtained as,

$$\ln\frac{v(t)}{V_{CO}} = -\frac{t}{RC} \tag{4.21}$$

Expressing the equation (4.21) in exponential form as,

$$v(t) = V_{CO}\, e^{-(t/RC)} \tag{4.22}$$

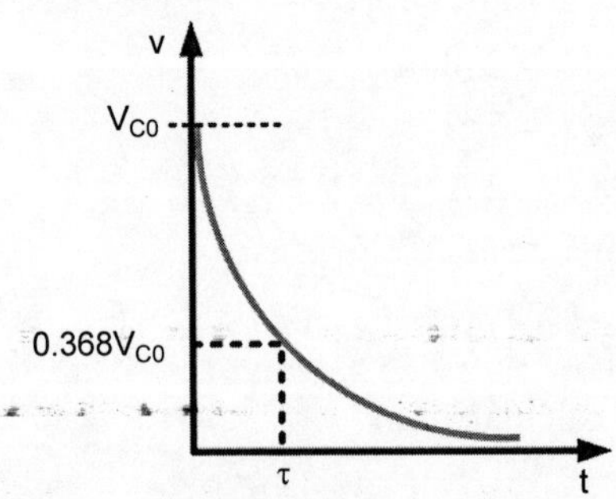

Fig. 4.4: Current response of source free series *R-C* circuit with $v(0) = V_{CO}$

The natural response of the series *R-C* circuit is given by equation (4.22) and it depicts that the voltage across the charged capacitor in a source free series *R-C* circuit decays exponentially from the initial voltage value (V_{CO}). This voltage is also known as transient voltage and it's value becomes zero as $t \to \infty$. The response is illustrated by Fig. 4.4.

If t is considered as in equation (4.22), then the value of $v(t)$ becomes,

$$v(t) = V_{C0}\, e^{-1} = 0.368 V_{C0} \tag{4.23}$$

RC is known as the time constant of the series *R-C* circuit. Smaller the value of time constant, faster the rate of decay in voltage across the capacitor. The time constant for a series *R-C* circuit is expressed as,

$$\tau = RC \tag{4.24}$$

Current through the resistor is given by,

$$i_R(t) = \frac{v(t)}{R} = \frac{V_{C0}}{R} e^{-(t/RC)} \tag{4.25}$$

Using equation (4.19), current through the capacitor is found out as,

$$i_c(t) = C\frac{dv(t)}{dt} = -i_R(t) = -\frac{V_{C0}}{R} e^{-(t/RC)} \tag{4.26}$$

Power dissipated in the resistor is given by,

$$p(t) = v(t)\, i_R(t) = \frac{V_{C0}^{\ 2}}{R} e^{-(2t/RC)} = \frac{V_{C0}^{\ 2}}{R} e^{-(2t/\tau)} \tag{4.27}$$

The energy absorbed by the resistor is given by,

$$w(t) = \int_0^t \frac{V_{C0}^{\ 2}}{R} e^{-(2t/RC)} dt = \frac{1}{2} C V_{C0}^{\ 2} \left(1 - e^{-(2t/\tau)}\right) \tag{4.28}$$

Equation (4.28) clearly shows that as $t \to \infty$, the energy absorbed by the resistor becomes the initial energy available in the charged capacitor and it is given by,

$$w(\infty) = \frac{1}{2} C V_0^{\ 2} \tag{4.29}$$

Example 4.3:

Find the current flowing through the capacitor for the circuit given below considering the initial voltage across the capacitor as 10 V. Also find the time constant of the circuit.

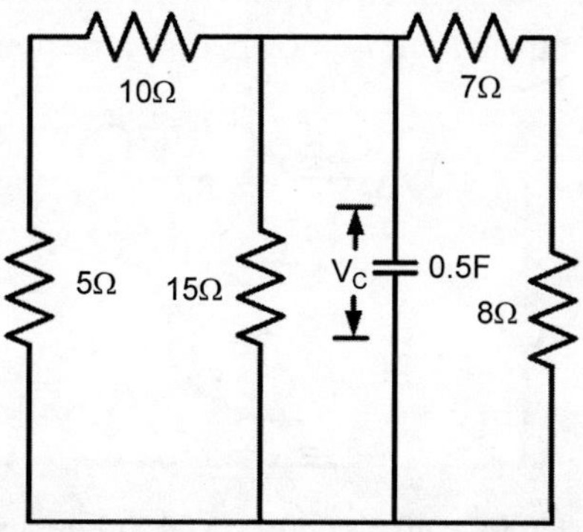

Solution:

$$C = 0.5\text{ F}$$

$$V_{CO} = 10\text{ V}$$

Equivalent resistance of the circuit is calculated as,

$$R_{eq} = (10+5)||15||(7+8) = \frac{15}{3} = 5\text{ohm}$$

Time constant of the circuit is calculated as,

$$\tau = R_{eq}C = 5 \times 0.5 = 2.5\text{ sec}$$

Current through the capacitor is given by,

$$i(t) = \frac{V_{CO}}{R_{eq}} e^{-\left(\frac{t}{\tau}\right)} = \frac{10}{5} e^{-\left(\frac{t}{2.5}\right)} = 2e^{-0.4t}\text{ A}$$

Example 4.4:

Find the current flowing through the 5 Ω resistor of the circuit mentioned in **Example 4.3** Also calculate the voltage across that resistor and power dissipated in that same resistor. Consider the same initial condition as mentioned in **Example 4.3.**

Solution:

As all the three parallel branches of resistors have equal resistance, then the current flowing through each branch or current through 5 Ω resistor is calculated as,

$$i_{5\Omega}(t) = \frac{i(t)}{3} = \frac{2}{3} e^{-0.4t} = 0.67e^{-0.4t}\text{ A}$$

Voltage across that resistor is calculated as,

$$v_{5\Omega}(t) = i_{5\Omega}(t) \times 5 = 2.22e^{-0.4t}\text{ V}$$

Power dissipated in 5 Ω resistor is calculated as,

$$p(t) = [i_{5\Omega}(t)]^2 R = [0.67e^{-0.4t}]^2 \times 5 = 3.33e^{-0.8t}\text{ W}$$

4.3.3 Step Response of Series *R-L* Circuit

When the voltage or current source of a circuit is applied suddenly and if the source can be modelled as step function then the response of the circuit is known as step response.

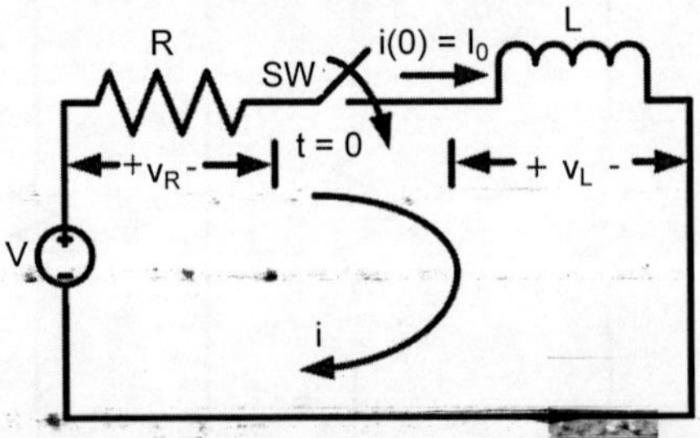

Fig. 4.5: Series *R-L* circuit with step input voltage and initial condition $i(0) = I_0$

A circuit with a series combination of pure resistor (R) and pure charged inductor (L) with an initial current of I_0 is excited by an independent dc voltage source as shown in Fig. 4.5. Initially the voltage source was disconnected from the circuit and then the dc voltage is suddenly applied to the circuit at t=0 by closing the switch (SW). The response of this circuit is considered as current through the circuit. The initial current through the circuit due to the charged inductor at t=0 is given by,

$$i(0) = I_0 \tag{4.30}$$

Applying KVL in the circuit shown in Fig. 4.5, the relation between voltage across the inductor and resistor is given by,

$$i(t)R + L\frac{di(t)}{dt} = Vu(t) \tag{4.31}$$

$u(t)$ represents the unit step function. Rearranging the equation (4.31), it is obtained as,

$$\frac{di(t)}{dt} + \frac{R}{L}i(t) = \frac{Vu(t)}{L} \tag{4.32}$$

The solution of equation (4.32) is obtained by comparing it with the solution of equation of (4.1) as mentioned in equation (4.5),

$$i(t) = Ke^{-(Rt/L)} + e^{-(Rt/L)}\int \frac{V}{L}e^{(Rt/L)}dt \tag{4.33}$$

The forced response part of equation (4.33) is given by,

$$i(t)_f = e^{-(Rt/L)}\int \frac{V}{L}e^{(Rt/L)}dt = \frac{V}{R} \tag{4.34}$$

Inductor does not allow sudden change in current through it as per the theory of electromagnetic induction. Hence, the current through the inductor just before the switching is same as the current through it just after switching as mentioned below,

$$i(0^-) = i(0^+) = I_0 \tag{4.35}$$

With this initial condition the equation (4.33) becomes,

$$I_0 = K + \frac{V}{R}$$

$$K = I \;\; -- \tag{4.36}$$

Substituting the value of K and in equation (4.33), the complete solution becomes,

$$i(t) = \left(I_0 - \frac{V}{R}\right)e^{-(Rt/L)} + \frac{V}{R} \tag{4.37}$$

The step response of the circuit can be expressed as,

$$i(t) = \begin{cases} I_0, & t < 0 \\ I_0 e^{-(t/\tau)} + \frac{V}{R}\left(1 - e^{-(t/\tau)}\right), & t > 0 \end{cases} \tag{4.37a}$$

Equation (4.37) depicts the step response of a series *R-L* circuit. The ratio of *L* and *R* is the time constant of this circuit. From this equation, it is clear that the current decays exponentially from an initial value of I_0 and settles at $\frac{V}{R}$ at $t \to \infty$ if the circuit is excited by a step voltage function. The response is demonstrated in Fig. 4.6

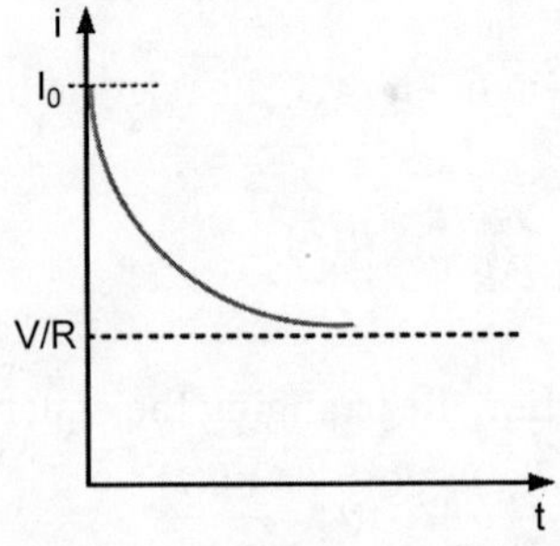

Fig. 4.6: Current response of series *R-L* circuit with step input voltage and initial condition $i(0) = I_0$

If $t \to \infty$ then the final response of the circuit, i.e., $i(\infty)$ will be the ratio of applied voltage and resistance. This is called the steady state response of the circuit. The equation (4.37) can be expressed as,

$$i(t) = [i(0) - i(\infty)]\, e^{-(t/\tau)} + i(\infty) \tag{4.38}$$

It is clear from the equation (4.38) that initial current value ($i(0)$, final current value $i(\infty)$ and the time constant of the circuit(τ) are required to find out the step response of a series R-L circuit.

Voltage across the resistor is given by,

$$v_R(t) = i(t)R = (I_0R - V)e^{-\left(\frac{Rt}{L}\right)} + V \tag{4.39}$$

It is clear from the equation (4.39) that the voltage across the resistor decays exponentially and it settles at *V* at steady state. Similarly the Using equation (4.7), voltage across the inductor is found out as,

$$v_L(t) = L\frac{di(t)}{dt} = (V - I_0R)e^{-(Rt/L)} \tag{4.40}$$

Instead of an initially charged inductor, if an uncharged inductor is used in the same circuit then the initial current value is given by,

$$i(0) = 0 \tag{4.41}$$

In that case, the step response of the circuit becomes,

$$i(t) = \begin{cases} 0, & t < 0 \\ \frac{V}{R}\left(1 - e^{-(t/\tau)}\right), & t > 0 \end{cases} \tag{4.42}$$

Voltage across the inductor and voltage across the resistor with zero initial current through the inductor is calculated as,

$$v_L(t) = L\frac{di(t)}{dt} = Ve^{-(t/\tau)} \tag{4.43}$$

$$v_R(t) = i(t)_R = V(1 - e^{-(t/\tau)}) \quad (4.44)$$

Equations (4.42) and (4.44) depict that the current through the inductor rises exponentially and voltage across it decays exponentially when a series *R-L* circuit with zero initial current through the inductor is excited by a step voltage function. The same is shown in Fig. 4.7.

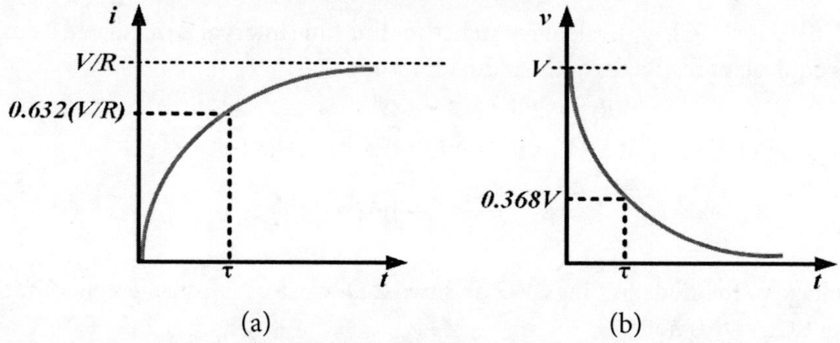

Fig. 4.7: Current through inductor and voltage across inductor of series *R-L* circuit with step input voltage and zero initial condition

Example 4.5:

For the circuit given below, SW1 is closed at t =0 and SW2 is closed at t = 2 sec. Find the current flowing through the circuit for t > 0. Also find the time constants of the circuit for different time intervals.

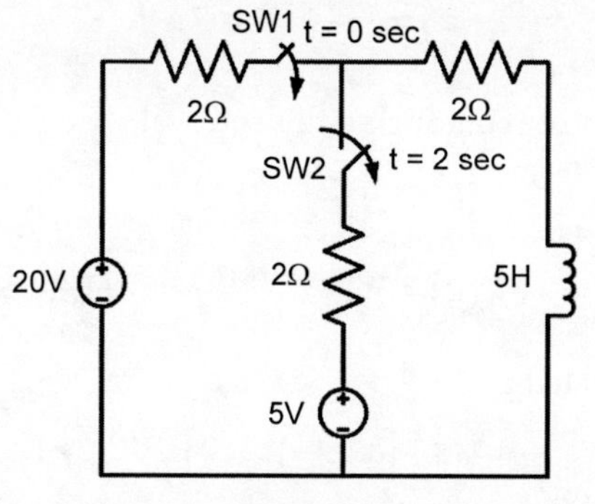

Solution:

For the time interval $0 \le t \le 2$, only SW1 is closed. During that time interval the equivalent resistance of the circuit is obtained as,

$$R_{eq1} = 2 + 2 = 4\ \Omega$$

Time constant of the circuit during this time interval is calculated as,

$$\tau = \frac{L}{R_{eq1}} = \frac{5}{4} = 1.25\,\text{sec}$$

The current through a series *R-L* circuit for step input voltage is given by,

$$i(t) = \left[i(0) - i(\infty)\right]e^{-(t/\tau)} + i(\infty)$$

Here,

$$i(0) = 0$$

$$i(\infty)=\frac{V}{R_{eq1}}=\frac{20}{4}=5\text{A}$$

$$i(t)=[0-5]e^{-(t/1.25)}+5=5(1-e^{-0.8t})\text{A}$$

For $t \geq 2$, SW2 is closed. Equivalent resistance for this time interval is calculated by using the concept of Thevenin's equivalent resistance at the inductor terminal as,

$$R_{eq2} = 2 \,||\, 2 + 2 = 3\ \Omega$$

Time constant of the circuit for this time interval is calculated as,

$$\tau_1=\frac{L}{R_{eq2}}=\frac{5}{3}=1.67\text{ sec}$$

If the voltage of the node joining SW2 and two 2 Ω resistors is considered as V_1, then to find out $i(\infty)$, applying KCL at this node as,

$$\frac{20-V_1}{2}+\frac{5-V_1}{2}=\frac{V_1}{2}$$

$$V_1=\frac{25}{3}=8.33\text{A}$$

$$i_1(\infty)=\frac{8.33}{2}=4.167\text{A}$$

Current flowing through the circuit during this interval,

$$i_1(t)=\left[i_1(2)-i_1(\infty)\right]e^{-(t/\tau)}+i_1(\infty)$$

$$i_1(2)=5\left(1-e^{-0.8\times 2}\right)=5\times\left(1-0.2\right)=3.99\text{A}$$

Hence the current is calculated as,

$$i_1(t)=\left[3.99-4.167\right]e^{-(t/1.67)}+4.167$$

$$i_1(t)=4.167-0.176e^{-0.6t}\text{A}$$

Example 4.6:

A coil has a resistance of 5 Ω and inductance of 2 H is connected to a dc voltage source of 100 V through a switch. The switch is closed at t = 0. Calculate the current flowing through the circuit and voltage across the resistor of the coil for t > 0. Also find the time constant of the circuit and steady state current through the circuit.

Solution:

Current through the circuit at $t = 0$ is given by,

$$i(0) = 0$$

$$R = 5\Omega$$

$$L = 2\text{H}$$

$$V = 100\text{V}$$

Time constant of the circuit is calculated as,

$$\tau = \frac{L}{R} = \frac{2}{5} = 0.4\,\text{sec}$$

Current through the circuit is calculated as,

$$i(t) = \frac{V}{R}\left(1 - e^{-(t/\tau)}\right) = \frac{100}{5}\left(1 - e^{-(t/0.4)}\right) = 20\left(1 - e^{-2.5t}\right)\text{A}$$

Voltage across the resistor for $t > 0$,

$$v_R(t) = V\left(1 - e^{-(t/\tau)}\right) = 100\left(1 - e^{-2.5t}\right)\text{V}$$

Steady state current is obtained as,

$$i_{steady\_state} = \frac{100}{5} = 20\text{A}$$

4.3.4 Step response of series *R-C* circuit

A circuit with a series combination of pure resistor (R) and pure charged capacitor (C) with an initial voltage of V_{C0} is excited by an independent dc voltage source as shown in Fig. 4.8. Initially the voltage source was not connected and then the dc voltage is suddenly connected at by closing the switch (SW).

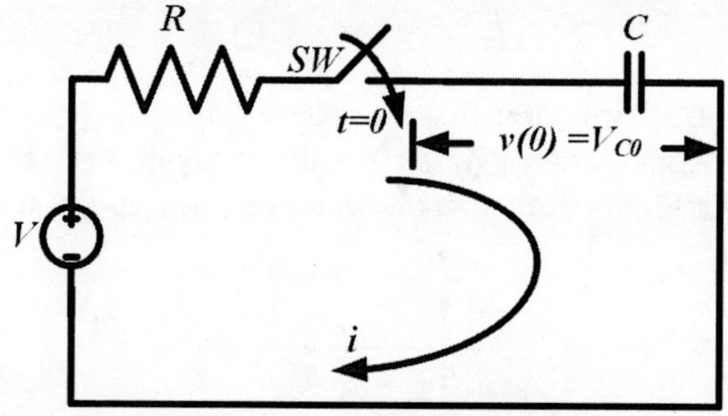

Fig. 4.8: Series *R-C* circuit with step input voltage with initial condition $v(0) = V_{CO}$

The response of this circuit is considered as current through the circuit. The initial voltage across the capacitor due to the charged capacitor at is given by,

$$v(0) = V_{C0} \tag{4.45}$$

As the voltage across the capacitor does not change suddenly, the voltage across the capacitor just before closing the switch is same as the voltage across it just after closing the switch. It is expressed as,

$$v(0^-) = v(0^+) = V_{C0} \tag{4.46}$$

Applying KCL in the circuit shown in Fig. 4.8, the relation between current through the capacitor and resistor is given by,

$$i_C(t) + i_R(t) = 0$$

$$C\frac{dv(t)}{dt} + \frac{v(t) - Vu(t)}{R} = 0$$

For $t > 0$, the above equation becomes,

$$C\frac{dv(t)}{dt}+\frac{v(t)-V}{R}=0 \tag{4.47}$$

Rearranging and integrating the equation (4.45), it is obtained as,

$$\int_{V_{CO}}^{v(t)}\frac{dv(t)}{v(t)-V}=-\int_0^t\frac{1}{RC}dt \tag{4.48}$$

The solution of equation (4.20) is obtained as,

$$\ln\frac{v(t)-V}{V_{CO}-V}=-\frac{t}{RC} \tag{4.49}$$

Expressing the equation (4.49) in exponential form and rearranging it as below,

$$v(t)=V+(V_{CO}-V)e^{-(t/RC)} \tag{4.50}$$

Equation (4.50) depicts the step response of a series *R-C* circuit and v(t) represents the voltage across the capacitor. $\frac{1}{RC}$ is the time constant of this circuit.

The step response of the voltage across the capacitor can also represented as,

$$v(t)=\begin{cases} V_{CO}, & t<0 \\ V+(V_{CO}-V)e^{-(t/\tau)}, & t>0 \end{cases} \tag{4.51}$$

From the equation (4.51), it is clear that the voltage across the capacitor increases exponentially from an initial value of V_{CO} and settles at V as $t \to \infty$ if this circuit is excited by a step voltage function with an initial condition of V_{CO} voltage across the capacitor. The response is demonstrated with the help of Fig. 4.9

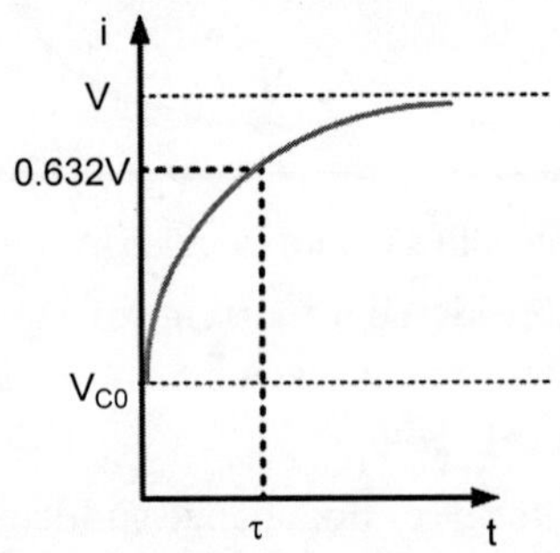

Fig. 4.9: Voltage response of series *R-C* circuit with step input voltage and initial condition $v(0) = V_{CO}$

The equation (4.50) can also be expressed as,

$$v(t)=\left[v(0)-v(\infty)\right]e^{-(t/\tau)}+v(\infty) \tag{4.52}$$

It is clear from the equation (4.52) that the initial value of voltage across capacitor, v(0), final value of voltage across capacitor, v(∞) and the time constant of the circuit(τ) are required to find out the step response of a series R-C circuit with an initially charged capacitor.

Current through the resistor is given by,

$$i_R(t)=\frac{v(t)-V}{R}=\frac{(V_{C0}-V)}{R}e^{-(t/RC)} \tag{4.53}$$

Using equation (4.19), current through the capacitor is found out as,

$$i_c(t)=C\frac{dv(t)}{dt}=-\frac{(V_{C0}-V)}{R}e^{-(t/RC)} \tag{4.54}$$

Both the equations (4.53) and (4.54) confirms that the current through this circuit is exponentially decaying from an initial value of and settles at zero as as .

Instead of an initially charged capacitor, if an uncharged capacitor is used in the same circuit then the initial voltage value across the capacitor is given by,

$$v(0) = 0 \tag{4.55}$$

In that case, the step response of the circuit becomes,

$$v(t)=\begin{cases} 0, & t<0 \\ V\left(1-e^{-(t/\tau)}\right), & t>0 \end{cases} \tag{4.56}$$

Current through the circuit with zero initial voltage across the capacitor is calculated as,

$$i(t)=\frac{V}{R}e^{-(t/RC)} \tag{4.57}$$

Equations (4.56) and (4.57) depicts that the voltage across capacitor rises exponentially from zero and settles at V and current through it decays exponentially from $\frac{V}{R}$ and settles at zero as $t \to \infty$, when a series *R-C* circuit with zero initial voltage across capacitor is excited by a step voltage function. The same is shown in Fig. 4.10.

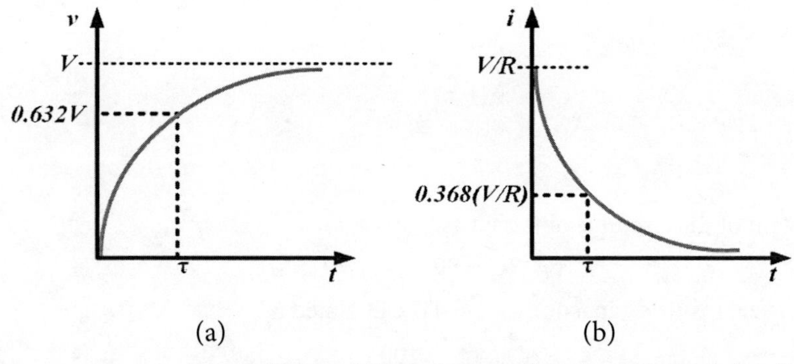

Fig. 4.10: Voltage across capacitor and current through capacitor of series *R-C* circuit with step input voltage and zero initial condition

Example 4.7:

A step input voltage source of 10 V is applied to a series *R-C* circuit of resistance of 10 Ω and capacitance of 0.1 F at t = 0. Calculate the steady state charge of the capacitor. Find the time taken to charge the capacitor up to 75% of its steady state value.

Solution:

$$R = 10\Omega$$
$$C = 0.1\text{F}$$
$$V = 10\text{V}$$

Time constant of the circuit is obtained as,

$$\tau = RC = 10 \times 0.1 = 1 \text{ sec}$$

Voltage across the capacitor for t > 0 is obtained as,

$$v(t) = V\left(1 - e^{-(t/\tau)}\right) = 10\left(1 - e^{-t}\right)\text{V}$$

Steady state voltage across the capacitor is calculated as,

$$V_{steady\ state} = 10 \text{ V}$$

Steady state charge in the capacitor is obtained as,

$$Q_{steady\ state} = C \times V_{steady\ state} = 0.1 \times 10 = 1 \text{ C}$$

Time taken by the circuit to charge the capacitor up to 75% of its steady state value is calculated as,

$$0.75 \times Q_{steady\ state} = Qs_{teady\ state}\ (1 - e^{-(t/\tau)})$$
$$0.75 = (1 - e^{-(t/1)}) = (1 - e^{-t})$$
$$e^{-t} = 0.25$$
$$t = 1.386 \text{ sec}$$

Example 4.8:

A step input voltage source of 500 V is applied to a series *R-C* circuit with resistance of 50 Ω and capacitance of 0.1 F at t = 0. The initial charge of the capacitor for t < 0 was 100 C. Calculate the current through the circuit for t > 0.

Solution:

$$R = 50\Omega$$
$$C = 0.01\text{F}$$
$$V = 500\text{V}$$
$$Q_0 = 100\text{C}$$

Time constant of the circuit is obtained as,

$$\tau = RC = 50 \times 0.01 = 0.5 \text{ sec}$$

Initial voltage across the capacitor for t < 0 is calculated as,

$$V_{C0} = \frac{Q_0}{C} = \frac{100}{0.1} = 1000\text{V}$$

Current through the circuit is calculated as,

$$i(t) = \frac{(V_{C0} - V)}{R} e^{-\left(\frac{t}{RC}\right)} = \frac{1000 - 500}{50} e^{-t/0.5} = 10e^{-2t} \text{ A}$$

4.3.5 Step Response of Parallel *R-L* Circuit

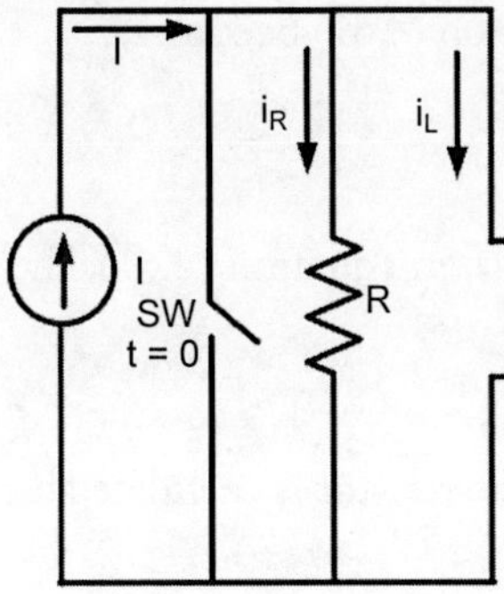

Fig. 4.11: Parallel *R-L* circuit with step input current

A circuit with a parallel combination of pure resistor (*R*) and pure inductor (*L*) connected across an independent dc current source as shown in Fig. 4.11 is considered for analysis. Initially the switch (*SW*) was closed and no current was flowing through the parallel combination of *R* and *L*. The switch is open at and current starts flowing through *R* and *L*. The response of this circuit is considered as current through the circuit i.e. parallel combination of *R* and *L*. The initial current through the parallel combination of *R* and *L* is zero as the inductor was not charged initially.

Applying KCL in the circuit shown in Fig. 4.11, the relation between current through inductor, resistor and the source current is given by,

$$i_R(t) + i_L(t) = I_u(t) \tag{4.58}$$

As the *R* and *L* are connected in parallel, the voltage across them are same and it is shown as,

$$i_R(t)R = L\frac{di_L(t)}{dt}$$

$$Iu(t)R = i_L(t)R + L\frac{di_L(t)}{dt} \tag{4.59}$$

Rearranging the equation (4.59), it is obtained for $t > 0$ as,

$$\frac{di_L(t)}{dt} + i_L(t)\frac{R}{L} = \frac{IR}{L} \tag{4.60}$$

The solution of equation (4.60) is obtained as,

$$i_L(t) = Ke^{-(Rt/L)} + e^{-(Rt/L)}\int \frac{IR}{L}e^{(Rt/L)}dt \tag{4.61}$$

The forced response part of equation (4.61) is given by,

$$i_L(t)_f = e^{-(Rt/L)}\int \frac{IR}{L}e^{(Rt/L)}dt = I \tag{4.62}$$

Inductor does not allow sudden change in current through it as per the theory of electromagnetic induction. Hence, the current through the inductor just before the switching is same as the current through it just after switching as mentioned below,

$$i_L(0^-) = i_L(0^+) = 0 \tag{4.63}$$

With this initial condition the equation (4.61) becomes,

$$0 = K + I$$

$$K = -I \tag{4.64}$$

Substituting the value of K and $i_L(t)_f$ in equation (4.61), the complete step response for inductor current becomes

$$i_L(t) = I(1 - e^{-(Rt/L)}) \tag{4.65}$$

Step response for current through the resistor is calculated as,

$$i_R(t) = I - i_L(t) = Ie^{-(Rt/L)} \tag{4.66}$$

Equations (4.65) and (4.66) depict that the current through the inductor rises exponentially and it will reach to a steady state value of I as $t \to \infty$. Similarly the current through the resistor decays exponentially from an initial value of I and it will become zero as $t \to \infty$. The same is illustrated in Fig.4.12.

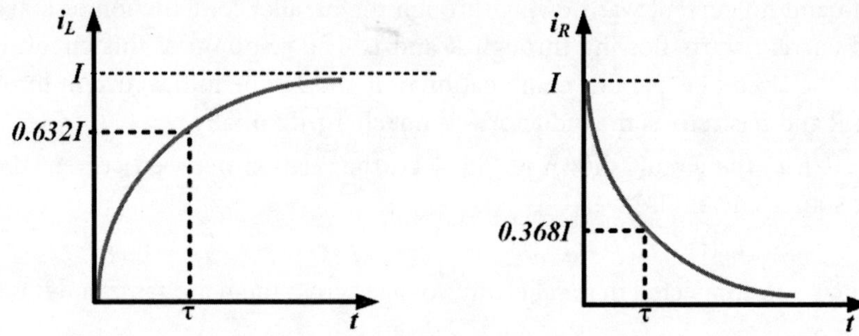

Fig. 4.12: Current through inductor and resistor of parallel *R-L* circuit with step input current and zero initial condition

If the switch (*SW*) is open for a long time, then the entire source current flows through the inductor and no current flows through the resistor.

Example 4.9:

A step input current source of 10 A is applied across a parallel combination of *R* and *L* with $R = 10\ \Omega$ and $L = 4$ H at $t = 0$. Find the time constant of the circuit, current through the inductor for $t > 0$ and for $t = 5$ times of time constant of the circuit. Find the steady state energy stored in the inductor.

Solution:

$$R = 10\Omega$$

$$L = 4\text{H}$$

$$I = 10\text{A}$$

Time constant of the circuit is obtained as,

$$\tau = \frac{L}{R} = \frac{4}{10} = 0.4\,\text{sec}$$

Current through the inductor for t > 0 is calculated as,

$$i_L(t) = I\left(1 - e^{-\left(\frac{Rt}{L}\right)}\right) = 10 \times \left(1 - e^{-\frac{10t}{4}}\right) = 10\left(1 - e^{-2.5t}\right)\text{A}$$

Current through the inductor for t = 5τ is calculated as,

$$i_L(t) = 10\left(1 - e^{-2.5\times5\times0.4}\right) = 10 \times 0.993\text{A} = 9.93\text{A}$$

Steady state current through the inductor is given by,

$$I_{L\text{-steady state}} = 10 \text{ A}$$

Energy stored in inductor at steady state is calculated as,

$$E = \frac{1}{2}L\left(I_{L\_steady\_state}\right)^2 = \frac{1}{2} \times 4 \times (10)^2 = 200\text{J}$$

4.3.6 Step response of parallel *R-C* circuit

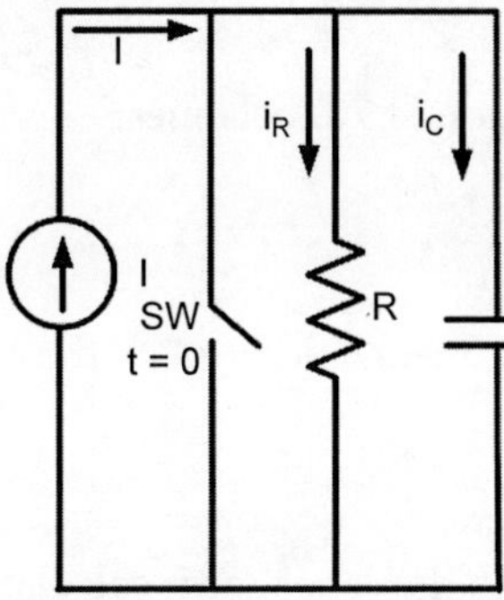

Fig. 4.13: Parallel *R-C* circuit with step input current

A circuit with a parallel combination of pure resistor (R) and pure capacitor (C) connected across an independent dc current source as shown in Fig. 4.13 is considered for analysis. Initially the switch (SW) was closed and no current was flowing through the parallel combination of R and C. The switch is open at *t*=0 and current starts flowing through R and C. The response of this circuit is considered as current through the circuit i.e. parallel combination of R and C. The initial voltage across the C is zero as the capacitor was not charged initially.

Applying KCL in the circuit shown in Fig. 4.13, the relation between current through capacitor, resistor and the source current is given by,

$$i_C(t) + i_R(t) = Iu(t) \tag{4.67}$$

As the *R* and *C* are connected in parallel, the voltage across them are same and it is shown as,

$$i_R(t)R = \frac{\int i_C(t)dt}{C} \tag{4.68}$$

Then equation (4.67) can be presented as,

$$C\frac{di_R(t)R}{dt}+i_R(t)=Iu(t)$$

$$\frac{di_R(t)}{dt}+\frac{1}{RC}i_R(t)=\frac{I}{RC}\ for\ t>0 \qquad (4.69)$$

The solution of equation (4.69) is obtained as,

$$i_R(t)=Ke^{-(t/RC)}+e^{-(t/RC)}\int\frac{I}{RC}e^{(t/RC)}dt \qquad (4.70)$$

The forced response part of equation (4.70) is given by,

$$i_R(t)_f=e^{-(t/RC)}\int\frac{I}{RC}e^{(t/RC)}dt=I \qquad (4.71)$$

Initially when the switch was closed then no current was flowing through the parallel combination of *R* and *C* as current flows through the minimum resistance path.

$$i_R(0)=0 \qquad (4.72)$$

With this initial condition the equation (4.70) becomes,

$$0=K+I$$

$$K=-I \qquad (4.73)$$

Substituting the value of K and $i_R(t)_f$ in equation (4.70), the complete step response for the current through resistor becomes,

$$i_R(t)=I(1-e^{-(t/RC)}) \qquad (4.74)$$

Step response for the current through the capacitor is calculated as,

$$i_C(t)=I-i_R(t)=Ie^{-(t/RC)} \qquad (4.75)$$

Equations (4.74) and (4.75) depict that the current through the resistor rises exponentially and it will reach to a steady state value of I as $t\to\infty$. Similarly the current through the capacitor decays exponentially from an initial value of I and it will become zero as $t\to\infty$. The same is illustrated in Fig.4.14.

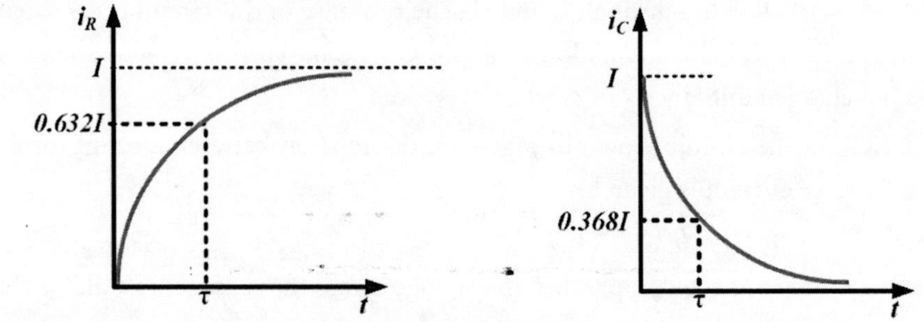

Fig. 4.14: Current through resistor and capacitor of parallel *R-C* circuit with step input current and zero initial condition

If the switch (*SW*) is open for a long time, then the entire source current flows through the resistor and the branch where capacitor is connected behaves as an open circuit as no current flows through the capacitor.

Example 4.9:

A step input current source of 10 A is applied across a parallel combination of *R* and *C* with $R = 20\ \Omega$ and $C = 0.1$ F at t = 0. Find the current through the capacitor and resistor for t > 0. Find the power dissipated in the resistor for t > 0.

Solution:

$$R = 20\Omega$$

$$C = 0.1\text{F}$$

$$I = 10\text{A}$$

Time constant of the circuit is obtained as,

$$\tau = RC = 20 \times 0.1 = 2 \text{ sec}$$

Current through resistor is given by,

$$i_R(t) = I\left(1 - e^{-(t/RC)}\right) = 10 \times (1 - e^{-(t/2)}) = 10(1 - e^{-0.5t})\text{A}$$

Current through the capacitor is calculated as,

$$i_C(t) = I - i_R(t) = Ie^{-(t/RC)} = 10e^{-0.5t}\text{ A}$$

Power dissipated in the resistor is given by,

$$p_R(t) = \left[i_R(t)\right]^2 R = \left[10(1 - e^{-0.5t})\right]^2 \times 20 = 2000(1 - e^{-0.5t})^2 \text{ W}$$

4.4 SECOND ORDER DIFFERENTIAL EQUATIONS

Second order differential equation is represented by,

$$a\frac{d^2y}{dx^2} + b\frac{dy}{dt} + cy = 0 \qquad (4.76)$$

Where *y* is the output and *a*, *b*, *c* are constant. The solution of equation of equation (4.76) is given by,

$$y = K_1 e^{\alpha_1 t} + K_2 e^{\alpha_2 t} \qquad (4.77)$$

Where K_1 and K_2 are constants and α_1 and α_2 are the roots of the characteristics equation given by,

$$ap^2 + bp + c = 0 \qquad (4.78)$$

The values of α_1 and α_2 are given by,

$$\alpha_1 = \frac{-b + \sqrt{b^2 - 4ac}}{2a} \qquad (4.79)$$

$$\alpha_2 = \frac{-b - \sqrt{b^2 - 4ac}}{2a} \qquad (4.80)$$

4.5 SECOND ORDER ELECTRIC CIRCUITS

The response of the series and parallel *R-L-C* circuits are governed by second order differential equations. Therefore, these circuits are together known as second order electric circuits. Normally, two energy storage elements and resistances form a second order electric circuit.

4.5.1 Step Response of Series *R-L-C* Circuit

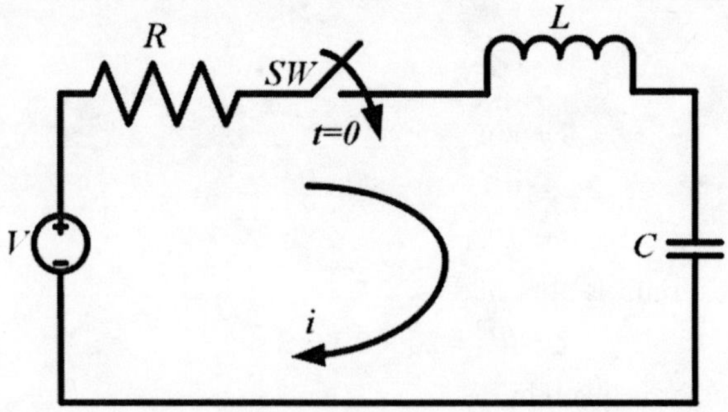

Fig. 4.15: Series *R-L-C* circuit with step input voltage

A circuit with a series combination of pure resistor (R), pure inductor (L) and pure capacitor (C) connected across an independent dc voltage source as shown in Fig. 4.15 is considered for analysis. Initially the switch (SW) was open and no current was flowing through the circuit. The switch (SW) is closed at t=0 and current starts flowing through the circuit. The response of this circuit is considered as current through the circuit.

Applying KVL in the circuit shown in Fig. 4.15, the relation between the voltage across resistor, inductor, capacitor and the source voltage is given by,

$$i(t)R+L\frac{di(t)}{dt}+\frac{1}{C}\int i(t)dt=Vu(t)$$

$$i(t)R+L\frac{di(t)}{dt}+\frac{1}{C}\int i(t)dt=V \; for \; t>0 \tag{4.81}$$

Taking differentiation in both side, it is obtained as,

$$L\frac{d\,i(t)}{dt}+R\frac{di(t)}{dt}+\frac{i}{C}=$$

$$\left(p^2+\frac{R}{L}p+\frac{1}{LC}\right)i=0 \tag{4.82}$$

Equation (4.82) is a linear, homogeneous, second order differential equation. The characteristic equation is given by,

$$p^2+\frac{R}{L}p+\frac{1}{LC}=0 \tag{4.83}$$

The roots of the characteristic equation are obtained by applying Shreedhara Acharya's formula as,

$$p_1, p_2 = \frac{-\frac{R}{L} \pm \sqrt{\left(\frac{R}{L}\right)^2 - \frac{4}{LC}}}{2} \tag{4.84}$$

Let us consider,

$$\beta = -\frac{R}{2L} \tag{4.85}$$

$$\gamma = \sqrt{\left(\frac{R}{2L}\right)^2 - \frac{1}{LC}} \tag{4.86}$$

$|\beta|$ is basically the neper frequency and $\frac{1}{LC}$ is the resonant frequency of the series R-L-C circuit.

Then p_1 and p_2 becomes

$$p_1 = \beta + \gamma \tag{4.87}$$

$$p_2 = \beta - \gamma \tag{4.88}$$

The solution of the differential equation mentioned in (4.82) is then given by,

$$i(t) = K_1 e^{p_1 t} + K2\, e^{p_2 t} \tag{4.89}$$

Where, K_1 and K_2 are constants.

The solution mentioned in (4.89) can be any one of the following based on the relation between neper frequency (β) and resonant frequency,

Condition 1:	$\left(\frac{R}{2L}\right)^2 > \frac{1}{LC}$	Over-damped response
Condition 2:	$\left(\frac{R}{2L}\right)^2 = \frac{1}{LC}$	Critically damped response
Condition 3:	$\left(\frac{R}{2L}\right)^2 < \frac{1}{LC}$	Under-damped response

Condition 1:

For this condition the value of γ is real and positive. Hence, the values of p_1, p_2 are real and unequal which is given by,

$$p_1 = \beta + \gamma$$

$$p_2 = \beta - \gamma$$

Hence the solution is obtained as,

$$i(t) = K_1 e^{(\beta+\gamma)t} + K_2 e^{(\beta-\gamma)t}$$

$$i(t) = e^{\beta t}(K_1 e^{\gamma t} + K_2 e^{-\gamma t}) \tag{4.90}$$

The over-damped current response is illustrated in Fig. 4.16. This figure shows that the response decays and approaches zero as time increases.

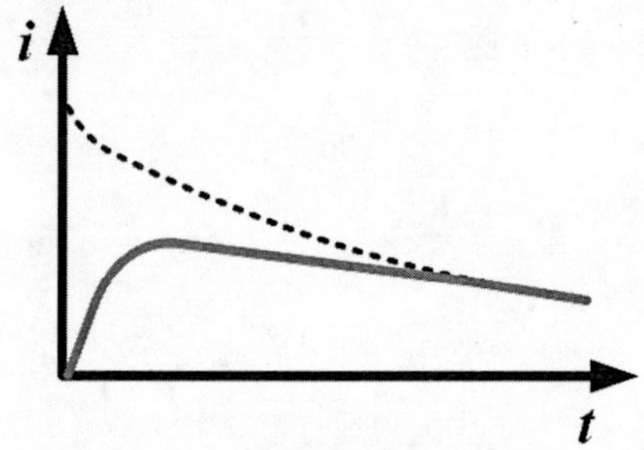

Fig. 4.16: Over-damped current response of series *R-L-C* circuit with step input voltage and zero initial condition

Condition 2:

For this condition the value of γ is zero. Hence, the values of p_1, p_2 are real and equal which is given by,

$$p_1 = p_1 = \beta$$

Hence the solution is obtained as,

$$i(t) = K_1\, e^{\beta t} + K_2\, te^{\beta t}$$

$$i(t) = e^{\beta t}\,(K_1 + K_2\, t) \tag{4.91}$$

The critically damped current response is illustrated in Fig. 4.17. The response reaches to its maximum value first at one time constant and then decays as time progresses.

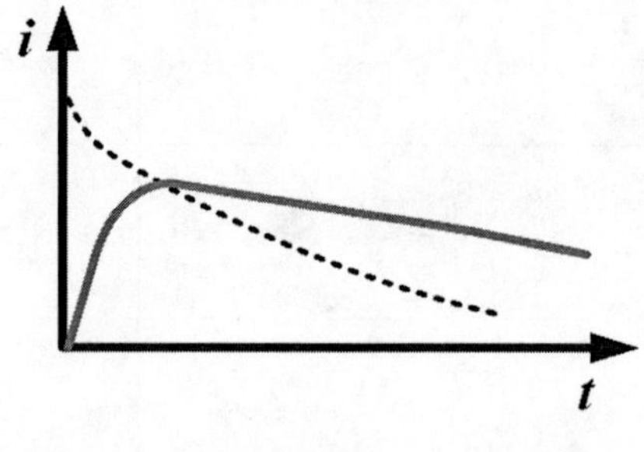

Fig. 4.17: Critically damped current response of series *R-L-C* circuit with step input voltage and zero initial condition

Condition 3:

For this condition the value of γ is imaginary. Hence, the values of p_1, p_2 are complex conjugates and they are presented as,

$$p_1 = \beta + j\gamma$$

$$p_2 = \beta - j\gamma$$

Hence the solution is obtained as,

$$i(t) = K_1\, e^{(\beta + j\gamma)t} + K_2\, e^{(\beta - j\gamma)t}$$

$$i(t) = e^{\beta t}\,(K_1\, e^{j\gamma t} + K_2\, e^{-j\gamma t}) \tag{4.92}$$

The under-damped current response is illustrated in Fig. 4.18. The response for this condition is oscillatory and exponentially damped in nature.

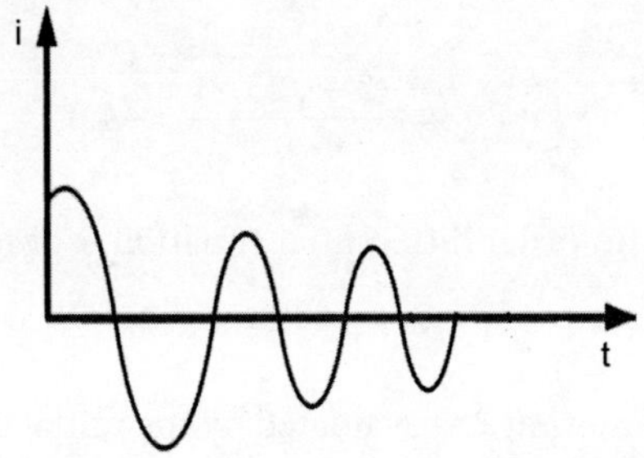

Fig. 4.18: Under-damped current response of series *R-L-C* circuit with step input voltage and zero initial condition

Example 4.10 A series *R-L-C* circuit with R = 10 Ω, L = 2 H and C = 2 F is excited by a dc step voltage of 40 V at t =0. Find the current flowing the circuit at t > 0.

Solution:

$$R = 10\Omega$$
$$L = 2\text{H}$$
$$C = 2\text{F}$$
$$V = 40\text{V}$$

For t > 0 the current through the circuit is mentioned through the relation mentioned below by,

$$i(t)R + L\frac{di(t)}{dt} + \frac{1}{C}\int i(t)dt = V$$

Differentiating the above equation it is obtained as,

$$L\frac{d^2i(t)}{dt^2} + R\frac{di(t)}{dt} + \frac{i}{C} = 0$$

$$\left(p^2 + \frac{R}{L}p + \frac{1}{LC}\right)i = 0$$

The characteristic equation is given by,

$$p^2 + \frac{R}{L}p + \frac{1}{LC} = 0$$

$$p^2 + \frac{10}{2}p + \frac{1}{2\times 2} = 0$$

$$p_2 + 5p + 0.25 = 0$$

The roots of the characteristic equation are obtained by applying Shreedhara Acharya's formula as,

$$p_1, p_2 = \frac{-5 \pm \sqrt{25-1}}{2}$$

$$p_1 = \frac{-5+\sqrt{25-1}}{2} = -0.05$$

$$p_2 = \frac{-5-\sqrt{25-1}}{2} = -4.95$$

The solution of the second order differential equation is obtained as,

$$i(t) = K_1 e^{-0.05t} + K_2 e^{-4.95t}$$

K_1 and K_2 are constant and they are evaluated by the initial conditions.

$i(0) = 0$ and $\frac{1}{C}\int i(0)dt = 0$

For t = 0 the current through the circuit is written through the relation mentioned below by,

$$i(0)\times 10 + 2\times\frac{di(0)}{dt} + \frac{1}{2}\int i(0)dt = 40$$

$$2\times\frac{di(\;)}{dt} = 40$$

$$\frac{di(0)}{dt} = 20\text{A / sec}$$

Substituting, $i(0) = 0$ in the solution it is obtained as,

$$i(0) = 0 = K_1\, e^0 + K_2\, e^0 = K_1 + K_2 \qquad \text{(a)}$$

Differentiating the solution it is obtained as,

$$\frac{di(t)}{dt} = -0.05K_1 e^{-0.05t} - 4.95K_2 e^{-4.95t}$$

It is already known as,

$$\frac{di(0)}{dt} = 20\text{A / sec}$$

Hence,

$$\frac{di(0)}{dt} = 20 = -0.05K_1 - 4.95K_2 \qquad \text{(b)}$$

Solving (a) and (b), the values of K_1 and K_2 are calculated as,

$$K_2 = -4.08 \text{ and } K_1 = 4.08$$

Current flowing through the circuit for t > 0 is then obtained as,

$$i(t) = 4.08e^{-0.05t} - 4.08\, e^{-4.95t}\ \text{A}$$

4.5.2 Step Response of Parallel *R-L-C* Circuit

A circuit with a parallel combination of pure resistor (R), pure inductor (L) and pure capacitor (C) connected across an independent dc current source as shown in Fig. 4.19 is considered for analysis. Initially the switch (SW) was closed and no current was flowing through the circuit. The switch (SW) is open at t=0 and current starts flowing through the parallel combination of R-L-C. The response of this circuit is considered as current through the circuit.

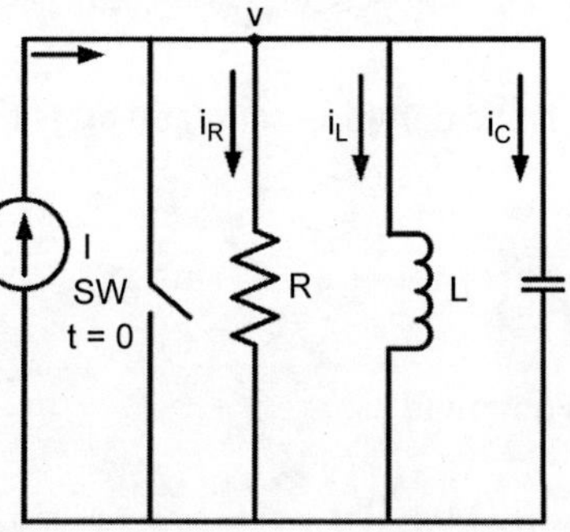

Fig. 4.19: Parallel *R-L-C* circuit with step input current

Applying KCL in the circuit shown in Fig. 4.19, the relation between the current through resistor, inductor, capacitor and the source current is given by,

$$i_R(t) + i_L(t) + i_C(t) = Iu(t) \tag{4.93}$$

As the R, L and C are connected in parallel, the voltage across them are same and it is considered as $v(t)$ shown as in Fig. 4.17. For $t > 0$ the equation (4.93) becomes,

$$\frac{v(t)}{R} + C\frac{dv(t)}{dt} + \frac{\int v(t)dt}{L} = I \tag{4.94}$$

Taking differentiation in both side of equation (4.94) and rearranging them, it is written as,

$$\frac{1}{R}\frac{dv(t)}{dt} + C\frac{d^2v(t)}{dt^2} + \frac{v(t)}{L} = 0$$

$$\frac{d^2v(t)}{dt^2} + \frac{1}{RC}\frac{dv(t)}{dt} + \frac{v(t)}{LC} = 0 \tag{4.95}$$

$$\left(p^2 + \frac{1}{RC}p + \frac{1}{LC}\right)v(t) = 0 \tag{4.96}$$

Equation (4.96) is a linear, homogeneous, second order differential equation. The characteristic equation is given by,

$$p^2 + \frac{1}{RC}p + \frac{1}{LC} = 0 \tag{4.97}$$

The roots of the characteristic equation are obtained by applying Shreedhara Acharya's formula as,

$$p_1, p_2 = \frac{-\frac{1}{RC} \pm \sqrt{\left(\frac{1}{RC}\right)^2 - \frac{4}{LC}}}{2} \tag{4.98}$$

It is assumed that,

$$\omega_0{}^2 = \frac{1}{LC}, 2\xi\omega_0 = \frac{1}{RC} and \omega_d = \sqrt{\omega_0{}^2 - (\xi\omega_0)^2}$$

ω_0 is the resonant frequency, ξ is the damping ratio and ω_d is the damping frequency. The solution of equation (4.96) is then given by,

$$v(t) = \frac{1}{\omega_d C} e^{-\xi\omega_0 t} \sin \omega_d t \tag{4.99}$$

Current through the resistor is obtained as,

$$i_R(t) = \frac{v(t)}{R} = \frac{1}{\omega_d CR} e^{-\xi\omega_0 t} \sin \omega_d t \tag{4.100}$$

Current through the inductor is given by,

$$i_L(t) = \frac{\int_0^t v(t)dt}{L} = \frac{1}{\omega_d LC} \frac{e^{-\xi\omega_0 t}}{\sqrt{(\xi\omega_0)^2 - \omega_d{}^2}} \sin\left(\omega_d t - \tan^{-1}\frac{\omega_d}{\xi\omega_0}\right) \tag{4.101}$$

Current through the capacitor is given by,

$$i_C(t) = C\frac{dv(t)}{dt} = \frac{e^{-\xi\omega_0 t}}{\omega_d}\left\{(\xi\omega_0)^2 + \omega_d{}^2\right\} \sin\left(\omega_d t - \tan^{-1}\frac{\omega_d}{\xi\omega_0}\right) \tag{4.102}$$

4.6 APPLICATION OF LAPLACE TRANSFORM IN ELECTRIC CIRCUITS

Laplace transform is an integral transformation of a function from time domain to frequency domain. It is a very useful technique to solve linear differential equation. It is understood from the previous sections that the transient response of first order and second order electric circuits are represented by linear differential equations. Hence, this technique is very useful for finding the transient response of first and second order electric circuits. Moreover, in Laplace transform the initial conditions (voltage or current in energy storage elements) are taken care of automatically unlike in the time domain. This technique involves three major steps to obtain the solution of a differential equation.

Step1: Transforming the differential equation in algebraic equation.

Step2: Obtaining the solution of the constituent algebraic equation.

Step3: Transforming back the solution in such a way that it represents the solution of the original differential equation.

4.6.1 Step Response of Series *R-L* Circuit using Laplace Transform

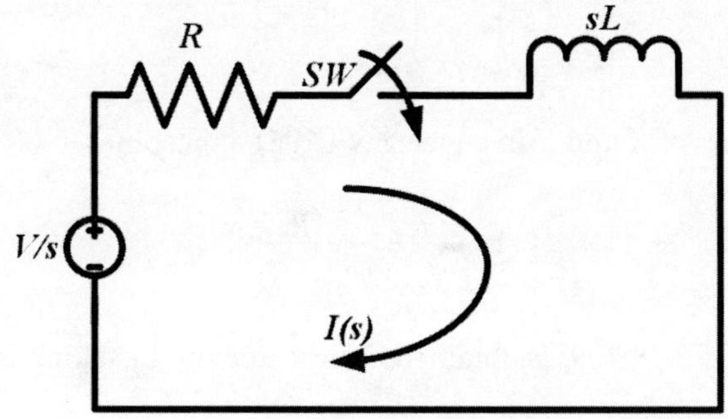

Fig. 4.20: Series *R-L* circuit with step input voltage in Laplace domain

A circuit with a series combination of pure resistor (R) and pure inductor (L) is energised by an independent dc voltage source as shown in Fig. 4.20. Initially the voltage source was disconnected from the circuit and then the dc voltage is suddenly applied to the circuit at t=0 by closing the switch (SW). The response of this circuit is considered as current through the circuit. The initial current through the circuit at t=0 is given by,

$$i(0) = 0 \tag{4.103}$$

Applying KVL in the circuit shown in Fig. 4.20, the relation between voltage across the inductor and resistor is given by,

$$i(t)R + L\frac{di(t)}{dt} = Vu(t) \tag{4.104}$$

The equation (4.104) is represented in Laplace domain as,

$$I(s)R + L[sI(s) - i(0^-)] = \frac{V}{s}$$

$$I(s)(R + sL) = \frac{V}{s}$$

$$I(s) = \frac{V/L}{s(R/L + s)} \tag{4.105}$$

Using the partial fraction concept of Laplace transform it can be written as,

$$I(s) = \frac{X}{s} + \frac{Y}{(R/L + s)} \tag{4.106}$$

Where the values of X and Y are given by,

$$X = \left[\frac{V/L}{(R/L + s)}\right]_{s=0} = \frac{V}{R} \tag{4.107}$$

$$Y = \left[\frac{V/L}{s} \right]_{s=-R/L} = -\frac{V}{R} \tag{4.108}$$

Substituting the values of X and Y in equation (4.106), it becomes,

$$I(s) = \frac{V/R}{s} - \frac{V/R}{(R/L + s)} \tag{4.109}$$

The solution of equation (4.109) is obtained in time domain by taking inverse Laplace transform of (4.109) for $t > 0$ as,

(4.109) for t > 0 as,

$$i(t) = \frac{V}{R}\left(1 - e^{-(Rt/L)}\right) \tag{4.110}$$

Equation (4.110) presents the current flowing through the circuit at $t > 0$.

4.6.2 Step Response of Parallel *R-C* Circuit using Laplace Transform

A circuit with a parallel combination of pure resistor (R) and pure capacitor (C) connected across an independent dc current source as shown in Fig. 4.21 is considered for analysis. Initially the switch (SW) was closed and no current was flowing through the parallel combination of R and C. The switch is open at t=0 and current starts flowing through R and C. The response of this circuit is considered as current through the circuit i.e parallel combination of R and C. The initial voltage across the C is zero as the capacitor was not charged initially.

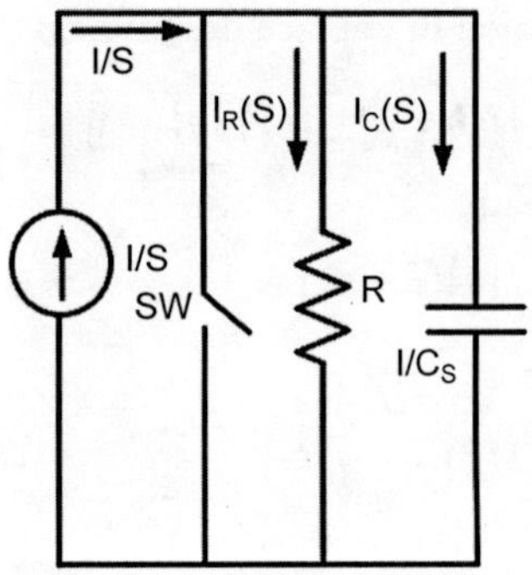

Fig. 4.21: Parallel *R-C* circuit with step input current in Laplace domain

Applying KCL in the circuit shown in Fig. 4.21, the relation between current through capacitor, resistor and the source current is given by,

$$i_C(t) + i_R(t) = Iu(t) \tag{4.111}$$

As the *R* and *C* are connected in parallel, the voltage across them are same and it is shown as,

$$i_R(t)R = \frac{\int i_C(t)dt}{C} \tag{4.112}$$

Multiplying both side by *R*, the equation (4.111) can be presented as,

$Ri_C(t) + Ri_R(t) = Iu(t)R$

$$Ri_C(t) + \frac{\int i_C(t)dt}{C} = Iu(t)R \tag{4.113}$$

Taking Laplace transform in both side of equation (4.113), it is obtained as,

$$RI_C(s) + \frac{I_C(s)}{Cs} - \frac{Q(0^-)}{Cs} = \frac{IR}{s}$$

$$RI_C(s) + \frac{I_C(s)}{Cs} = \frac{IR}{s} \tag{4.114}$$

Rearranging the terms of equation (4.114), it is represented as,

$$I_C(s)\left[R + \frac{1}{Cs}\right] = \frac{IR}{s}$$

$$I_C(s) = \frac{I}{\left(s + \frac{1}{RC}\right)} \tag{4.115}$$

The solution of equation (4.115) is obtained in time domain by taking inverse Laplace transform of (4.115) for $t > 0$ as,

$$i_C(t) = Ie^{-t/RC} \tag{4.116}$$

Equation (4.116) presents the current flowing through the capacitor at $t > 0$. Current flowing through the resistor is given by,

$$i_R(t) = I - i_C(t) = I(1 - e^{-t/RC}) \tag{4.117}$$

4.6.3 Step Response of Series *R-L-C* Circuit using Laplace Transform

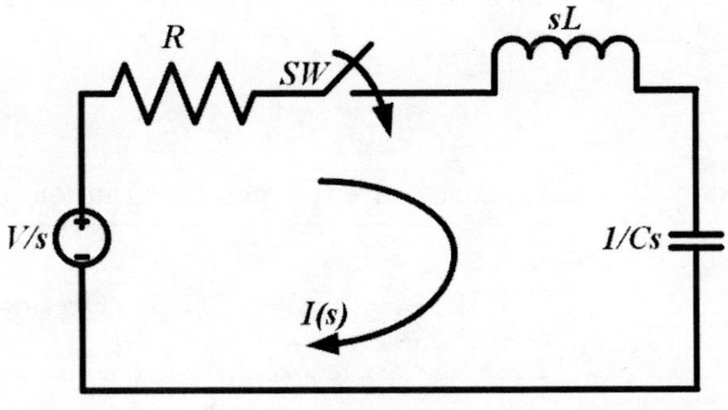

Fig. 4.22: Series *R-L-C* circuit with step input voltage in Laplace domain

A circuit with a series combination of pure resistor (R), pure inductor (L) and pure capacitor (C) connected across an independent dc voltage source as shown in Fig. 4.22 is considered for analysis. Initially the switch (SW) was open and no current was flowing through the circuit. The switch (SW) is closed at t=0 and current starts flowing through the circuit. The response of this circuit is considered as current through the circuit.

Applying KVL in the circuit shown in Fig. 4.22, the relation between the voltage across resistor, inductor, capacitor and the source voltage is given by,

$$i(t)R + L\frac{di(t)}{dt} + \frac{1}{C}\int i(t)dt = Vu(t)$$

$$i(t)R + L\frac{di(t)}{dt} + \frac{1}{C}\int i(t)dt = V \text{ for } t > 0 \tag{4.118}$$

Taking Laplace Transform in both side, it is obtained as,

$$I(s)R + L[I(s)s - i(0^-)] + \frac{1}{C}\left[\frac{I(s)}{s} - \frac{Q(0^-)}{s}\right] = \frac{V}{s} \tag{4.119}$$

Taking the initial conditions as zero, equation (4.119) can be written as,

$$I(s) = \frac{V}{\left(Ls^2 + Rs + \frac{1}{C}\right)}$$

$$I(s) = \frac{V}{\left(s^2 + \frac{R}{L}s + \frac{1}{LC}\right)} \tag{4.120}$$

Equation (4.120) can be written as,

$$I(s) = \frac{\frac{V}{Ls}}{(s+\alpha)(s+\beta)} \tag{4.121}$$

Where the values of α and β by applying Shreedhara Acharya's formula as,

$$\alpha, \beta = -\frac{R}{2L} \pm \sqrt{\left(\frac{R}{2L}\right)^2 - \frac{1}{LC}} \tag{4.122}$$

The equation mentioned in (4.121) can be solved by considering any one of the following conditions,

Condition 1:	$\left(\frac{R}{2L}\right)^2 > \frac{1}{LC}$	Over-damped response
Condition 2:	$\left(\frac{R}{2L}\right)^2 = \frac{1}{LC}$	Critically damped response
Condition 3:	$\left(\frac{R}{2L}\right)^2 < \frac{1}{LC}$	Under-damped response

Condition 1:

During this condition both α and β are real and unequal. Using the partial fraction concept of Laplace transform the equation (4.121) can be written as,

$$I(s)=\frac{X}{(s+\beta)}+\frac{Y}{(s+\beta)} \tag{4.123}$$

Where X and Y are given by,

$$X=\left[\frac{V/L}{(s+\beta)(s+\alpha)}\right]_{s=-\alpha}=\frac{V}{L(\beta-\alpha)} \tag{4.124a}$$

$$Y=\left[\frac{V/L}{(s+\alpha)(s+\beta)}\right]_{s=-\beta}=\frac{V}{L(\alpha-\beta)} \tag{4.124a}$$

Equation (4.123) can be written as,

$$I(s)=\frac{V/L}{(\beta-\alpha)(s+\beta)}+\frac{V/L}{(\alpha-\beta)(s+\beta)} \tag{4.125}$$

The solution of equation (4.125) is obtained in time domain by taking inverse Laplace transform of (4.125) for $t > 0$ as,

$$i(t)=\frac{V/L}{(\beta-\alpha)}e^{-\beta t}+\frac{V/L}{(\alpha-\beta)}e^{-\alpha t} \tag{4.126}$$

Condition 2:

For this condition the values of α and β are real and equal which is given by,

$$\alpha=\beta=K$$

Hence the equation (4.121) can be written as,

$$I(s)=\frac{V/L}{(s+K)^2} \tag{4.127}$$

The solution of equation (4.127) is obtained in time domain by taking inverse Laplace transform of (4.127) for $t > 0$ as,

$$i(t)=\frac{V}{L}te^{-Kt} \tag{4.128}$$

Condition 3:

For this condition the value of γ is imaginary. Hence, the values of α,β are complex conjugates and they are presented as,

$$\alpha=-X+jY$$
$$\beta=-X-jY$$

Equation (4.121) can be written as,

$$I(s)=\frac{V/L}{(s+X+jY)(s+X-jY)}$$

$$I(s)=\frac{V/L}{\left(s+X^2\right)+Y^2}$$

$$I(s)=\frac{V}{LY}\frac{Y}{\left(s+X^2\right)+Y^2} \tag{4.129}$$

The solution of equation (4.129) is obtained in time domain by taking inverse Laplace transform of (4.129) for $t > 0$ as,

$$i(t)=\frac{V}{LY}te^{-Xt}\sin Yt \tag{4.130}$$

Example 4.11:

A capacitor of C = 2 μF with initial voltage across the capacitor as 20 V is connected to a resistor of 100 Ω through a switch at t = 0. Find the current through the circuit using Laplace Transform for t > 0.

Solution:

$$R = 100\Omega$$
$$C = 2\mu F$$
$$v(0)=\frac{Q_0}{C}=10V$$

The circuit is drawn as,

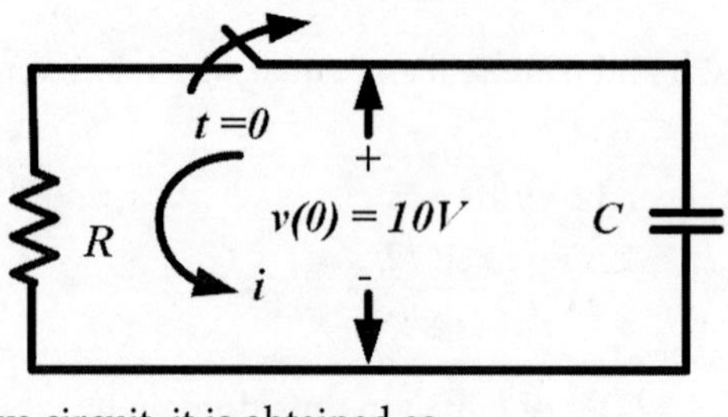

Applying KVL in the above circuit, it is obtained as,

$$i_R(t)R+\frac{\int i_C(t)dt}{C}=0$$

Taking Laplace Transform of the above equation, it is written as,

$$I(s)R+\frac{I(s)}{Cs}+\frac{Q_0}{Cs}=0$$

$$I(s)\left(R+\frac{1}{Cs}\right)=-\frac{Q_0}{Cs}$$

$$I(s)=-\frac{Q_0/Cs}{\left(R+\frac{1}{Cs}\right)}=-\frac{Q_0}{CR\left(s+\frac{1}{RC}\right)}$$

$$I(s) = -\frac{10}{100\left(s + \frac{1}{100 \times 2 \times 10^{-6}}\right)}$$

Taking inverse Laplace Transform of the above equation,

$$i(t) = -0.1e^{-5000t} \text{ A}$$

Example 4.12:

For the circuit given below, the dc voltage of 100 V was applied to the circuit for a long time when the switch was open. The switch is closed at t = 0. Find the current through the circuit using Laplace Transform for t > 0.

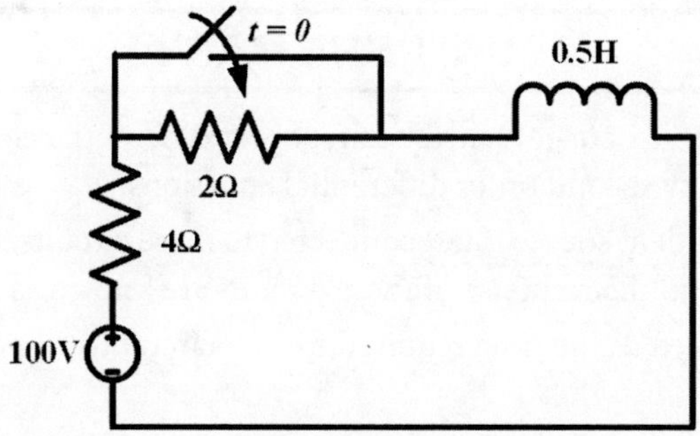

Solution:

When the switch was open current was flowing through both the resistances and the inductance for a long time. Current through inductor does not change suddenly. The current through the circuit during that time (t < 0) is given by,

$$i(0^+) = i(0^-) = i(0) = \frac{100}{6} = 16.67\text{A}$$

When the switch is closed at t = 0, applying KVL in the circuit, it is obtained as,

$$i(t) \times 4 + 0.5 \times \frac{di(t)}{dt} = 100u(t)$$

The above equation is represented in Laplace domain as,

$$I(s) \times 4 + 0.5 \times [sI(s) - i(0)] = \frac{100}{s}$$

$$I(s) = \frac{100 + 0.5si(0)}{s(4 + 0.5s)} = \frac{200 + si(0)}{s(8 + s)}$$

Using partial fraction, the above equation can be written as,

$$I(s) = \frac{X}{s} + \frac{Y}{(8 + s)}$$

$$X = \left[\frac{200 + si(0)}{(8 + s)}\right]_{s=0} = \frac{200}{8} = 25$$

$$Y = \left[\frac{200 + si(0)}{s}\right]_{s=-8} = \frac{200 - 8i(0)}{-8} = \frac{200 - 8 \times 16.67}{-8} = -8.33$$

Hence the value of I(s) is written as,

$$I(s) = \frac{25}{s} - \frac{8.33}{(8+s)}$$

Taking inverse Laplace Transform in the above equation, the current through the circuit is obtained as,

$$i(t) = 25 - 8.33e^{-8t} \text{ A}$$

UNIT SUMMARY

1. First order electric circuits are governed by first order differential equations whereas second order circuits are governed by second order differential equations.
2. If the external independent source is not connected to these circuits, then they are known as source free electric circuits and the response of these circuits are known as natural response.
3. Time constant is defined as the time required for the circuit to respond to a value of 36.8% of the initial value.
4. For a source free *R-L* circuit, the energy stored in the inductor decays exponentially and it dissipates across the resistor.
5. For a source free *R-C* circuit, the charge stored in the capacitor decays exponentially and it dissipates across the resistor.
6. When the voltage or current source of a circuit is applied suddenly and if those sources can be modelled as step function then the response of the circuit is known as step response.
7. Current through the inductor or voltage across the capacitor rises exponentially when an independent voltage or current source is connected to the circuit.
8. Laplace Transform is an important mathematical tool to find out the transient response of electric circuits. It is an integral transformation of a function from time domain to complex frequency domain.

EXERCISES

Multiple Choice Questions

1. Time constant of RC circuit is

(a) 1/RC (b) RC
(c) R/C (d) C/R

2. Time constant of RL circuit is

(a) 1/RL (b) RL
(c) R/L (d) L/R

3. Time dependent voltage source, $V = f(t)$ can be represented in Laplace domain as,

(a) V/s (b) V(s)
(c) Vs (d) 0

4. Inductance (L) in Laplace domain is represented as,

(a) L (b) sL
(c) L/s (d) s^2L

5. The final value of current through the circuit in a series R-L circuit with step input voltage of V and initial current of I_0 through the inductor is given by

(a) V (b) V/R
(c) I_0 (d) V/L

6. The final value of voltage across C in a series R-C circuit with step input voltage of V and initial voltage across C of V_{C0} is given by

(a) V_{C0} (b) 0
(c) ∞ (d) V

7. If the value of voltage source is 0 at $t < t_0$ and V at $t > t_0$ then it can be expressed as

(a) v(t) = 0 (b) $v(t) = Vu(t + t_0)$
(c) $v(t) = Vu(t - t_0)$ (d) v(t) = V

8. In a RC circuit with R=1 Ω and C=1 F, the time required for the capacitor voltage to reach 40% of its steady state value is

(a) 0.51 sec (b) 0.41 sec
(c) 0.61 sec (d) 0.71 sec

9. In a RL circuit with R=1 Ω and L=1 H, the time required for the inductor current to reach 80% of its steady state value is

(a) 1.41 sec (b) 1.51 sec
(c) 1.61 sec (d) 1.71 sec

10. Inverse of time constant of a circuit is known as

(a) resonance (b) damping ratio
(c) frequency (d) neper frequency

ANSWERS															
1.	(b)	2.	(d)	3.	(b)	4.	(b)	5.	(b)	6.	(d)	7.	(c)	8.	(a)
9.	(c)	10.	(b)												

Short Answer Questions

1. What is the meaning of time constant of the circuit?
2. Prove that at t= 5 of times of time constant of the circuit, the transient response of the circuit is equal to its steady state value.
3. Define damping ratio.
4. What is the significance of the smaller time constant of the circuit?
5. What do you mean by first order electric circuit?

Long Answer Questions

1. Prove that as $t \to \infty$, the energy absorbed by the resistor in a source free R-L circuit is equal to the initial energy available in the charged inductor.
2. Prove that the current flowing through the capacitor becomes zero as $t \to \infty$ in a parallel R-C circuit excited by a step dc current.
3. Derive the expression for current through the resistor for the circuit mentioned in Fig. 4.19 using Laplace Transform.
4. Derive the expression for current through the circuit mentioned in Fig. 4.8 using Laplace Transform.

Numerical Problems

1. When a series *R-L* circuit with R = 100 ohm and L = 10 H is connected to a dc voltage source at t = 0, the current through the circuit rises at the rate of 10A/sec. Find the applied voltage and final energy stored in the inductor.
2. For the circuit shown below, initially the switch was in contact with position 1 for 20 sec. The switch changes its contact from position from 1 to 2 at t =0. Determine the voltage across the capacitor at t=0 and current through the circuit at t > 0.

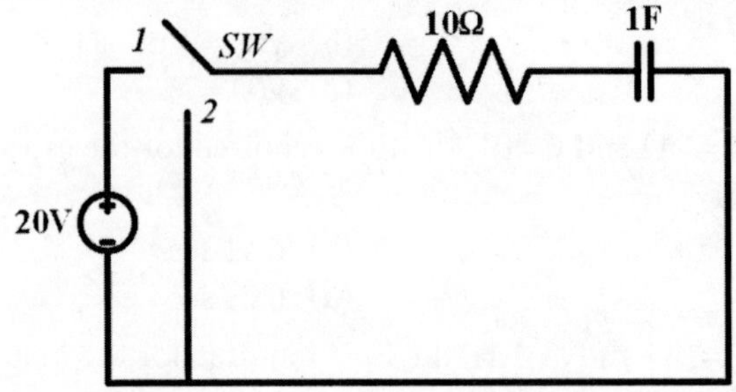

3. A constant voltage of 100 V is applied at t=0 to a series *R-L-C* circuit with R = 10 ohm, L =2 H and C =100 μF. Find the current flowing through the circuit for t > 0.
4. A constant current of 1 A is applied at t =0 across a parallel combination of *R-L* circuit with R = 1 ohm and L = 0.5 H. Find the current through the resistor and inductor for t > 0.
5. For the circuit given below, the switch was closed for a long time. At t =0, the switch becomes open then find the current through the circuit for t > 0 and voltage across the switch at the time of opening.

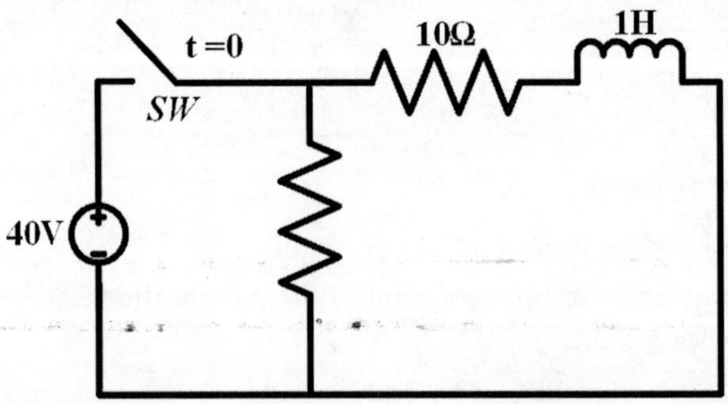

6. For the Fig. 4.19 consider I =10 A, R = 1 ohm, L =1 H and C =1 F. The switch becomes open at t = 0. Find out the current through the resistor for t > 0 using Laplace Transform.

7. A step voltage of 100 V is applied at t = 0 to a series combination of *R-C* circuit where *R* = 5 ohm, *C* =2 F. Initial voltage across the capacitor is zero. Find the current through the circuit for t > 0 using Laplace Transform.
8. A step voltage of 10 V is applied at t = 0 to a series combination of *R-C* circuit where *R* = 5 ohm, *C* =2 F. Initial charge on the capacitor is 4C. Find the current through the circuit for t > 0 using Laplace Transform.

PRACTICAL

1. Use LTspice to determine the current through a series R-L-C circuit with R=10 ohm, L= 2 H and C=50 **μF when connected to a step voltage input of 200V DC.**
2. Use LTspice to determine the current through a series R-L circuit with R=100 ohm and L= 4 H with zero initial current through L at t < 0 when connected to a step voltage input of 10V DC.

KNOW MORE

1. Find the current flowing through the capacitor for the circuit given in Example 4.3 considering the initial voltage across the capacitor as 10 V using LTspice.

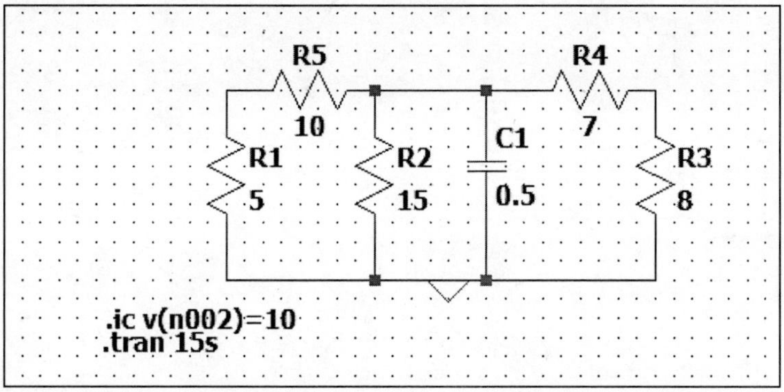

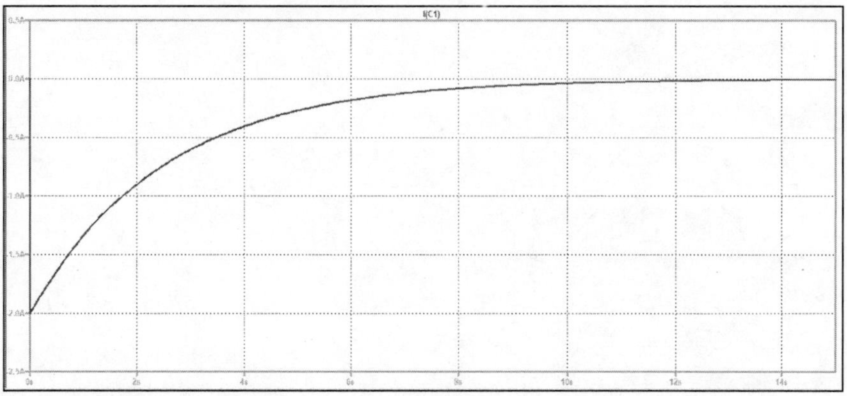

Fig. 4.23: Current through capacitor for the circuit given in Example 4.3 using LTspice Current through capacitor is initially 2 A and it decays as time progresses.

REFERENCES AND SUGGESTED READINGS

1. Joel L. Schiff, "The Laplace Transform Theory and Applications", Springer Science + Business Media, New York, 1999.
2. Charles K. Alexander and Matthew N. O. Sadiku, "Fundamentals of Electric Circuits", McGraw Hill Publishers, 7th Edition, 2021.

https://archive.nptel.ac.in/courses/111/106/111106139/

Dynamic QR Code for Further Reading

5 Graph Theory

UNIT SPECIFICS

Through this unit we have discussed the following aspects:

- Network topology
- Tie-set matrix
- Cut-set matrix
- Two-port networks
- Impedance and Admittance parameters
- ABCD and h parameters
- Relations between the two-port parameters
- Interconnection of the electric networks

RATIONALE

An electric network consists of voltage sources, current sources and passive elements (R, L, C) in series and parallel combinations. In order to analyse the network, many network theorems are applied in the earlier chapters. In this chapter, such circuits are solved with the help of the graph theory concepts. Hence, the networks are transformed into the oriented graph. This graph is useful for the formation of the tie-set matrix and the cut-set matrix. These matrices are useful for solving such networks without doing many calculations. Most of the networks have two-ports namely input port and output ports. Hence these networks are denoted as two-port networks. The two-port network parameters are impedance parameters, admittance parameters, ABCD or transmission parameters and hybrid parameters. These two-port networks allow one to find the response of the network for the given input signal without calculating or finding the internal voltages and currents of the network. Also, the relations between these two-port networks are derived. . Many networks are interconnected in series, parallel or cascade mode and discussed the method of finding the two-port parameters of such interconnected networks.

PRE-REQUISITES

Basic graph theory

Network theorems

UNIT OUTCOMES

List of outcomes of this unit is as follows:

U5-O1: Understand the basics of the network topology and its definitions

U5-O2: Formulate the tie-set matrix and cut-set matrix for the network

U5-O3: Comprehend the two-port network parameters namely Z, Y, ABCD and h parameters

U5-O4: Realize the relations between the two-port parameters

U5-O5: Analyse and Understand the interconnection of the networks

Unit-5 Outcome	EXPECTED MAPPING WITH COURSE OUTCOMES (1 - Weak Correlation; 2 - Medium Correlation; 3 - Strong Correlation)				
	CO-1	CO-2	CO-3	CO-4	CO-5
U5-O1	1	1	-	2	-
U5-O2	-	-	-	3	-
U5-O3	1	1	-	3	-
U5-O4	-	-	-	3	-
U5-O5	-	-	-	2	-

5.1 NETWORK GRAPHS

An electric network consists of voltage sources, current sources, resistors, inductors and capacitors. The network graphs are the graphical representation of the electric networks. The nodes and lines are considered for making the graph. A single point in which two or more electrical elements are connected together is denoted as a node. The lines are used to represent the branch in a circuit. A circuit branch is an electrical element or multiple electrical elements that are connected between two nodes. These circuit branches are replaced by the lines in a network graph. A sample electric network is shown in Fig.5.1 which consists of a voltage source and resistors R_1 to R_5. As per the definition of the node and branch, the four nodes including reference node, 5 branches are shown in Fig.5.2. Basic terminologies in the network graphs are listed with explanation.

Basic terminologies used in network graph:

Branch: A branch is a line segment that represents an electrical element or multiple electrical elements connected between the nodes. In Fig. 5.2, the line segment named as 1 to 5 are the branches which will not inform about the type of the electrical component. It provides information about a component or an element present on that segment.

Node: A node is the end points of the line segment. Each line segment has two nodes. The line segment 1 shown in Fig.5.2 has two end points that is named as Nodes N1 and N2 respectively. It has totally four nodes N1, N2, N3 and N4.

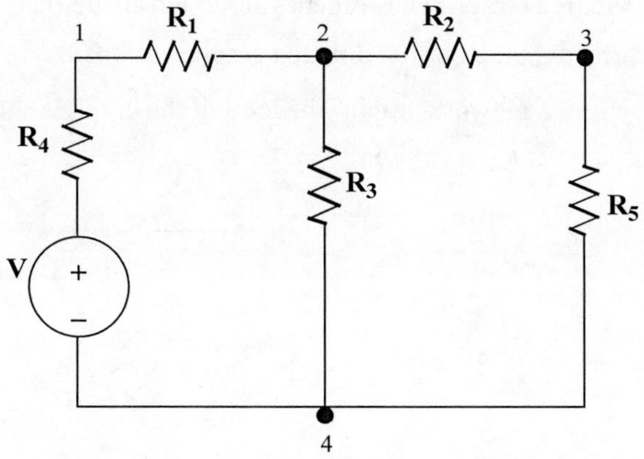

Fig. 5.1: Sample electric circuit for explaining concept of network graph

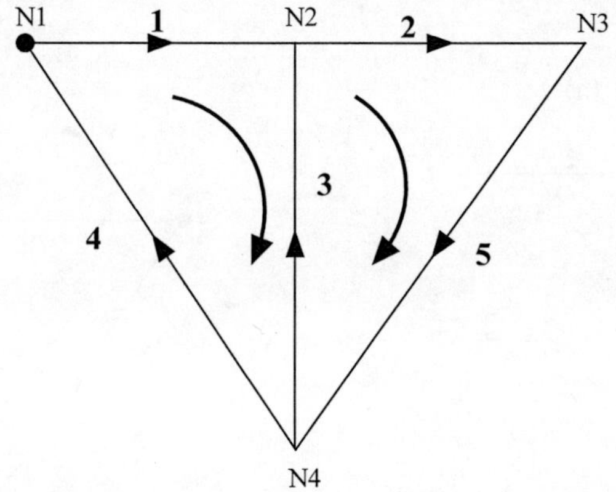

Fig. 5.2: Network graph of the electric circuit shown in Fig.5.1

Degree of a node: The number of branches or line segments connected to that node is called the degree of a node. In Fig. 5.2, the degree of the node for node N2 is 3, since the line segments 1, 2 and 3 are connected to it.

Tree: It is an interconnected and undirected sub graph of a network which covers all the nodes in a network. There is no closed loop in the tree or it has an open set of branches. A graph may have multiple numbers of trees and that is shown in Fig.5.3.

The connected subgraph is known as the tree of the graph if the subgraph has all nodes of the graph without containing any loop.

Twigs: It is known as the branches in a tree. Or Tree branches are called Twigs. If the number of nodes is n in a graph. The number of twigs in a tree is (n-1).

Links or Co-tree: It consists of the branches which are not in the tree of a network graph. In simple way, by combining tree branches and tree links, all branches in a graph could be obtained. The trees of the graph shown in Fig. 5.2 are shown in Fig.5.3, the corresponding tree link is shown in Fig.5.4 in the same order. For example, the tree link of the Tree of Fig.5.3 (a) is presented in Fig.5.4 (a) in a dotted line.

Directed Graph or Oriented Graph: The branches of a graph are marked with the current direction. Such a graph is called an oriented graph or the directed graph.

Loop: It is a closed path in a network graph. The loop of the graph is shown in Fig. 5.2. It consists of two independent loops.

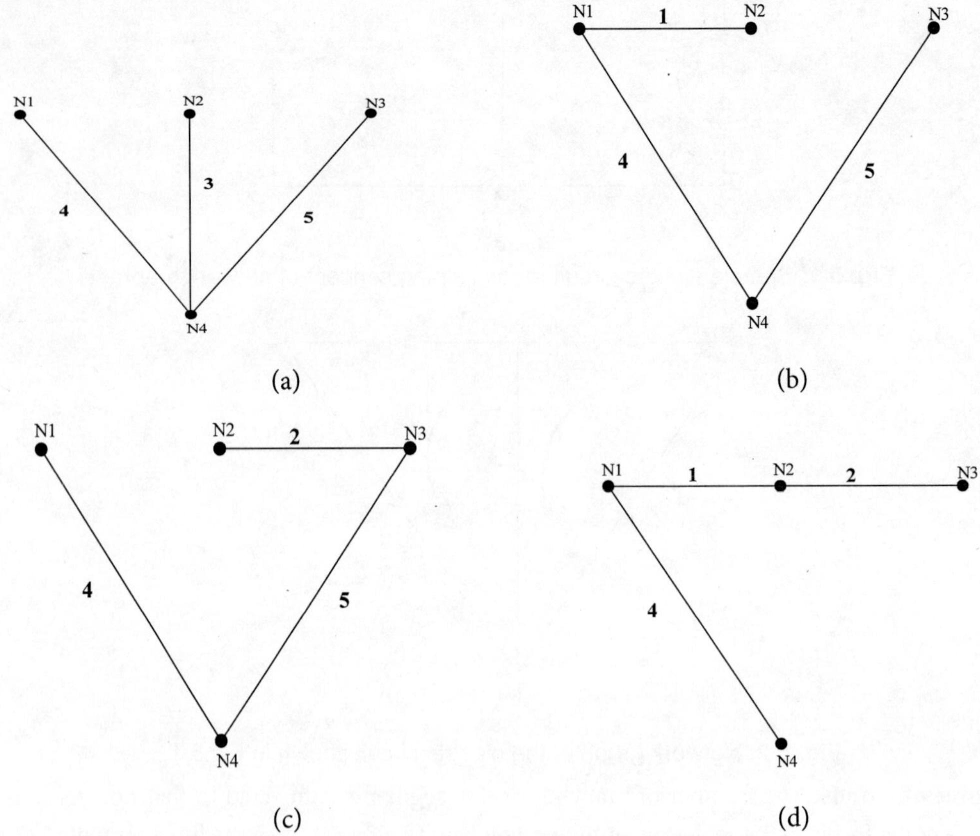

Fig. 5.3: Trees of the network graph of the electric circuit shown in Fig.5.2

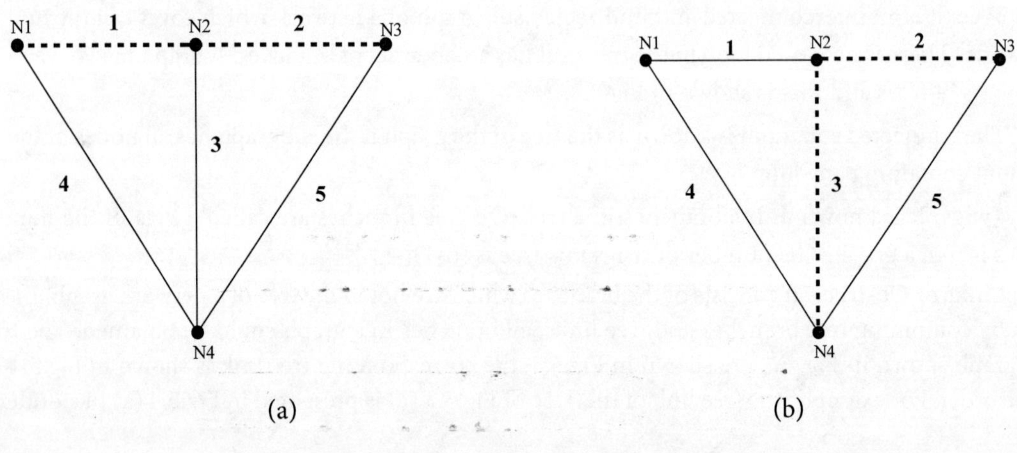

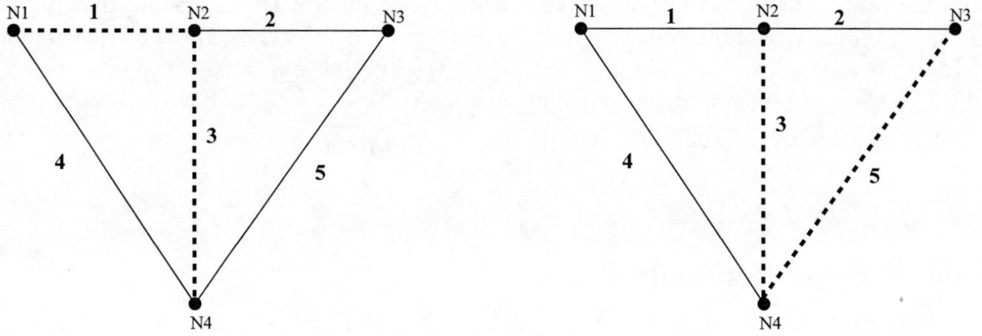

(c) (d)

Fig. 5.4: Tree links of the trees shown in Fig.5.3

Incidence Matrix using KCL: For the circuit shown in Fig.5.5, the number of nodes is 4 and the number of branches is 6 that are numbered in closed brackets. Let us assume that the branch currents i_b1, i_b2,i_b3, …..i_b6flows in the corresponding branches. By applying the Kirchoff's current law at each node, the following equation 5.1 is obtained. This equation is represented as matrix form with the branch currents and incidence matrix in equation (5.2).

$$i_{b1}-i_{b2}-i_{b6}=0-i_{b1}+i_{b3}+i_{b4}=0i_{b2}-i_{b3}-i_{b5}=0-i_{b4}+i_{b5}+i_{b6}=0\}\tag{5.1}$$

$$[1-1000-1-10110001-10-10000-111][i_{b1}i_{b2}i_{b3}i_{b4}i_{b5}i_{b6}]=0\tag{5.2a}$$

$$A_a\, I_b = 0\tag{5.2b}$$

Where, A_a is the complete incidence matrix of the graph and I_b is known as the branch currents of the graph.

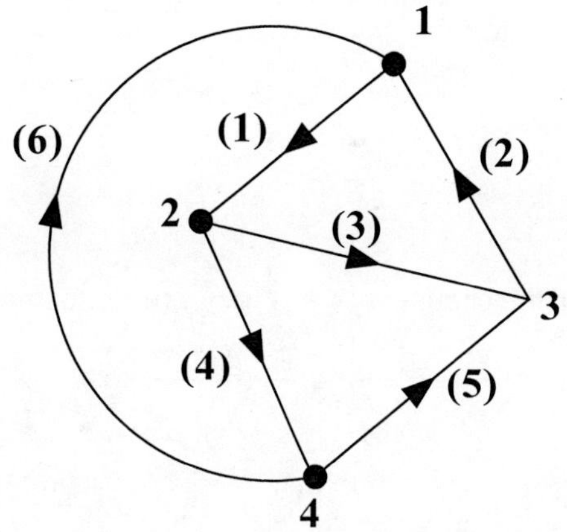

Fig. 5.5: Oriented graph presenting for the incidence matrix

Reduced Incidence Matrix: A reduced incidence matrix (A) is obtained by eliminating one row from the actual incidence matrix. The eliminated row is corresponding to the reference node or datum node. In this example, node 4 is taken as a reference node. Hence the obtained reduced incidence matrix is given below.

A = [1 –1000 –1 – 10110001 –10 –10]

Incidence Matrix using KVL: For the circuit shown in Fig.5.5, let us assume that the branch voltages $v_{b1}, v_{b2}, v_{b3}, \ldots v_{b6}$ are the voltage across the corresponding branches. These branch voltages are presented as node voltages. $v_{b1} = v_{n1} - v_{n2}$; $v_{b2} = v_{n1} - v_{n3}$; $v_{b3} = v_{n2} - v_{n3}$; $v_{b4} = v_{n2} - v_{n4}$; $v_{b5} = v_{n4} - v_{n3}$; $v_{b6} = v_{n4} - v_{n1}$. These are presented in matrix form in equation (5.3).

$$[1\ -10010\ -1001\ -10010\ -100\ -11\ -\ 1001][v_{n1}\ v_{n2}\ v_{n3}\ v_{n4}] = [v_{b1}\ v_{b2}\ v_{b3}\ v_{b4}\ v_{b5}\ v_{b6}] \quad (5.3a)$$

OR

$$[A_a^T][v_n] = [v_b] \quad (5.3b)$$

Properties of the incidence matrix:

- The sum of the entries in any row is zero.
- The determinant of the incidence matrix of a closed loop is zero.
- The rank of the incidence matrix of an oriented graph is $(n-1)$. Where n is the number of nodes in a graph.

5.2 TIE SET ANALYSIS

A tie-set is a set of branches in a loop. Where the loop contains one link and the remaining are the tree branches. The figure shown in Fig. 5.6 is the oriented graph of a circuit. The tree of this graph is shown in Fig. 5.7. The possible fundamental loops of that tree is shown in Fig.5.8. In this graph, there are three loops available. The Fundamental loop branches and links are [(1,2), (6)], [(1,3), (4)] and [(2,3), (5)]. These set of branches form the three tie-sets.

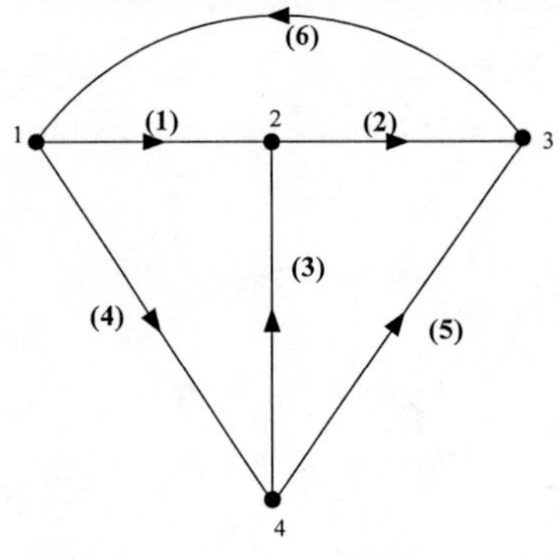

Fig. 5.6: Oriented graph of the circuit

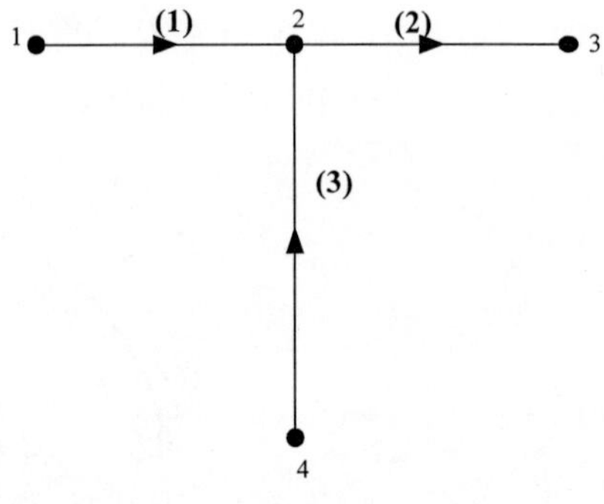

Fig. 5.7: Tree for the oriented graph shown in Fig.5.6

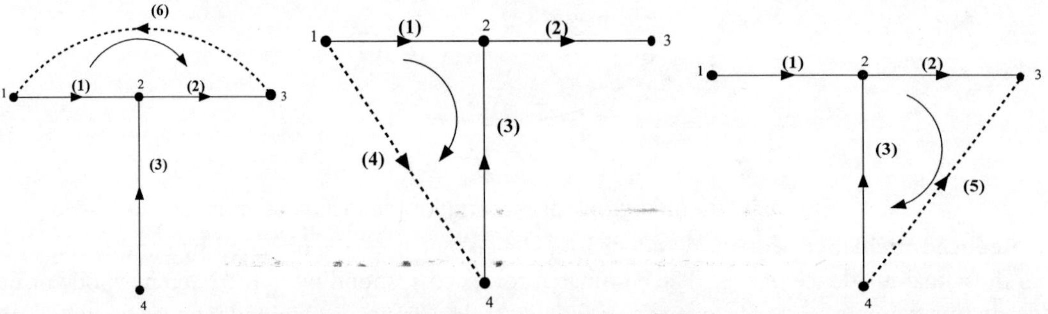

(a) Fundamental loop1 (b) Fundamental loop2 (c) Fundamental loop 3

Fig. 5.8: Fundamental loops of the oriented graph shown in Fig.5.5 and the tree of Fig.5.6

Tie-set matrix: A tie-set matrix (B) is a rectangular matrix whose rows are the links of the fundamental loops and the columns are the branches in an oriented graph. Hence the tie-set matrix for the circuit is shown in equation 5.5. It has three fundamental loops and six branches. The tie-set matrix elements are filled with the equation (5.4). The tie-set matrix of the oriented graph of Fig. 5.5 is shown in equation 5.5. The rows are the links of the fundamental loops 1, 2 and 3. Similarly the columns are corresponding to its branches.

Bij = {+1 if the fundamental loop direction is same as the tree/link branch direction-1 if the fundamental loop direction is opposite as the tree/link branch direction if the tree/link branch direction is not in the fundamental loop direction (5.4)

$$\text{B} = [-1\ -1000\ -110\ -1\ -1000110\ -10] \quad (5.5)$$

Tie-set matrix with KVL: By applying the KVL to the fundamental loops 1, the concern equation is obtained in equation (5.6).

$$-V_{b1} - V_{b2} - V_{b6} = 0 \quad (5.6)$$

This concept is applied to other fundamental loops and the concern equations are presented in equation (5.7) and (5.8) respectively.

$$V_{b1} - V_{b3} - V_{b4} = 0 \quad (5.7)$$

$$V_{b2} - V_{b3} - V_{b5} = 0 \quad (5.8)$$

These equations (5.6) to (5.8) are formed as matrix format that is shown in equation (5.9). Where V_b represents the branch voltages. This equation could be written as like in the equation (5.10).

$$[-1\ -1000\ -110\ -1\ -1000110\ -10][V_{b1}\ V_{b2}\ V_{b3}\ V_{b4}\ V_{b5}\ V_{b6}] = 0 \quad (5.9)$$

$$[B][V_b] = 0 \quad (5.10)$$

Tie-set matrix with KCL: Let assume that the fundamental loop currents are I_{L1}, I_{L2} and I_{L3} for the fundamental loops shown in Fig. 5.7. The branch current of the branch (1) is denoted as (I_{b1}). Based on this, the branch currents are represented in terms of the loop currents that are shown in equation (5.11).

$$I_{b1} = -I_{L1} + I_{L2};\ I_{b2} = -I_{L1} + I_{L3};\ I_{b3} = -I_{L2} + I_{L3};\ I_{b4} = -I_{L2};\ I_{b5} = -I_{L3};\ I_{b6} = -I_{L1} \quad (5.11)$$

This equation is presented in the matrix form as shown in equation (5.12) and this equation could be written as like in the equation (5.13).

$$[I_{b1}\ I_{b2}\ I_{b3}\ I_{b4}\ I_{b5}\ I_{b6}] = [-110 - 1010\ -110\ -1000\ -1\ -100][I_{L1}\ I_{L2}\ I_{L3}] \quad (5.12)$$

$$[I_b] = [\text{B}]^{\text{T}}\ [I_L] \quad (5.13)$$

5.3 CUT-SET ANALYSIS

A cut-set is a set of branches or the elements that are removed from the connected graph, the graph divides into the two groups of elements. A cut-set is the minimum number of branches in a graph such that the removal of these branches reduces the rank of the graph by one. In simple words, the removal of these branches results in the unconnected two graphs.

The figure shown in Fig. 5.6 is the oriented graph of a circuit. The tree of this graph is shown in Fig. 5.7. The links or the twigs of this graph are the branches (4), (5) and (6). The graph is shown with its twigs and links in Fig.5.9. The removal of the branches (1), (4) and (6) make the unconnected graph. Node 1 forms one graph and Node 2,3 and 4 forms another graph.

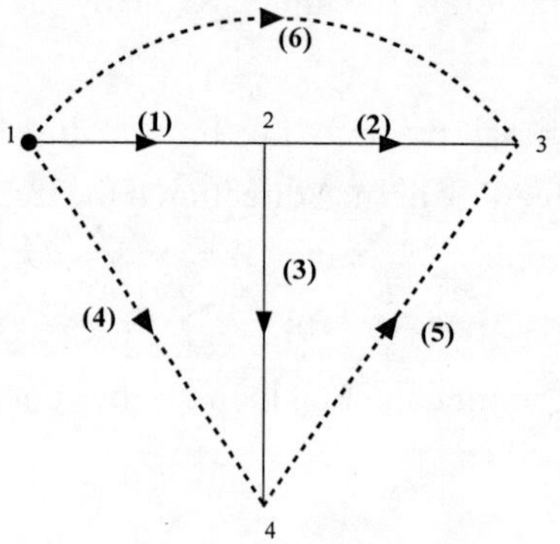

Fig. 5.9: Oriented graph of the circuit with its tree the branches and link branches

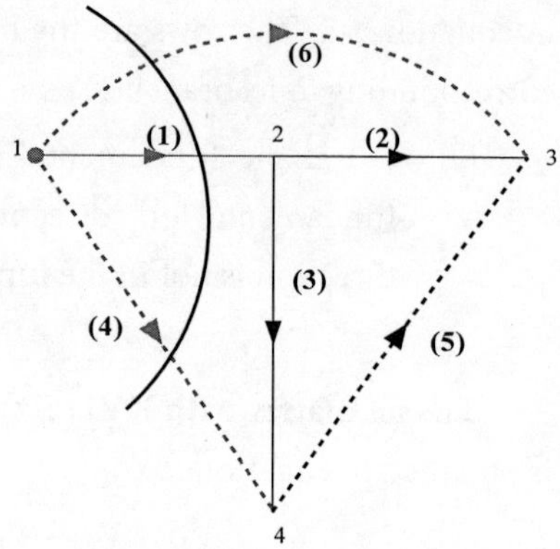

Fig. 5.10: The resultant graph with the removal of (1), (4) and (6)

Fundamental cut-set: It is a cut-set that cuts the graph into two parts. It contains only one tree branch. Hence the number of cut-sets in an oriented graph or a connected graph is equal to the number of tree branches.

Finding the fundamental cut-set:

- The oriented graph of the circuit is drawn (See Fig.5.6). Find the tree of the graph (See Fig.5.7).
- Focus on the tree branches one by one (See Fig.5.9).
- Check whether by removing this tree branch (Consider the tree branch (1) in Fig.5.9) from the tree of the oriented graph separates the graph into two parts (Node 1 and Node 2, 3 & 4).
- All the links from this tree branch connected node to other nodes of the graph are noted down. These links [Link (4) and (6)] and the concerned tree branch [tree (1)] forms the fundamental cut-set.
- Hence the fundamental cut-set of the oriented graph is

  Fundamental cut-set 1 : [1,4,6]

  Fundamental cut-set 2 : [2,5,6]

  Fundamental cut-set 3 : [3,4,5]

  - The fundamental cut-set divides the graph into two parts.
  - Each cut-set consists of one tree branch.
  - The cut-set direction is the direction of the tree branch of that cut-set.
- Fundamental cut-sets of the graph shown in Fig.5.9 are mentioned above the concern graphs are shown in Fig. 5.11. The cut-set direction is indicated in that figure.

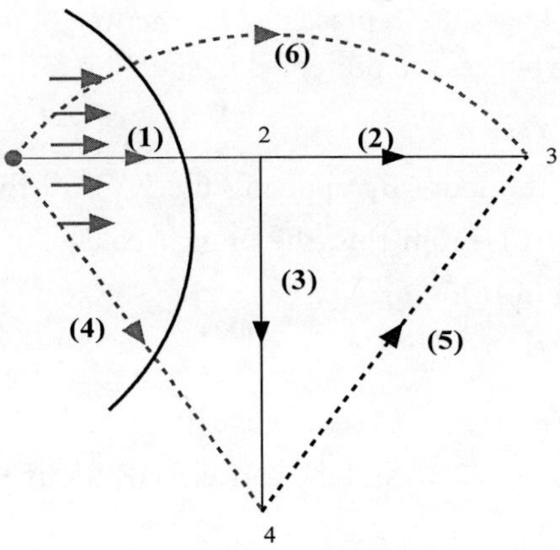

(a) Fundamental cut-set1

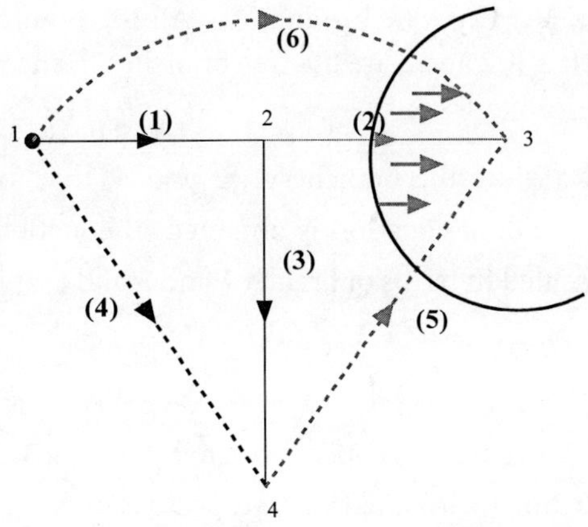

(b) Fundamental cut-set2

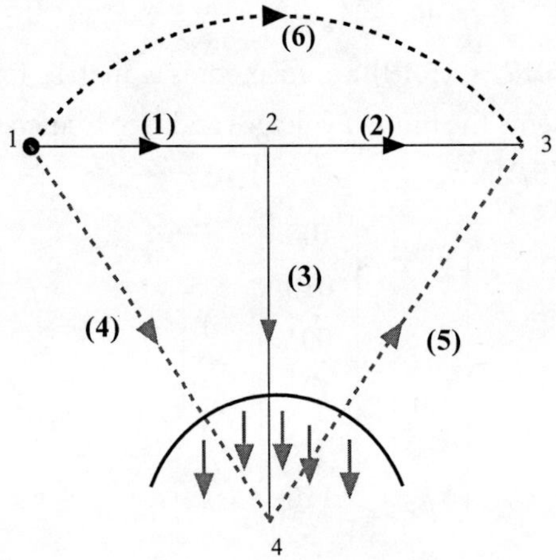

(c) Fundamental cut-set3

Fig. 5.11: Fundamental cut-sets and its direction of the oriented graph shown in Fig.5.9

Cut-set matrix: A cut-set matrix (Q_C) is a rectangular matrix whose rows are the cut-sets and the columns are the branches in an oriented graph. Hence the cut-set matrix for the circuit is shown in equation 5.14. It has three fundamental cut-sets and six branches. The cut-set matrix elements are filled as follows.

$$Q_C^{ij} = \{ \begin{matrix} +1 \text{ if the branch j in the cut – set i and thier orientation coincide} \\ -1 \text{ if the branch j in the cut – set i and thier orientation oppose} \\ 0 \text{ if the branch j is not in the cut – set i} \end{matrix}$$

The cut-set matrix of the oriented graph of Fig. 5.9 is shown in equation 5.14. The rows are the cut-sets. Similarly, the columns are corresponding to its branches.

$$Q_C = [1001010100110011 -10] \quad (5.14)$$

Cut-set matrix with KVL: TThe oriented graph is shown in Fig.5.9 and the corresponding tree is shown in Fig.5.10. The tree branches are branches 1, 2 and 3. Similarly the links are branch 4, 5 and 6. Let us assume that the tree branch voltages are V_{t1}, V_{t2} and V_{t3}. Similarly, the branch voltages are denoted as V_{b1}, V_{b2}, V_{b3}, V_{b4}, V_{b5} and V_{b6}. All the branch voltages are represented by the tree branch voltages. Branches 1, 2 and 3 are the tree branches. Hence its voltages are presented in equation (5.15).

$$V_{b1} = V_{t1}\ V_{b2} = V_{t2}\ V_{b3} = V_{t3}\} \tag{5.15}$$

Consider the branches 1, 3 and 4, that form the loop. By applying the KVL to that loop, the corresponding equation is obtained in equation (5.16). From this, the branch voltage of branch 4 is represented in terms of branch 1 and 3 and that is in (5.16).

$$V_{b1} + V_{b3} - V_{b4} = 0 \tag{5.16}$$

$$V_{b4} = V_{b1} + V_{b3} = V_{t1} + V_{t3} \tag{5.17}$$

This concept is applied to other loops (2, 3, 5) and (1, 2, 6) and the concern equations are presented in equation (5.18) and (5.19) respectively.

$$V_{b5} = V_{t2} - V_{t3} \tag{5.18}$$

$$V_{b6} = V_{t1} + V_{t2} \tag{5.19}$$

These equations (5.15), (5.17) – (5.19) are formed as a matrix format that is shown in equation (5.20). Where V_b and V_t represents the branch voltages and tree branch voltages. This equation could be written as like in the equation (5.21).

$$\begin{bmatrix} V_{b1} \\ V_{b2} \\ V_{b3} \\ V_{b4} \\ V_{b5} \\ V_{b6} \end{bmatrix} = \begin{bmatrix} 100 \\ 010 \\ 001 \\ 101 \\ 01-1 \\ 110 \end{bmatrix} \begin{bmatrix} V_{t1} \\ V_{t2} \\ V_{t3} \end{bmatrix} \tag{5.20}$$

$$[V_b] = [Q_C^T][V_t] \tag{5.21}$$

Cut-set matrix with KCL: The branch current of the branch (1) is denoted as (I_{b1}). Apply the KCL at the nodes where the links are connected, the equation is presented in equation (5.22)

$$\left.\begin{aligned} I_{b1} + I_{b4} + I_{b6} &= 0 \\ -I_{b2} - I_{b5} - I_{b6} &= 0 \\ -I_{b3} - I_{b4} + I_{b5} &= 0 \end{aligned}\right\} \tag{5.22}$$

The equation (5.22) is modified as follows

$$\left.\begin{aligned} I_{b1} + I_{b4} + I_{b6} &= 0 \\ +I_{b2} + I_{b5} + I_{b6} &= 0 \\ +I_{b3} + I_{b4} - I_{b5} &= 0 \end{aligned}\right\}$$

This equation is presented in the matrix form as shown in equation (5.23) and this equation could be written as like in the equation (5.24).

$$\begin{bmatrix} 100101 \\ 010011 \\ 0011-10 \end{bmatrix} \begin{bmatrix} I_{b1} \\ I_{b2} \\ I_{b3} \\ I_{b4} \\ I_{b5} \\ I_{b6} \end{bmatrix} = 0 \quad (5.23)$$

$$[Q_C][I_b] = 0 \quad (5.24)$$

The summarization of the three matrices (incidence matrix, tie-set matrix and cut-set matrix) are presented in Table 5.1.

Table 5.1: Summarization of incidence matrix, tie-set matrix and cut-set matrix

Matrix	KVL	KCL
Incidence matrix	$\left[A_a^T\right][v_n] = [v_b]$	Aa Ib = 0
Tie-set matrix	$[B][V_b] = 0$	$[Ib] = [B]^T [IL]$
Cut-set matrix	$[V_b] = \left[Q_C^T\right][V_t]$	$[Q_C][I_b] = 0$

5.4 NETWORK EQUILIBRIUM EQUATIONS

The network equilibrium equations are the set of equations which describe the state of the network at any instant of time. It is represented with either current variables or voltage variables. These equations are unique if the number of independent variables is equal to the number of independent equations.

Number of independent variables or equations = {(b – (n – 1) for loop method of analysis@(n – 1) for node method of analysis)

The network equations could be formed in any of the following methods.

- Based on Kirchhoff's voltage law, the loop equations are formed in which the current is the independent variable.
- Based on Kirchhoff's current law, the node equations are formed in which the voltage is the independent variable.

Formation of Loop equations based on KVL: The branch k having an impedance Z_K and a voltage source V_{SK} in series and the current flowing in the branch is i_K. The branch voltage (v_K) is written in equation (5.25). This equation should be in the matrix format in equation (5.26).

$$v_k = Z_K\, i_K + V_{SK} \quad (5.25)$$

$$[V_b] = [Z_b][I_b] + [V_S] \quad (5.26)$$

Z_B - Branch impedance

I_K - Branch current

V_S - Source voltage

V_B - Branch voltage

$[Z_B]$ - Branch impedance matrix

$[I_B]$ - Branch current matrix which is column vector

$[V_S]$ - Source voltage matrix which is column vector

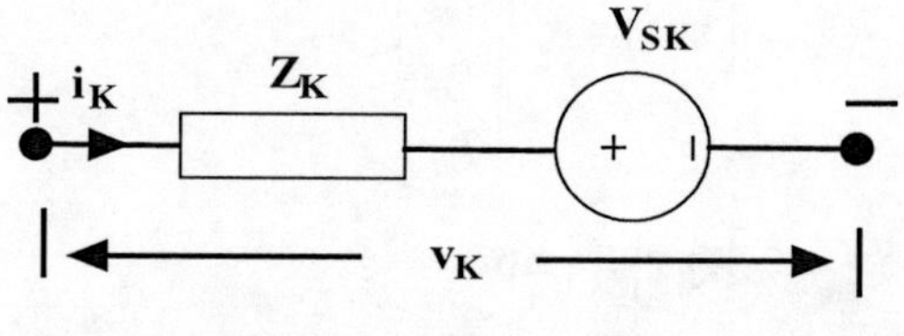

Fig. 5.12: A voltage source in series with an impedance

The equation (5.26) is rewritten based on as per the equation (5.10), that is written in equation (5.27). The equation (5.27) is modified as the equation (5.28). The equation (5.28) is written by applying the equation (5.13), which is presented in equation (5.29).

$$[B][V_b] = [B][Z_b][I_b] + [B][V_S] \tag{5.27}$$

$$[B][V_b] = 0 \text{ (as per equation (5.10))}$$

$$[B][Z_b][I_b] + [B][V_S] = 0$$

$$[B][Z_b][I_b] = -[B][V_S] \tag{5.28}$$

$$[B][Z_b]\ [B]^T\ [I_L] = -[B][V_S] \tag{5.29}$$

$$[Z][I_L] = -[E] \tag{5.30}$$

$$[E] = [B][V_S]$$

Where, $[Z]$ is the loop impedance matrix which is written in equation (5.31).

$$[Z] = [B][Z_b]\ [B]^T \tag{5.31}$$

Formation of Node equations based on KCL: The branch k having an admittance y_k and a current source i_SK in parallel and the voltage across the branch k is v_K. The current (i_K) is written in equation (5.32). This equation should be in the matrix format in equation (5.33).

$$i_k = y_K\, v_K - i_{SK} \tag{5.32}$$

$$[I_b] = [Y_b][V_b] - [I_S] \tag{5.33}$$

Y_B - Branch admittance

I_B - Branch current

I_S - Source current

V_B - Branch voltage

$[Y_B]$ - Branch admittance matrix

$[I_B]$ - Branch current matrix which is column vector

$[I_S]$ - Source current matrix which is column vector

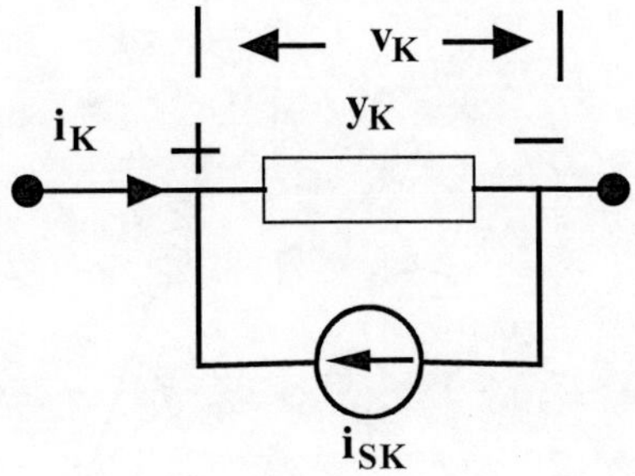

Fig. 5.13: A current source in parallel with an admittance

The equation (5.33) is rewritten based on as per the equation (5.2), that is written in equation (5.34). The equation (5.34) is modified as the equation (5.35). The equation (5.35) is written by applying the equation (5.3), which is presented in equation (5.36).

$$[A_a][I_b] = [A_a][Y_b][V_b] - [A_a][I_S] \tag{5.34}$$

$$[A_a][I_b] = 0 \text{ (as per equation (5.2))}$$

$$[A_a][Y_b][V_b] - [A_a][I_S] = 0$$

$$[A_a][Y_b][V_b] = [A_a][I_S] \tag{5.35}$$

$$[A_a][Y_b]\,[A_a]^T\,[V_n] = [A_a][I_S] \tag{5.36}$$

$$[Y][V_n] = [F] \tag{5.37}$$

$$[F] = [A_a][I_S]$$

Where, [Y] is the node admittance matrix which is written in equation (5.38).

$$[Y] = [A_a][Y_b]\,[A_a]^T \tag{5.38}$$

5.5 ANALYSIS OF RESISTIVE NETWORK

The network consists of the passive elements and active sources like voltage sources and current sources. Here, the network which consists of the resistors are only discussed in this chapter. The voltage across the branch elements and the current through the branch elements are to be found using the loop impedance matrix or node admittance matrix or in other words which is found by the topological application of KVL and KCL concepts. These electrical networks are also solved with network theorems and basic network solving methods like loop analysis and node analysis. The response of the network could be verified using KVL and KCL.

Example 5.1:

Determine the loop currents and branch currents using tie-set concepts for the circuit shown in Fig. 5.14.

Solutions:

The oriented graph of the circuit shown in Fig.5.14 is presented in Fig.5.15. One of the trees of the oriented graph is shown in Fig. 5.16. The loop current direction is shown in Fig.5.17. The tie-set matrix of the circuit is shown in equation (5.39). The dimension of the tie-set matrix is (number of loops X number of branches).

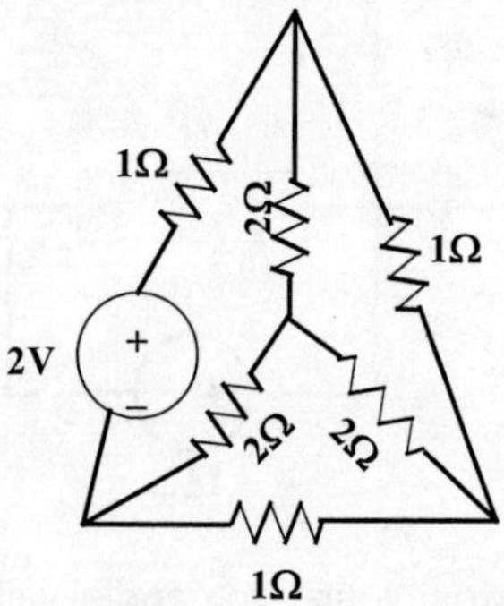

Fig. 5.14: Circuit for Example 5.1

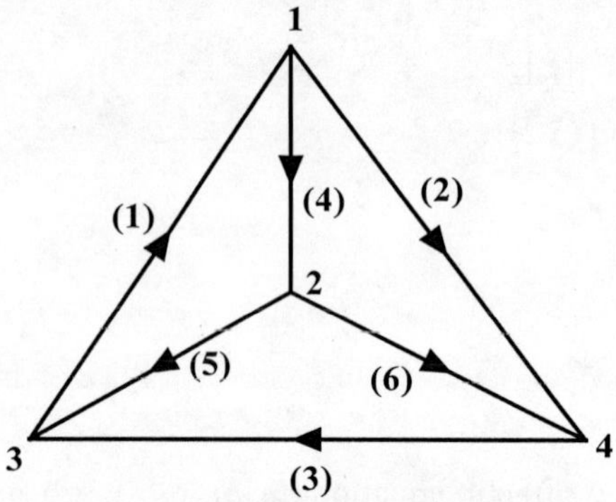

Fig. 5.15: Oriented graph of Example 5.1 and the circuit of Fig.5.14

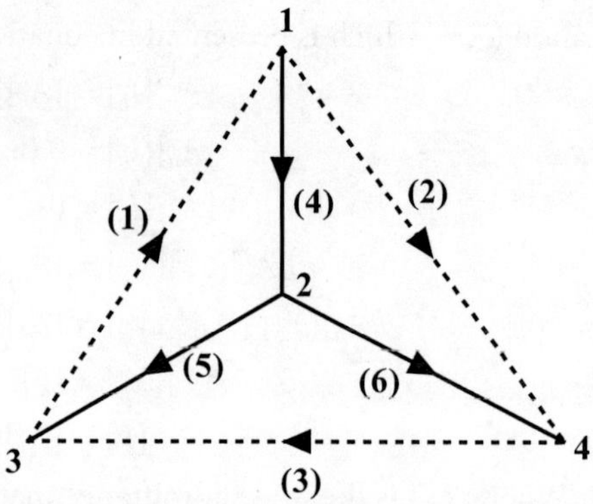

Fig. 5.16: Tree of the oriented graph of Fig.5.15

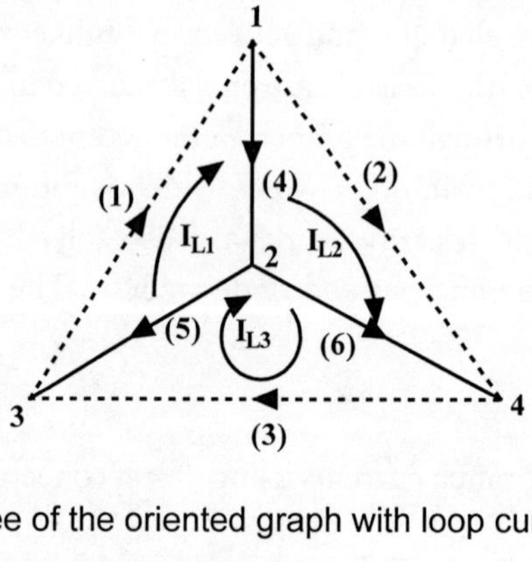

Fig. 5.17: Tree of the oriented graph with loop current direction

$$[B] = \begin{bmatrix} 100110 \\ 010-10-1 \\ 0010-11 \end{bmatrix} \tag{5.39}$$

The KVL equation in matrix is presented in (5.40) that is developed based on the equation (5.29).

$$[B][V] = \begin{bmatrix} 100110 \\ 010-10-1 \\ 0010\ \ 11 \end{bmatrix} \begin{bmatrix} \\ \\ \\ \\ \\ \\ \end{bmatrix} \tag{5.40}$$

$$[B][V_S] = \begin{bmatrix} 2 \\ 0 \\ 0 \end{bmatrix}$$

The branch impedance matrix $[Z_b]$ is presented in equation (5.41). The matrix $[B][Z_b]$ $[B]^T$ is calculated in equation (5.42).

$$[Z_b] = \begin{bmatrix} 1 & 0 & 0 & 0 & 0 & 0 \\ 0 & 1 & 0 & 0 & 0 & 0 \\ 0 & 0 & 1 & 0 & 0 & 0 \\ 0 & 0 & 0 & 2 & 0 & 0 \\ 0 & 0 & 0 & 0 & 2 & 0 \\ 0 & 0 & 0 & 0 & 0 & 2 \end{bmatrix} \tag{5.41}$$

$$[B][Z_b] = \begin{bmatrix} 1 & 0 & 0 & 1 & 1 & 0 \\ 0 & 1 & 0 & -1 & 0 & -1 \\ 0 & 0 & 1 & 0 & -1 & 1 \end{bmatrix} \begin{bmatrix} 1 & 0 & 0 & 0 & 0 & 0 \\ 0 & 1 & 0 & 0 & 0 & 0 \\ 0 & 0 & 1 & 0 & 0 & 0 \\ 0 & 0 & 0 & 2 & 0 & 0 \\ 0 & 0 & 0 & 0 & 2 & 0 \\ 0 & 0 & 0 & 0 & 0 & 2 \end{bmatrix}$$

$$[B][Z_b] = \begin{bmatrix} 1 & 0 & 0 & 2 & 2 & 0 \\ 0 & 1 & 0 & -2 & 0 & -2 \\ 0 & 0 & 1 & 0 & -2 & 2 \end{bmatrix}$$

$$[B][Z\ ][B] = \begin{bmatrix} 1 & 0 & 0 & 2 & 2 & 0 \\ 0 & 1 & 0 & -2 & 0 & -2 \\ 0 & 0 & 1 & 0 & 2 & 2 \end{bmatrix} \begin{bmatrix} 1 & 0 & 0 \\ 0 & 1 & 0 \\ 0 & 0 & 1 \\ 1 & 1 & 0 \\ 1 & 0 & 1 \\ 0 & 1 & 1 \end{bmatrix}$$

$$[B][Z_b][B]^T = \begin{bmatrix} 5 & -2 & -2 \\ -2 & 5 & -2 \\ -2 & -2 & 5 \end{bmatrix} \tag{5.42}$$

Based on the equation (5.29), the values are substituted using (5.40), (5.41) and (5.42) and that are presented in equation (5.43). The solution of the equation or the loop current value is $I_{L1} = -0.8571$A, $I_{L2} = -0.5714$A and $I_{L3} = -0.5714$.

$$-\begin{bmatrix} 2 \\ 0 \\ 0 \end{bmatrix} = \begin{bmatrix} 5 & -2 & -2 \\ -2 & 5 & -2 \\ -2 & -2 & 5 \end{bmatrix} \begin{bmatrix} I_{L1} \\ I_{L2} \\ I_{L3} \end{bmatrix} \tag{5.43}$$

The branch currents are calculated with the equation $[I_b] = [B]^T [I_L]$.

$$[I_b] = \begin{bmatrix} i_1 \\ i_2 \\ i_3 \\ i_4 \\ i_5 \\ i_6 \end{bmatrix} = \begin{bmatrix} 1 & 0 & 0 \\ 0 & 1 & 0 \\ 0 & 0 & 1 \\ 1 & -1 & 0 \\ 1 & 0 & -1 \\ 0 & -1 & 1 \end{bmatrix} \begin{bmatrix} -0.8571 \\ -0.5714 \\ -0.5714 \end{bmatrix}.$$

$$\begin{bmatrix} i_1 \\ i_2 \\ i_3 \\ i_4 \\ i_5 \\ i_6 \end{bmatrix} = \begin{bmatrix} -0.8571 \\ -0.5714 \\ -0.5714 \\ -0.2857 \\ -0.2857 \\ 0 \end{bmatrix}$$

Example 5.2:

(a) For the network shown in Fig. 5.18, draw the graph and tie-set schedule. Using the tie-set matrix, obtain the loop equations and find the branch currents.

(b) For the network shown in Fig. 5.18, draw the graph and cut-set schedule. Using the cut-set matrix, obtain the node equations and find the branch currents.

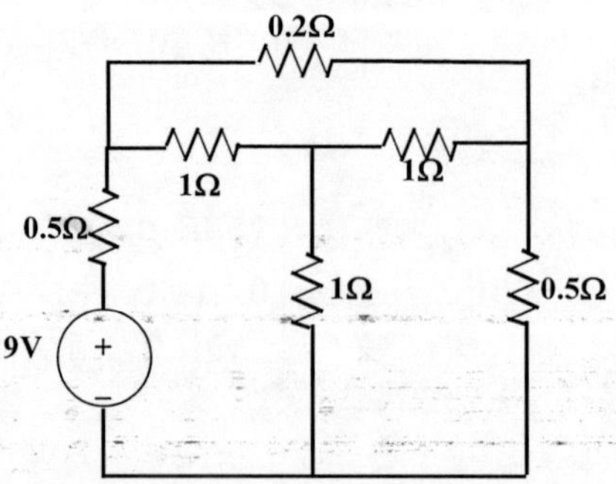

Fig. 5.18: Circuit for Example 5.2

Solutions:

The problem could be solved by two methods namely tie-set matrix and cut-set matrix.

Tie-set Matrix method: The oriented graph of the circuit shown in Fig.5.18 is presented in Fig.5.19. One of the trees of the oriented graph is shown in Fig. 5.20. The loop current direction is shown in Fig.5.20. The tie-set matrix of the circuit is shown in equation (5.44). The dimension of the tie-set matrix is (number of loops X number of branches).

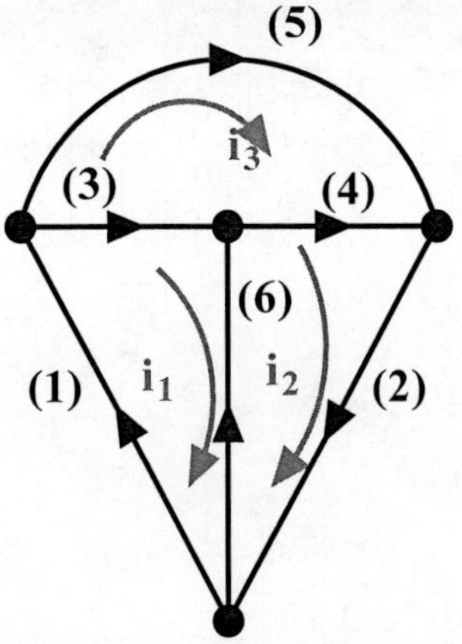

Fig. 5.19: Oriented graph of the circuit for Example 5.2

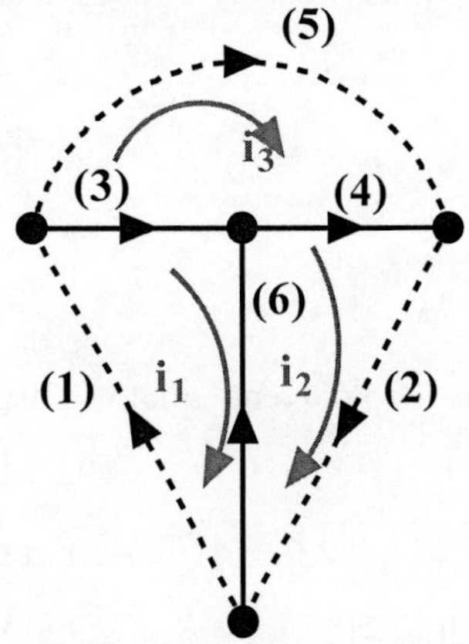

Fig. 5.20: Tree of the circuit for Example 5.2

$$[B]=\begin{bmatrix}1 & 0 & 1 & 0 & 0 & -1\\ 0 & 1 & 0 & 1 & 0 & 1\\ 0 & 0 & -1 & -1 & 1 & 0\end{bmatrix} \tag{5.44}$$

The branch impedance matrix is given below.

$$[Z_b]=\begin{bmatrix}0.5 & 0 & 0 & 0 & 0 & 0\\ 0 & 0.5 & 0 & 0 & 0 & 0\\ 0 & 0 & 1 & 0 & 0 & 0\\ 0 & 0 & 0 & 1 & 0 & 0\\ 0 & 0 & 0 & 0 & 0.2 & 0\\ 0 & 0 & 0 & 0 & 0 & 1\end{bmatrix}$$

The loop impedance matrix is calculated using $[B][Z_b]\,[B]^T$.

$$[B][Z_b][B]^T=\begin{bmatrix}1 & 0 & 1 & 0 & 0 & -1\\ 0 & 1 & 0 & 1 & 0 & 1\\ 0 & 0 & -1 & -1 & 1 & 0\end{bmatrix}\begin{bmatrix}0.5 & 0 & 0 & 0 & 0 & 0\\ 0 & 0.5 & 0 & 0 & 0 & 0\\ 0 & 0 & 1 & 0 & 0 & 0\\ 0 & 0 & 0 & 1 & 0 & 0\\ 0 & 0 & 0 & 0 & 0.2 & 0\\ 0 & 0 & 0 & 0 & 0 & 1\end{bmatrix}\begin{bmatrix}1 & 0 & 1 & 0 & 0 & -1\\ 0 & 1 & 0 & 1 & 0 & 1\\ 0 & 0 & -1 & -1 & 1 & 0\end{bmatrix}$$

$$[B][Z_b][B]^T = \begin{bmatrix} 2.5 & -1 & -1 \\ -1 & -2.5 & -1 \\ -1 & -1 & 2.2 \end{bmatrix}$$

Calculate the matrix -$[B][V_S]$.

$$-[B][V_S] = -\begin{bmatrix} 1 & 0 & 1 & 0 & 0 & -1 \\ 0 & 1 & 0 & 1 & 0 & 1 \\ 0 & 0 & -1 & -1 & 1 & 0 \end{bmatrix} \begin{bmatrix} -9 \\ 0 \\ 0 \\ 0 \\ 0 \\ 0 \end{bmatrix} = \begin{bmatrix} 9 \\ 0 \\ 0 \end{bmatrix}$$

Now the loop current equations are $[B][Z_b]\ [B]^T\ [I_L]=[Z][I_L] = -[B][V_S]$

$$[I_L] = [Z]^{-1} * (-[B][V_S])$$

$$[I_L] = \begin{bmatrix} i_{L1} \\ i_{L2} \\ i_{L3} \end{bmatrix} = \begin{bmatrix} 2.5 & -1 & -1 \\ -1 & -2.5 & -1 \\ -1 & -1 & 2.2 \end{bmatrix}^{-1} \begin{bmatrix} 9 \\ 0 \\ 0 \end{bmatrix} = \begin{bmatrix} 0.9890 & 0.7033 & 0.7692 \\ 0.7033 & 0.9890 & 0.7692 \\ 0.7692 & 0.7692 & 1.1538 \end{bmatrix} \begin{bmatrix} 9 \\ 0 \\ 0 \end{bmatrix}$$

$$[I_L] = \begin{bmatrix} i_{L1} \\ i_{L2} \\ i_{L3} \end{bmatrix} = \begin{bmatrix} 8.9 \\ 6.33 \\ 6.92 \end{bmatrix} \text{A}$$

The branch currents are calculated with the equation $[I_b]=[B]^T\ [I_L]$.

$$[I_b] = \begin{bmatrix} i_1 \\ i_2 \\ i_3 \\ i_4 \\ i_5 \\ i_6 \end{bmatrix} = \begin{bmatrix} 1 & 0 & 1 & 0 & 0 & -1 \\ 0 & 1 & 0 & 1 & 0 & 1 \\ 0 & 0 & -1 & -1 & 1 & 0 \end{bmatrix}^T \begin{bmatrix} 8.9 \\ 6.33 \\ 6.92 \end{bmatrix} = \begin{bmatrix} 8.90 \\ 6.33 \\ 1.98 \\ -0.59 \\ 6.92 \\ -2.57 \end{bmatrix} \text{A}$$

Cut-set Matrix method: The oriented graph of the circuit shown in Fig.5.18 is presented in Fig.5.21. One of the trees of the oriented graph with a cut set and its directions are shown in Fig. 5.22. The node potential (V_3, V_4, V_6) of the cut-set is shown in Fig.5.22. The cut-set matrix of the circuit is shown in equation (5.45). The dimension of the cut-set matrix is (number of cut-set nodes X number of branches).

$$[Q_C] = \begin{bmatrix} -1 & 0 & 1 & 0 & 1 & 0 \\ 0 & -1 & 0 & 1 & 1 & 0 \\ 1 & -1 & 0 & 0 & 0 & 1 \end{bmatrix} \tag{5.45}$$

From the cut-set matrix or by applying the KCL at the cut-set nodes, the equation (5.46) is obtained with the branch currents. Also, the branch voltage (v_1, v_2, v_3, v_4, v_5 & v_6) is found from the matrix which is written in equation (5.47).

$$\left.\begin{aligned}-i_1+i_3+i_5&=0\\-i_2+i_4+i_5&=0\\i_1-i_2+i_6&=0\end{aligned}\right\} \tag{5.45a}$$

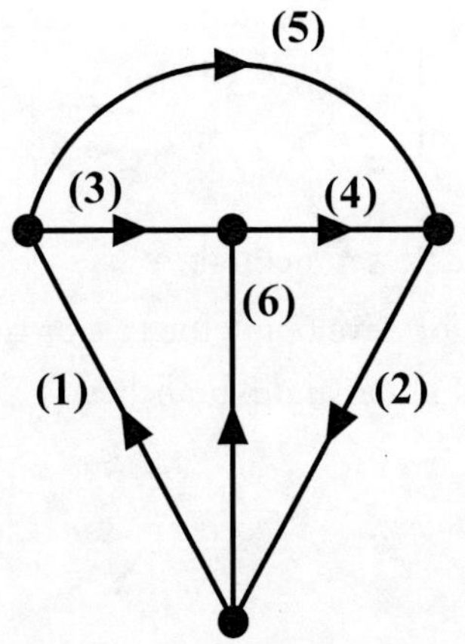

Fig. 5.21: Oriented graph of the circuit for Example 5.2

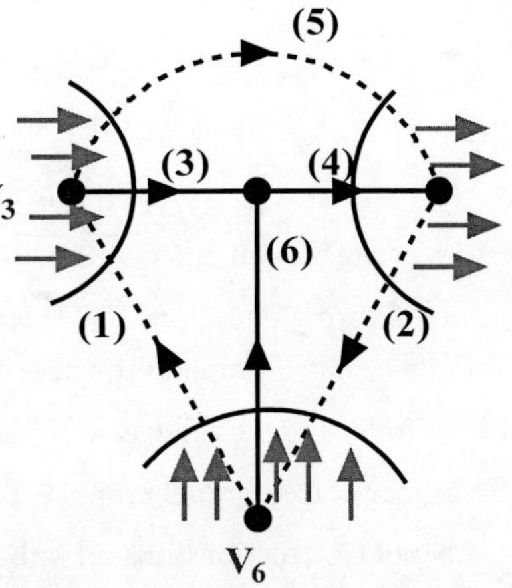

Fig. 5.22: Tree of the circuit with cut-set and its direction

$$\left.\begin{aligned}v_1&=V_6-V_3\\v_2&=-V_4-V_6\\v_3&=V_3\\v_4&=V_4\\v_5&=V_3+V_4\\v_6&=V_6\end{aligned}\right\} \tag{5.46}$$

The node current equations presented in (5.45a) are written in terms of the node voltages using (5.46). The branch 1 consists of the voltage source, hence the branch current $i_1=\frac{1}{0.5}(V_6-V_3+9)$. Similarly other branch currents are substituted in equation (5.45). The following equations are obtained.

$$8V_3+5V_4-2V_6=18$$

$$5V^3+8V_4+2V_6=0$$

$$-2V_3+2V_4+5V_6=-18$$

By solving these equations, the node potential is V_3 = 1.978V; V_4 = −0.5934V; V_6 = −2.57V. From this, the branch currents are evaluated in the following ways.

$$\left.\begin{aligned}
i_1 &= \frac{1}{0.5}(V_6 - V_3 + 9) = 8.9A \\
i_2 &= \frac{1}{0.5}(-V_4 - V_6) = 6.326A \\
i_3 &= \frac{1}{1}(V_3) = 1.978A \\
i_4 &= \frac{1}{1}(V_4) = -0.5934A \\
i_5 &= \frac{1}{0.2}(V_3 + V_4) = 6.923A \\
i_6 &= \frac{1}{1}(V_6) = -2.57A
\end{aligned}\right\}$$

The values of the branch current remain the same from the tie-set method.

Note: The following equations (5.47) to (5.50) are useful for developing the tree or twigs/links or co-trees from the oriented graph of the network consisting of N nodes and B branches.

$$\text{Number of independent voltages} = N - 1 \tag{5.47}$$

$$\text{Number of tree-branches} = N - 1 \tag{5.48}$$

$$\text{No of Co-trees or links} = L = B - (N - 1) \tag{5.49}$$

$$\text{Total number of branches (B)} = L + (N - 1) \tag{5.50}$$

5.6 TWO PORT NETWORKS

A network is classified as one port network or multi-port networks. One port network has only input terminals. If a network has two sets of terminals such as input terminals and output terminals. It is named as two port networks. The pictorial representation of these networks are shown in Fig. 5.23 (a) and 5.23(b). Example for one port network is shown in Fig. 5.24. It has the input source as voltage source. In general, an electric network has two pairs of terminals that are named as "input terminals" and "output terminals". The electrical signal or electrical energy enters into the terminals that are denoted as the input terminals. Similarly, the electrical signal or energy received at the terminals which are named as the output terminals. The circuit is analysed based on the given input and its response is observed at the output terminals in the form of voltages/ currents or any electrical parameters. These output terminals are well connected to the input terminals of other networks. In a two-port network shown in Fig. 5.23 (b), terminals "ab" are denoted as input terminals and terminals "cd" are denoted as output terminals. Common representations of the input terminals are 1 and 1' and the output terminals as 2 and 2' that is shown in Fig. 5.25. The current I_1 and I_2 enter at terminals 1 and 2 respectively. Similarly, current I_1 and I_2 leave at terminals 1' and 2' respectively. The voltage V_1 and V_2 are the voltage across the input and output terminals respectively. Some two-port networks and most of the electronic circuits have the additional property that terminals 1' and 2' are in common.

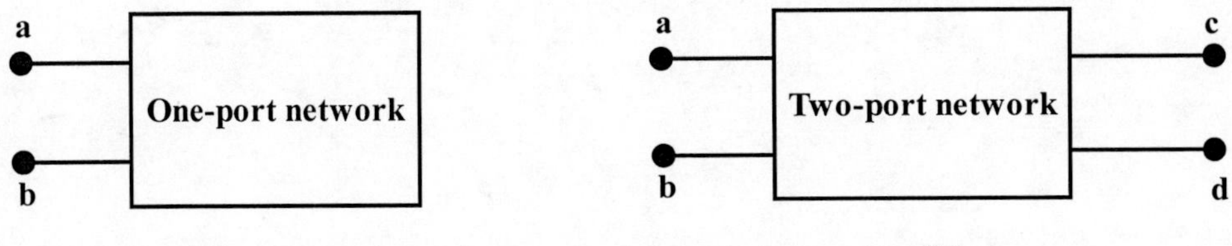

(a) (b)

Fig. 5.23: Block diagram of one-port and two-port networks

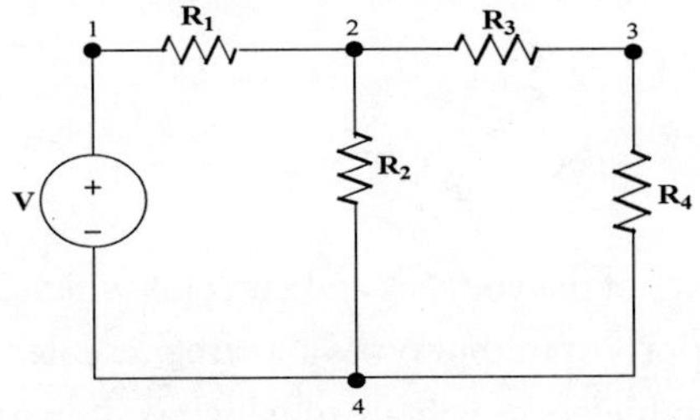

Fig. 5.24: Sample network for one-port network (Node 1 and 4 are the input terminals)

Fig. 5.25: Typical two-port network with input terminals 1 & 1' and output terminals 2 & 2'

These two-port networks are described with four variables: input port / output port currents and voltages. Among those variables, two variables are independent and the remaining two variables are dependent variables. Since the network is linear, the equations are formed by a set of linear equations that relates the port voltages and currents in the network. These relating parameters are denoted as two-port parameters. These two-port parameters are classified as impedance parameters, admittance parameters, hybrid parameters and transmission parameters based on the combination of the input and output port parameters or variables. These parameters are extremely helpful for analysing the electronic circuits and electrical power systems.

5.6.1 OPEN CIRCUIT IMPEDANCE PARAMETERS

The open circuit impedance parameters are called impedance parameters or Z parameters. Any electrical networks are driven by either voltage sources or current sources which are represented as two-port networks in Fig. 5.26 and Fig. 5.27 respectively. V_1, I_1 and V_2, I_2 are denoted as input port and output port terminal voltages and terminal currents respectively. Both input and output port voltages are linearly related with the input/output port currents sequentially by the equations (5.51) and (5.52). These equations could be arranged in matrix form in equation (5.53).

$$V_1 = Z_{11} I_1 + Z_{12} I_2 \tag{5.51}$$

$$V_2 = Z_{21} I_1 + Z_{22} I_2 \tag{5.52}$$

$$[V_1\ V_2] = [Z_{11}\ Z_{12}\ Z_{21}\ Z_{22}\ ][I_1\ I_2] \tag{5.53}$$

$$Z_{11} = \frac{V_1}{I_1}|_{I_2=0} \tag{5.54}$$

$$Z_{12} = \frac{V_1}{I_2}|_{I_1=0} \tag{5.55}$$

$$Z_{21} = \frac{V_2}{I_1}|_{I_2=0} \tag{5.56}$$

$$Z_{22} = \frac{V_2}{I_2}|_{I_1=0} \tag{5.57}$$

The Z parameters are the ratio of the voltages and currents which are obtained by making terminals open circuited with either input ports or output ports and are presented in Fig. 5.26 and 5.27 respectively. Based on that the admittance parameters are obtained which is shown in equations (5.63) to (5.66).

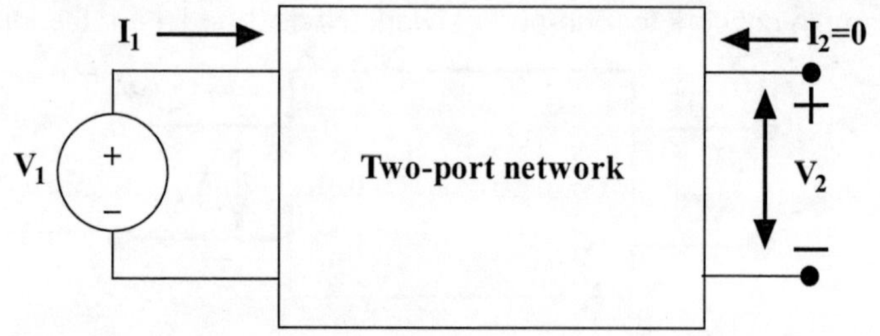

Fig. 5.26: Two-port network with output terminals 2 & 2' are to be opened (I2=0)

Z_{11} = Open-circuit input impedance

Z_{21} = Open-circuit transfer impedance from port 2 (output port) to port 1 (input port)

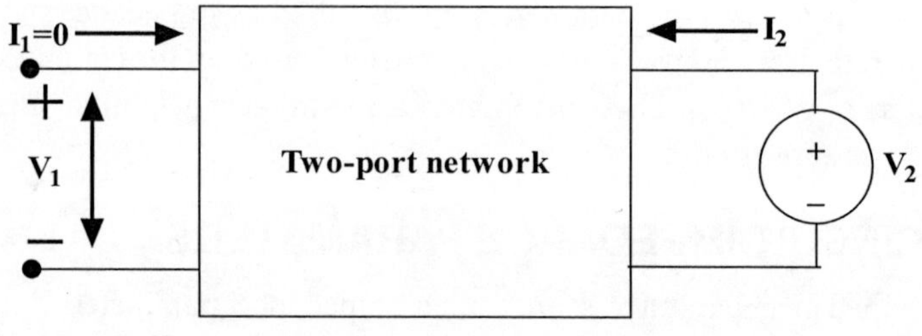

Fig. 5.27: Two-port network with input terminals 1 & 1' are to be opened (I_1=0)

Z_{12} = Open-circuit transfer impedance from port 1 (input port) to port 2 (output port)

Z_{22} = Open-circuit output impedance

Where Z_{11} and Z_{22} are also known as the driving point impedances and Z_{12} and Z_{21} are known as transfer impedances. These parameters are obtained when the circuit is driven by voltage sources or current sources.

When the two-port network is linear and there is no dependent sources, the transfer impedances remains the same ($Z_{12} = Z_{21}$).

Dependent sources

- Voltage dependent voltage sources

- Current dependent voltage sources

- Voltage dependent current sources

- Current dependent current sources

Where and are the voltage across any circuit elements and current flowing in any circuit elements in a particular circuit respectively.

α and δ are the constant parameters.

β is presented in (V/A) unit

γ is presented in (A/V) unit

Example 5.11:

Determine the Z parameters pf the circuit shown in Fig.5.28.

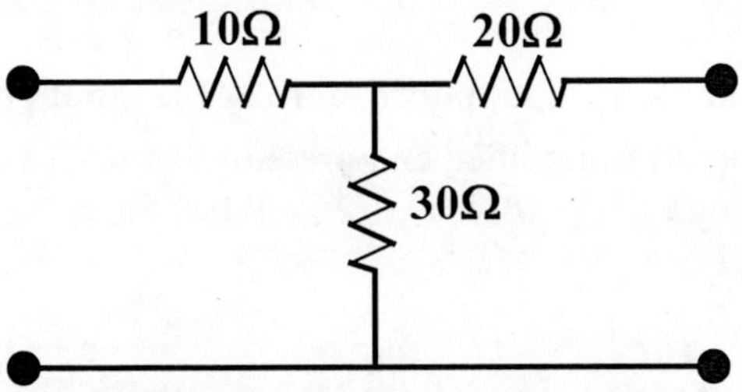

Fig. 5.28: Circuit for Example 5.11

Solutions:

The circuit shown in Fig.5.56 is represented as a two-port network model as shown in Fig.5.29. Solve this problem by applying the current entered into the input and output ports as zero.

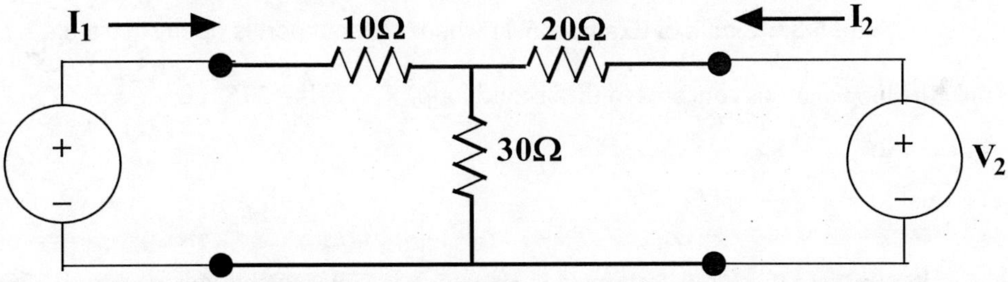

Fig. 5.29: Circuit of Example 5.11 rearranged as two-port network pattern

Step1: Setting the current at the output port as zero or the output port is open circuited.

With this concept, the circuit is modified and presented in Fig.5.30. By observing the circuit, it is understood that no current flows in the 20Ω resistor and the input current I_1 flows through 10Ω and 30Ω resistors only.

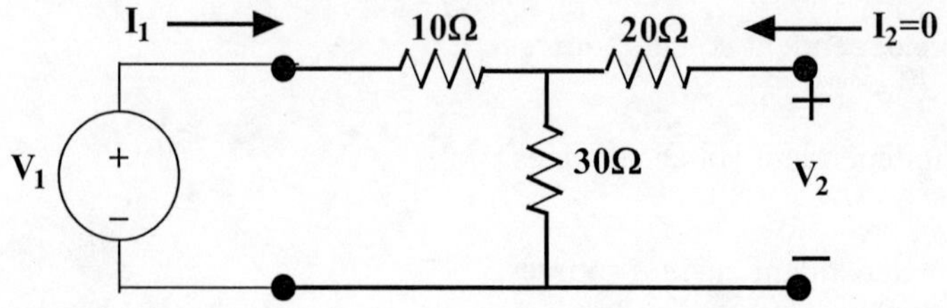

Fig. 5.30: Circuit of Example 5.11 when the output port is open circuited

Apply the loop analysis concepts to the first loop, $V_1 - 10I_1 - 30I_1 = 0$; $\frac{V_1}{I_1} = 40\Omega$

It is equal to $Z_{11} = \frac{V_1}{I_1} = 40\Omega$

The voltage across the output terminals is V_2 which is equal to the voltage across 30Ω. Since the current flows in the 20Ω resistor is zero.

Hence, $V_2 = 30I_1$. From this, $Z_{21} = \frac{V_2}{I_1} = 30\Omega$ is obtained.

Step2: Setting the current at the input port as zero or the input port is open circuited.

With this concept, the circuit is modified and presented in Fig.5.31. By observing the circuit, it is understood that no current flows in the 10Ω resistor and the output current I_2 flows through 20Ω and 30Ω resistors only.

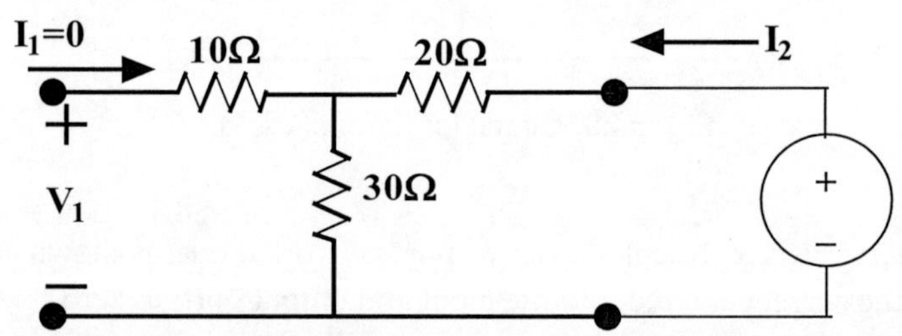

Fig. 5.31: Circuit of Example 5.11 when the input port is open circuited

Apply the loop analysis concepts to the second loop, $V_2 - 20I_2 - 30I_2 = 0$; $\frac{V_2}{I_2} = 50\Omega$

It is equal to $Z_{22} = \frac{V_2}{I_2} = 50\Omega$

The voltage across the input terminals is V_1 which is equal to the voltage across 30Ω. Since the current flows in the 10Ω resistor is zero.

Hence, V_1=30I_2. From this, $Z_{12} = \frac{V_1}{I_2} = 30\Omega$ is obtained.

Z parameters of the given circuit is $\begin{bmatrix} Z_{11} & Z_{12} \\ Z_{21} & Z_{22} \end{bmatrix} = \begin{bmatrix} 40 & 30 \\ 30 & 50 \end{bmatrix}$

Example 5.12:

Determine the Z parameters of the circuit shown in Fig.5.32.

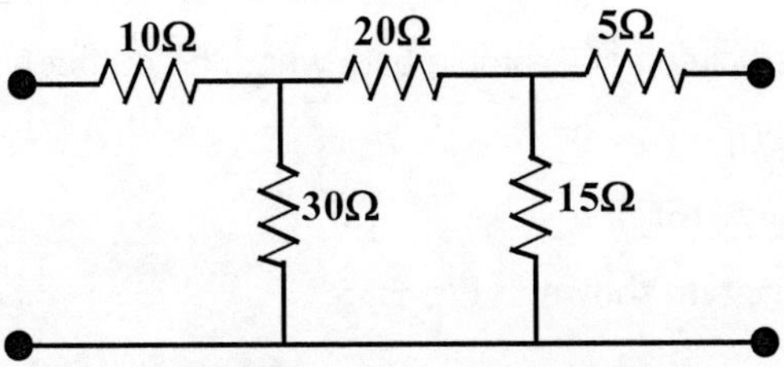

Fig. 5.32: Circuit for Example 5.12

Solutions:

The circuit shown in Fig.5.32 is represented as two-port network model as shown in Fig.5.33. Solve this problem by applying the current entered into the input and output ports as zero.

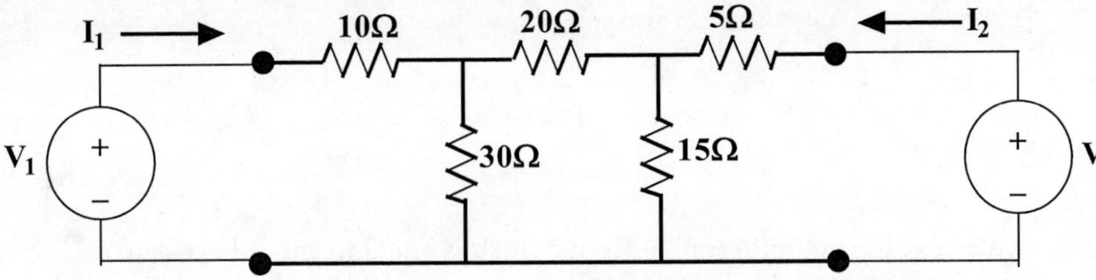

Fig. 5.33: Circuit of Example 5.12 rearranged as two-port network pattern

Step1: Setting the current at the output port as zero or the output port is open circuited.

With this concept, the circuit is modified and presented in Fig.5.34. By observing the circuit, it is understood that there is no current flows in the 5Ω resistor and the input current I1 flows through 10Ω and then divided to the parallel circuits. Hence the circuit is redrawn as shown in Fig.5.35. Let us assume that the two loops' currents i_1 and i_2 are flowing as shown in the figure. By applying the loop analysis concept, the KVL equations are formed and presented as equation (5.58) and (5.59).

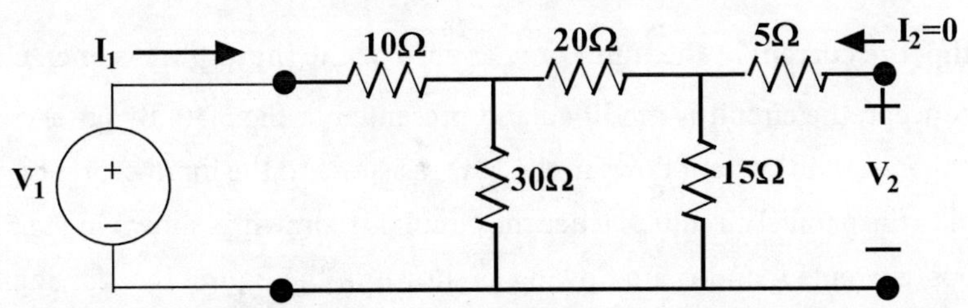

Fig. 5.34: Circuit of Example 5.12 when the output port is open circuited

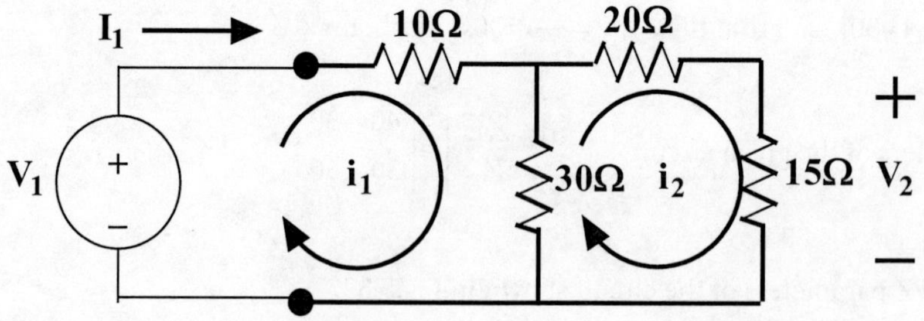

Fig. 5.35: Modified circuit of Example 5.12 when the output port is open circuited

$$V_1 - 10i_1 - 30(i_1 - i_2) = 0 \quad (5.58)$$

$$-30(i_2 - i_1) - 20i_2 - 15i_2 = 0 \quad (5.59)$$

$i_1 = I_1$ is as per circuit diagram shown in Fig. 5.35.

Based on equation (5.59), $i_2 = \frac{30}{65}i_1$ which is substituted in equation (5.58).

Hence, $V_1 = 26.15i_1$.

$i_1 = I_1$ is as per circuit diagram shown in Fig. 5.35. $V_1 = 26.15I_1$.

$$\frac{V_1}{I_1} = 26.15\Omega$$

It is equal to $Z_{11} = \frac{V_1}{I_1} = 26.15\Omega$

The voltage across the output terminals is V_2 which is equal to the voltage across 15Ω. Since the current flows in the 5Ω resistor is zero.

Hence, $V_2 = 15i_2$.

$$V_2 = 15i_2 = 15 * \left(\frac{30}{65}\right)i_1 = \frac{90}{13}i_1 = \frac{90}{13}I_1$$

From this, $Z_{21} = \frac{V_2}{I_1} = \frac{90}{13} = 6.92\Omega$ is obtained.

Step2: Setting the current at the input port as zero or the input port is open circuited.

With this concept, the circuit is modified and presented in Fig.5.36. By observing the circuit, it is understood that there are no current flows in the 10Ω resistor and the input current I_2 flows through 5Ω and then divided to the parallel circuits. Hence the circuit is redrawn as shown in Fig.5.37. Let us assume that the two loops' currents i_3 and i_4 are flowing as shown in the figure. By applying the mesh analysis concept, the KVL equations are formed and presented as equations (5.60) and (5.61).

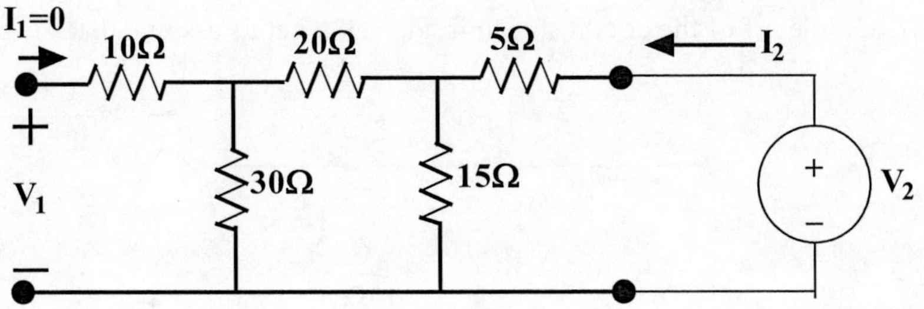

Fig. 5.36: Circuit of Example 5.12 when the input port is open circuited

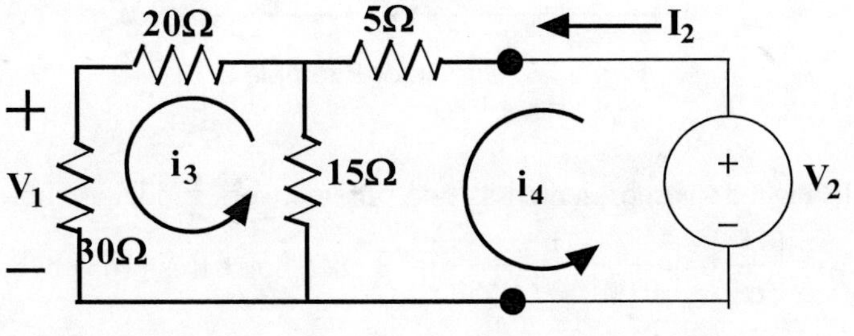

Fig. 5.37: Modified circuit of Example 5.12 when the output port is open circuited

$$V2 - 5i4 - 15(i_4 - i_3) = 0 \tag{5.60}$$

$-30i_3 - 20i_3 - 15(i3 - i_4) = 0$ (5.61)

Based on equation (5.61), $i_3 = \frac{15}{65} i_4$ which is substituted in equation (5.60).

Hence, $V_2 = 16.53i_4$.

$i_4 = I_2$ is as per circuit diagram shown in Fig. 5.37. $V_2 = 16.53I_2$.

$$\frac{V_2}{I_2} = 16.53\Omega$$

It is equal to $Z_{22} = \frac{V_2}{I_2} = 16.53\Omega$

The voltage across the input terminals is V_1 which is equal to the voltage across 30Ω. Since the current flows in the 10Ω resistor is zero.

Hence, $V_1 = 30i_3$.

$$V_1 = 30i_3 = 30 * \left(\frac{15}{65}\right) i_4 = \frac{90}{13} i_4 = \frac{90}{13} I_2$$

From this, $Z_{12} = \frac{V_1}{I_2} = \frac{90}{13} = 6.92\Omega$ is obtained.

Z parameters of the given circuit is $\begin{bmatrix} Z_{11} & Z_{12} \\ Z_{21} & Z_{22} \end{bmatrix} = \begin{bmatrix} 26.15 & 6.92 \\ 6.92 & 16.53 \end{bmatrix}$

Example 5.13:

Determine the Z parameters of the circuit shown in Fig.5.38. Let us assume that the source frequency is 50Hz.

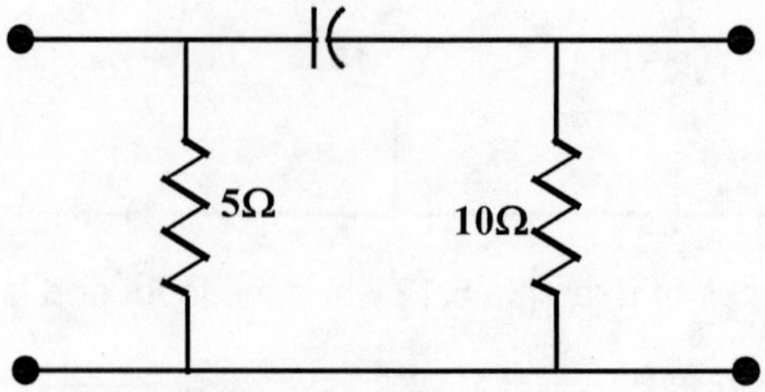

Fig. 5.38: Circuit for Example 5.13

Solutions:

The circuit shown in Fig.5.38 is represented as two-port network model as shown in Fig.5.39 with all impedances in Ω. $X_C = \left(\frac{1}{c\omega}\right) = \left(\frac{1}{0.001*2*\pi*50}\right) = 3.18\Omega$. Solve this problem by applying the current entered into the input and current in the output port as zero.

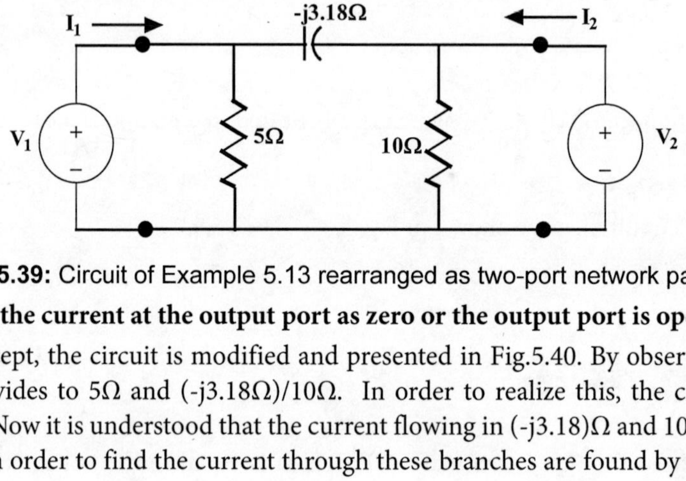

Fig. 5.39: Circuit of Example 5.13 rearranged as two-port network pattern

Step1: Setting the current at the output port as zero or the output port is open circuited.

With this concept, the circuit is modified and presented in Fig.5.40. By observing the circuit, the input current I_1 divides to 5Ω and (-j3.18Ω)/10Ω. In order to realize this, the circuit is redrawn as shown in Fig.5.41. Now it is understood that the current flowing in (-j3.18)Ω and 10Ω at the output side remain the same. In order to find the current through these branches are found by the current division technique.

$$I_{3.18\Omega\ or 10\Omega} = I_1 * \left(\frac{5}{5 - j3.18 + 10}\right) = I_1 * \left(\frac{5}{15 - j3.18}\right) = \left(\frac{5}{15.33\angle -11.96^0}\right) I_1 = 0.326\angle 11.96^0\, I_1$$

Voltage across the 10Ω resistor is V_2, which is equal to $I_{10\Omega}$*10. Hence, $V_2 = 3.26\angle 11.96^0\ I_1$

$$\frac{V_2}{I_1} = 3.26\angle 11.96^0\,\Omega$$

It is equal to $Z_{21} = \frac{V_2}{I_1} = 3.26\angle 11.96^0\,\Omega$

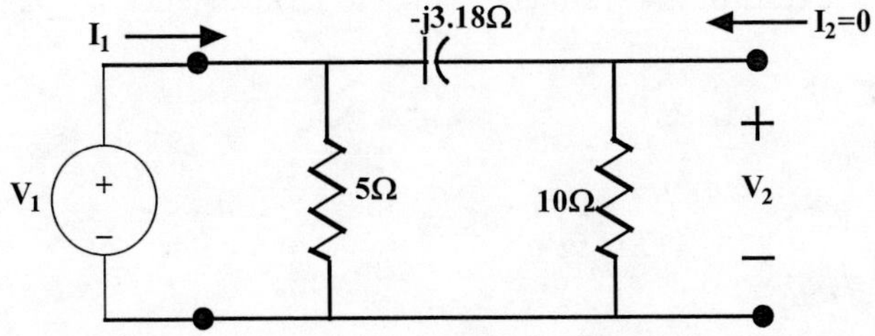

Fig. 5.40: Circuit of Example 5.13 when the output port is open circuited

In order to find Z_11, the circuit is simplified as shown in Fig.5.42. The impedances (-j3.18) and 10 are in series and the obtained result is (10-j3.18)Ω, then it is parallel to 5Ω resistor.

$$\text{Hence, } (10\text{-}j3.18\,||5) = \frac{(10-j3.18)(5)}{15-j3.18} = \frac{5*10.49\angle -17.64^0}{15.33\angle -11.96^0} = 3.42\angle -5.68^0\ \Omega.$$

From this, $Z_{11} = \frac{V_1}{I_1} = 3.42\angle -5.68^0\,\Omega$ is obtained.

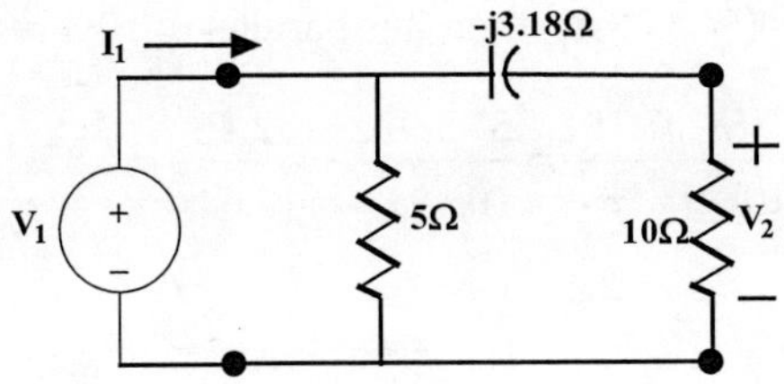

Fig. 5.41: Modified circuit of Example 5.13 when the output port is open circuited

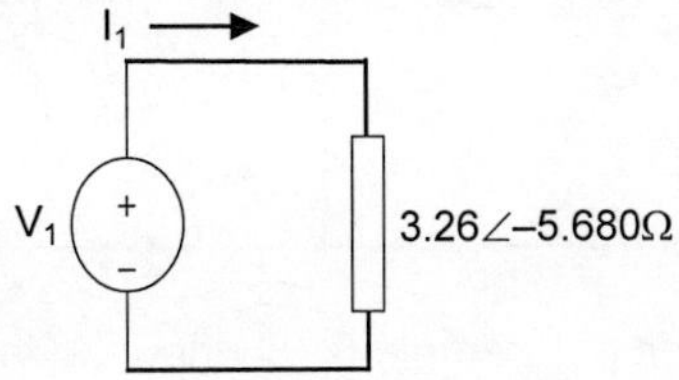

Fig. 5.42: Modified circuit of Example 5.13 for calculating Z_{11} when the output port is open circuited

Step2: Setting the current at the input port as zero or the input port is open circuited.

With this concept, the circuit is modified and presented in Fig.5.43. By observing the circuit, the input current I_2 divides to 10Ω and (-j3.18)Ω/5Ω. In order to realize this, the circuit is redrawn as shown in Fig.5.44. Now it is understood that the current flowing in (-j3.18)Ω and 5Ω at the input side remain the same. In order to find the current through these branches are found by the current division technique.

$$I_{3.18\Omega\ or 5\Omega} = I_2 * \left(\frac{10}{5-j3.18+10}\right) = I_2 * \left(\frac{10}{15-j3.18}\right) = \left(\frac{10}{15.33\angle -11.96^0}\right) I_2 = 0.652\angle 11.96^0\, I_2$$

Voltage across the 5Ω resistor is V_1, which is equal to $I_{5\Omega}$*5. Hence, $V_1 = 3.26\angle 11.96^0\, I_2$

$$\frac{V_1}{I_2} = 3.26\angle 11.96^0\,\Omega$$

It is equal to $Z_{12} = \dfrac{V_1}{I_2} = 3.26\angle 11.96^0\,\Omega$

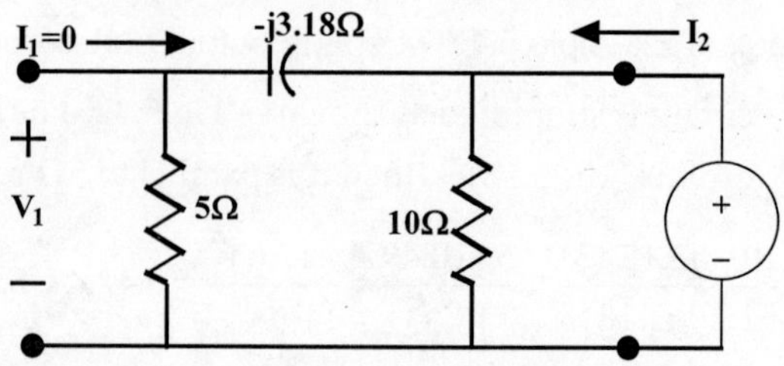

Fig. 5.42: Circuit of Example 5.13 when the input port is open circuited

In order to find , the circuit is simplified as shown in Fig.5.42. The impedances (-j3.18) and 5 are in series and the obtained result is (5-j3.18)Ω, then it is parallel to 10Ω resistor.

Hence, (5-j3.18) ||(5) $= \dfrac{5*(5-j3.18)}{10-j3.18} = \dfrac{5*5.92\angle -32.45^0}{10.49\angle --17.64^0} = 2.85\angle -14.81^0\,\Omega$

From this, is obtained.

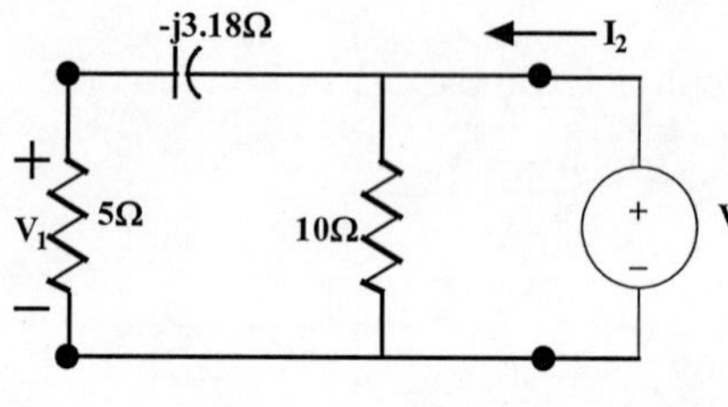

Fig. 5.43: Modified circuit of Example 5.13 when the input port is open circuited

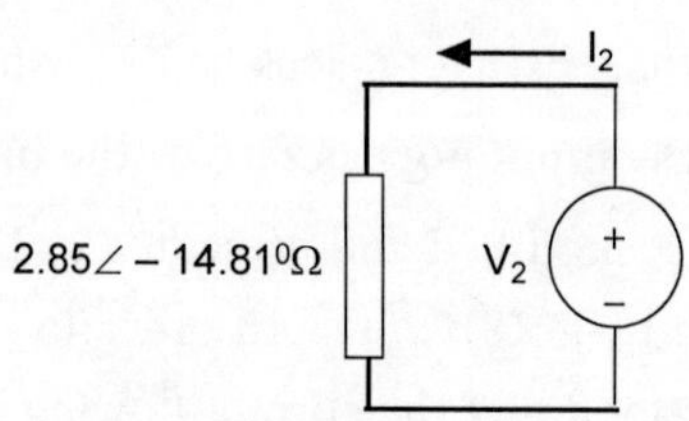

Fig. 5.44: Modified circuit of Example 5.13 for calculating Z_{11} when the input port is open circuited

Z parameters of the given circuit is $\begin{bmatrix} Z_{11} & Z_{12} \\ Z_{21} & Z_{22} \end{bmatrix} = \begin{bmatrix} 3.42\angle -5.68^0 & 3.26\angle 11.96^0 \\ 3.26\angle 11.96^0 & 2.85\angle -14.81^0 \end{bmatrix}$

5.6.2 Short Circuit Admittance Parameters

The short circuit admittance parameters are called as admittance parameters or Y parameters that are to be obtained for the two-port network shown in Fig.5.45. Both input and output port currents are linearly related with the input/output port voltages sequentially by the equations (5.60) and (5.61). These equations could be arranged in matrix form in equation (5.62).

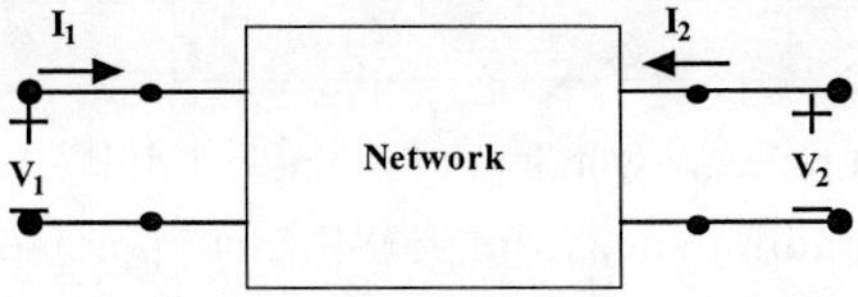

Fig. 5.45: Two-port network with input and output terminals

$$I_1 = Y_{11}\, V_1 + Y_{12}\, V_2 \tag{5.60}$$

$$I_2 = Y_{21}\, V_1 + Y_{22}\, V_2 \tag{5.61}$$

$$\begin{bmatrix} I_1 \\ I_2 \end{bmatrix} = \begin{bmatrix} Y_{11} & Y_{12} \\ Y_{21} & Y_{22} \end{bmatrix} \begin{bmatrix} V_1 \\ V_2 \end{bmatrix} \tag{5.62}$$

$$Y_{11} = \frac{I_1}{V_1}\Big|_{V_2=0} \tag{5.63}$$

$$Y_{12} = \frac{I_1}{V_2}\Big|_{V_1=0} \tag{5.64}$$

$$Y_{21} = \frac{I_2}{V_1}\Big|_{V_2=0} \tag{5.65}$$

$$Y_{22} = \frac{I_2}{V_2}\Big|_{V_1=0} \tag{5.66}$$

The Y parameters are the ratio of the currents and voltages which are obtained by short circuiting the input ports or output ports which is presented as shown in Fig. 5.46 and 5.47. Based on that the admittance parameters are obtained which are shown in equations (5.63) to (5.66).

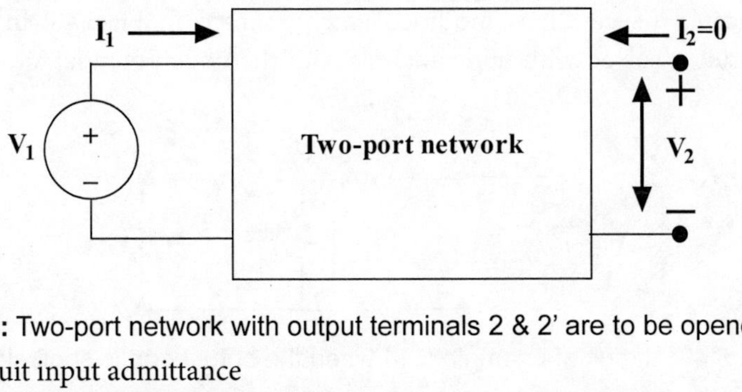

Fig. 5.46: Two-port network with output terminals 2 & 2' are to be opened (I_2=0)

Y_{11} = Short-circuit input admittance

Y_{21} = Short-circuit transfer admittance from port 2 (output port) to port 1 (input port)

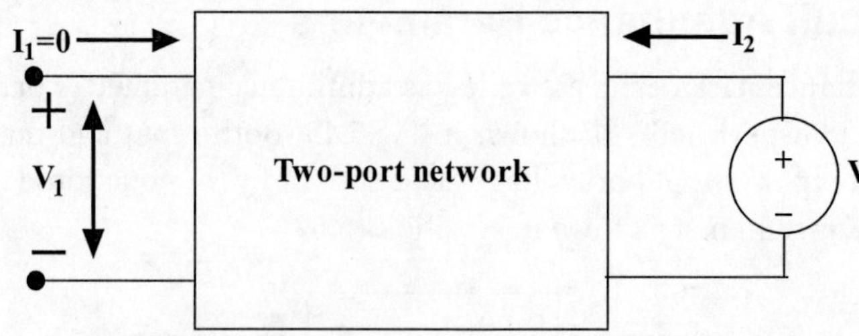

Fig. 5.47: Two-port network with input terminals 1 & 1' are to be opened (I_1=0)

Y_{12} = Short-circuit transfer admittance from port 1 (input port) to port 2 (output port)

Y_{22} = Short-circuit output admittance

Where Y_{11} and Y_{22} are also known as the driving point admittances and Y_{12} and Y_{21} are known as transfer admittances. These parameters are obtained when the circuit is driven by voltage sources or current sources.

When the two-port network is linear and there are no dependent sources, the transfer admittances remain the same ($Y_{12} = Y_{21}$)

Example 5.14:

Determine the Y parameters of the circuit shown in Fig.5.28.

Solutions:

The circuit shown in Fig.5.28 is represented as a two-port network model as shown in Fig.5.48. Solve this problem by short circuiting both the input and output ports sequentially.

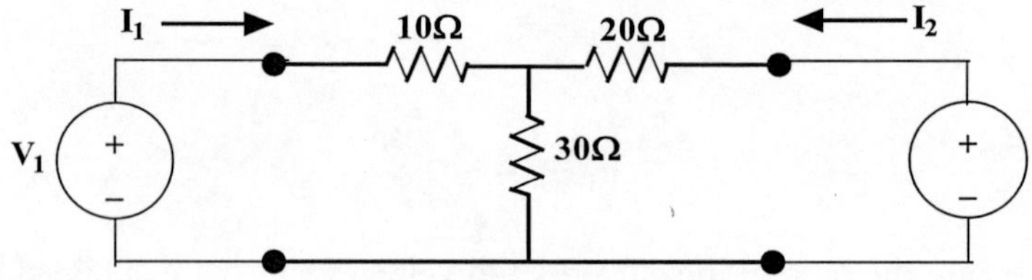

Fig. 5.48: Circuit of Example 5.14 rearranged as two-port network pattern

Step1: Setting the voltage at the output port as zero or the output port is short circuited.

With this concept, the circuit is modified and presented in Fig.5.48. In order to find the Y parameters, the circuit is solved with node analysis with the node potential V_A, the concern circuit is drawn in Fig.5.49.

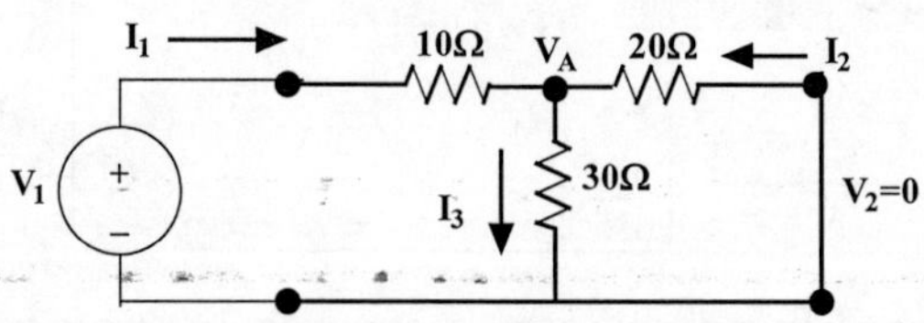

Fig. 5.49: Circuit of Example 5.14 when the output port is short circuited

Apply the node analysis concepts at VA node, $I_1 + I_2 = I_3$ (5.67)

The equation (5.67) is written as, $\frac{V_1 - V_A}{10} + \frac{0 - V_A}{20} = \frac{V_A}{30}$

Hence, VA = 0.546V_1.

Whereas the current flowing in 10Ω resistor is $I_1 = \frac{V_1 - V_A}{10} = \frac{V_1 - 0.546V_1}{10} = 0.0454V_1$

With this,

The current flowing in 20Ω resistor is $I_2 = \frac{0 - V_A}{20} = \frac{-0.546V_1}{20} = -0.0273V_1$

With this, $Y_{11} = \frac{I_1}{V_1} = 0.0454mho$

Step2: Setting the voltage at the input port as zero or the input port is short circuited.

With this concept, the circuit is modified and presented in Fig.5.50. In order to find the Y parameters, the circuit is solved with node analysis with the node potential V_B, the concern circuit is drawn in Fig.5.50.

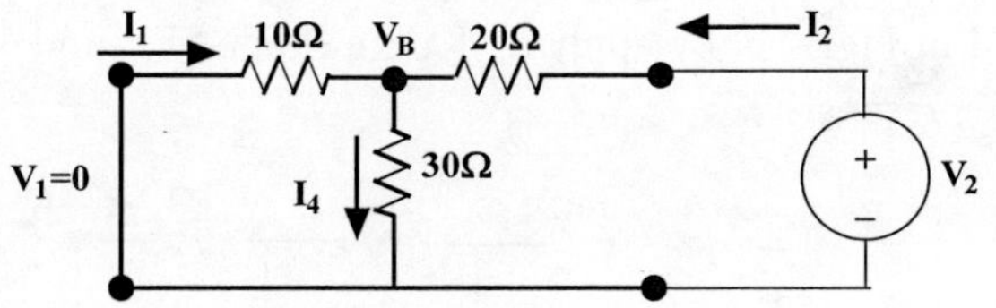

Fig. 5.50: Circuit of Example 5.14 when the input port is short circuited

Apply the node analysis concepts at V_B node, $I_1 + I_2 = I_4$ (5.68)

The equation (5.68) is written as, $\frac{0 - V_B}{10} + \frac{V_2 - V_B}{20} = \frac{V_B}{30}$

Hence, $V_B = 0.272\ V_2$.

Whereas the current flowing in 20Ω resistor is $I_2 = \frac{V_2 - V_B}{20} = \frac{V_2 - 0.272V_2}{20} = 0.0364V_2$

With this, $Y_{22} = \frac{I_2}{V_2} = 0.0364mho$

The current flowing in 10Ω resistor is $I_1 = \frac{0 - V_B}{10} = \frac{-0.272V_2}{10} = -0.0272V_2$

With this, $Y_{21} = \frac{I_2}{V_1} = -0.0272mho$

Y parameters of the given circuit is $\begin{bmatrix} Y_{11} & Y_{12} \\ Y_{21} & Y_{22} \end{bmatrix} = \begin{bmatrix} 0.0454 & -0.0272 \\ -0.0272 & 0.0364 \end{bmatrix}$

(Hint: Find the inverse of the Y parameter matrix, Z parameter matrix will be obtained.)

Example 5.15:

Determine the Y parameters of the circuit shown in Fig.5.32.

Solutions:

The circuit shown in Fig.5.60 is represented as a two-port network model as shown in Fig.5.51. Solve this problem by short circuiting the input and output ports sequentially.

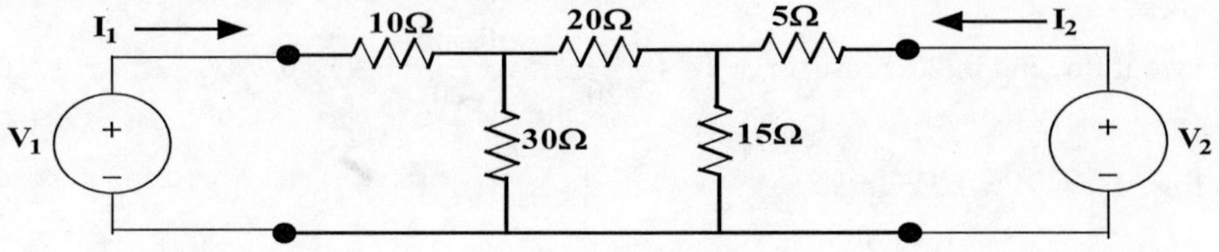

Fig. 5.51: Circuit of Example 5.15 rearranged as two-port network pattern

Step1: Setting the voltage at the output port as zero or the output port is short circuited.

With this concept, the circuit is modified and presented in Fig.5.52. In order to find the admittance parameters, the circuit is solved by the node analysis. For analysis, the branch current and node potentials are identified and marked in Fig.5.53. By applying KCL at nodes V_A and V_B, the linear equations are formed in (5.69) and (5.70) respectively.

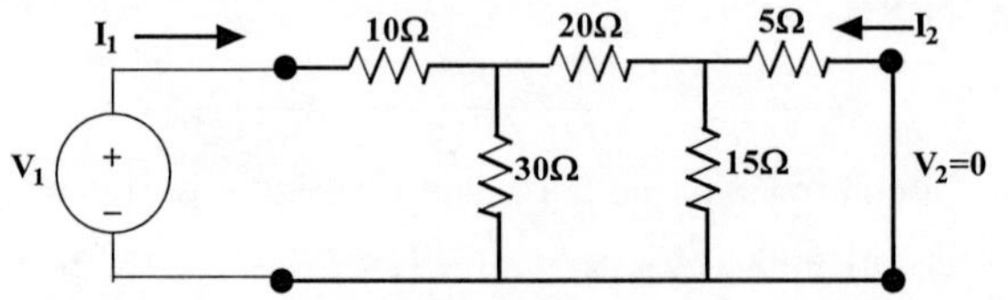

Fig. 5.52: Circuit of Example 5.15 when the output port is short circuited

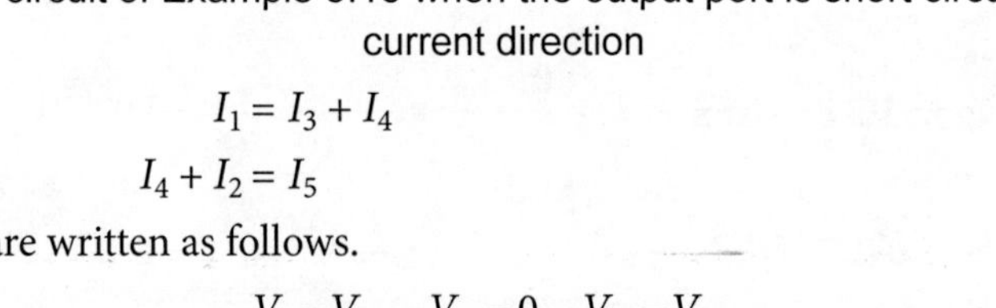

Fig. 5.53: Modified circuit of Example 5.15 when the output port is short circuited with the branch current direction

$$I_1 = I_3 + I_4 \tag{5.69}$$

$$I_4 + I_2 = I_5 \tag{5.70}$$

These equations are written as follows.

$$\frac{V_1 - V_A}{10} = \frac{V_A - 0}{30} + \frac{V_A - V_B}{20} \tag{5.71}$$

$$\frac{V_A - V_B}{20} + \frac{0 - V_B}{5} = \frac{V_B}{15} \tag{5.72}$$

$$\frac{V_A}{20} = (\frac{1}{20} + \frac{1}{15} + \frac{1}{5})V_B$$

$$V_A = 6.33V_B \tag{5.73}$$

By substituting (5.73) in equation (5.71), $\frac{V_1}{10} = \left(\frac{1}{30} + \frac{1}{20} + \frac{1}{10}\right)V_A - \frac{V_B}{20}$

$$\frac{V_1}{10} = \left(\frac{1}{30} + \frac{1}{20} + \frac{1}{10}\right)(6.33V_B) - \frac{V_B}{20}$$

$$V_1 = 11.105\ V_B$$

By the Fig. (5.82), the current and are written as follows.

$$I_1 = \frac{V_1 - V_A}{10}$$

$$I_2 = \frac{0 - V_B}{5}$$

Hence, $I_2 = \frac{0 - V_B}{5} = \frac{0 - (\frac{1}{11.105})V_1}{5} = -0.018V_1$

$$Y_{21} = \frac{I_2}{V_1} = -0.018\upsilon$$

$$I_1 = \frac{V_1 - V_A}{10} = \frac{V_1 - (6.33V_B)}{10} = \frac{V_1 - 6.33\left(\frac{1}{11.105}\right)V_1}{10} = 0.043V_1$$

$$Y_{11} = \frac{I_1}{V_1} = 0.043\upsilon$$

Step2: Setting the voltage at the input port as zero or the input port is short circuited.

With this concept, the circuit is modified and presented in Fig.5.54. In order to find the admittance parameters, the circuit is solved by the node analysis. For analysis, the branch current and node potentials are identified and marked in Fig.5.55. By applying KCL at nodes V_C and V_D, the linear equations are formed in (5.73) and (5.74) respectively.

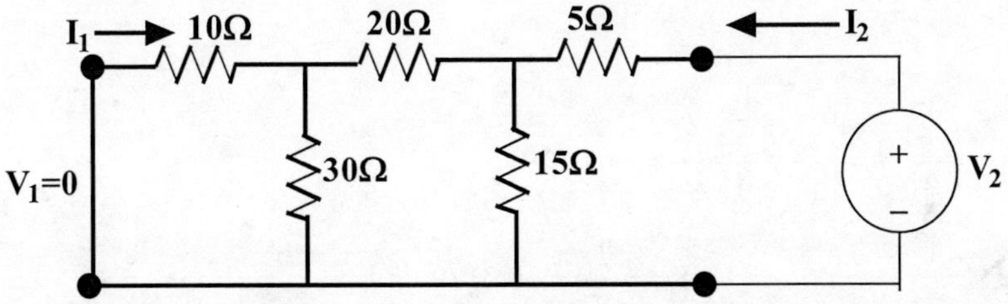

Fig. 5.54: Circuit of Example 5.15 when the input port is short circuited

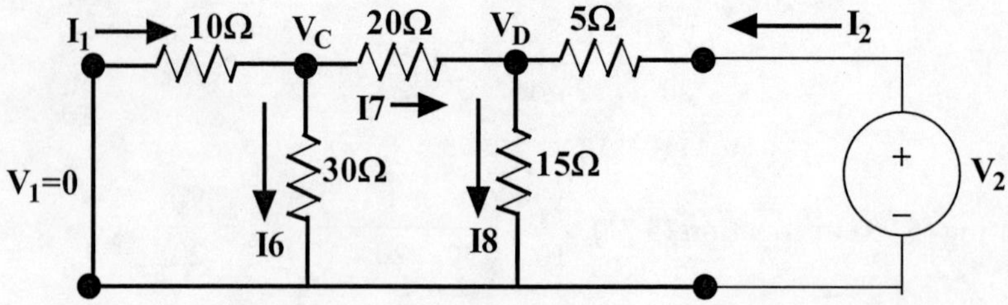

Fig. 5.55: Modified circuit of Example 5.15 when the input port is short circuited with the branch current direction

$$I_1 = I_6 + I_7 \tag{5.74}$$

$$I_7 + I_2 = I_8 \tag{5.74a}$$

These equations are written as follows.

$$\frac{0-V_C}{10} = \frac{V_c - 0}{30} + \frac{V_C - V_D}{20} \tag{5.75}$$

$$\frac{V_C - V_D}{20} + \frac{V_2 - V_D}{5} = \frac{V_D}{15} \tag{5.76}$$

$$\frac{V_D}{20} = \left(\frac{1}{20} + \frac{1}{10} + \frac{1}{30}\right)V_C \tag{5.77}$$

By substituting (5.77) in equation (5.76), $\frac{V_C - V_D}{20} + \frac{V_2 - V_D}{5} = \frac{V_D}{15}$

$$V_2 = 5.56V_C \tag{5.78}$$

$$\frac{V_2}{5} = \left(\frac{1}{15} + \frac{1}{5} + \frac{1}{20}\right)V_D - \frac{1}{20}V_C$$

By the Fig. (5.84), the current I_1 and I_2 are written as follows.

$$\frac{V_2}{5} = \left(\frac{1}{15} + \frac{1}{5} + \frac{1}{20}\right)V_D - \frac{1}{20}V_C$$

$$I_1 = \frac{0 - V_C}{10}$$

$$I_2 = \frac{V_2 - V_D}{5}$$

Hence, $I_1 = \frac{0 - V_C}{10} = \frac{0 - (\frac{1}{5.56})V_2}{10} = -0.0179V_2$

$$Y_{12} = \frac{I_1}{V_2} = -0.0179mho$$

$$I_2 = \frac{V_2 - V_D}{5} = \frac{V_2 - (3.67V_C)}{5} = \frac{V_2 - 3.67(\frac{1}{5.56})V_2}{5} = \frac{V_2(0.339)}{5} = 0.0678V_2$$

$$Y_{22} = \frac{I_2}{V_2} = 0.0678mho$$

Y parameters of the given circuit is $\begin{bmatrix} Y_{11} & Y_{12} \\ Y_{21} & Y_{22} \end{bmatrix} = \begin{bmatrix} 0.043 & -0.018 \\ -0.018 & 0.678 \end{bmatrix}$

Example 5.16:

Determine the Y parameters of the circuit shown in Fig.5.38. Let us assume that the source frequency is 50Hz.

Solutions:

The circuit shown in Fig.5.38 is represented as two-port network model as shown in Fig.5.39 with all impedances in Ω. $X_C = \left(\frac{1}{c\omega}\right) = \left(\frac{1}{0.001 * 2 * \pi * 50}\right) = 3.18\Omega$. Solve this problem by applying the current entered into the input and output ports as zero.

Step1: Setting the voltage at the output port as zero or the output port is short circuited.

With this concept, the circuit is modified and presented in Fig.5.56. By observing the circuit, the output port short circuited which makes the 10Ω resistor in the circuit absurd or meaningless. Hence the circuit is modified as shown in Fig.5.57. In order to find the Y parameters, it is assumed that the current I_1 is divided into current I_3 and I_4. By applying current division technique, these currents are found as follows.

$$I_4 = I_1 * \frac{5}{5 - j3.18} = I_1 * \frac{5}{5.93\angle -32.45^0} = 0.843\angle 32.45^0 I_1$$

$$I_3 = I_1 * \frac{-j3.18}{5 - j3.18} = I_1 * \frac{3.18\angle -90^0}{5.93\angle -32.45^0} = 0.536\angle -57.55^0 I_1$$

I2 = $-0.843\angle 32.45^0 I_1 = 0.843\angle -147.55^0 I_1$

By observing this circuit, it is understood that the current I_4 is opposing the current I_2. Hence $I_2 = -I_4$. Therefore,

$I_2 = -0.843\angle 32.45^0 I_1 = 0.843\angle -147.55^0 I_1$

Voltage across the 5Ω resistor is V_1, which is equal to I_3 *5. Hence, $V_1 = 2.68\angle -57.55^0 I_1$

$$\frac{I_1}{V_1} = \frac{1}{2.68\angle -57.55^0} = 0.373\angle 57.55^0$$

It is equal to $Y_{11} = \frac{I_1}{V_1} = 0.373\angle 57.55^0 \upsilon$

$$V_1 = 2.68\angle -57.55^0\, I_1 = 2.68\angle -57.55^0 \left(\frac{1}{0.843\angle -147.55^0}\right) I_2 = 3.179\angle 90^0\, I_2$$

$$Y_{21} = \frac{I_2}{V_1} = 0.314\angle -90^0\, \upsilon$$

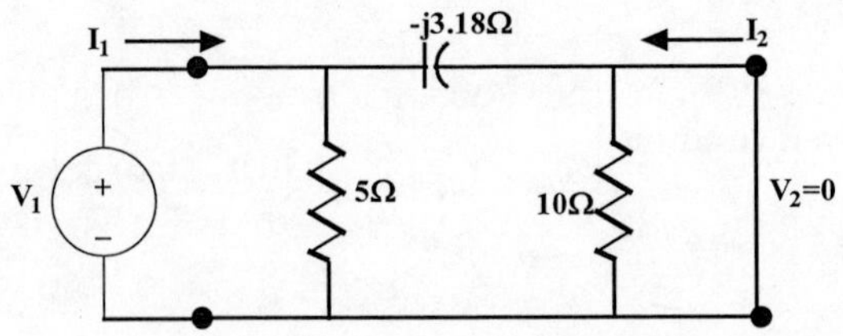

Fig. 5.56: Circuit of Example 5.16 when the output port is short circuited

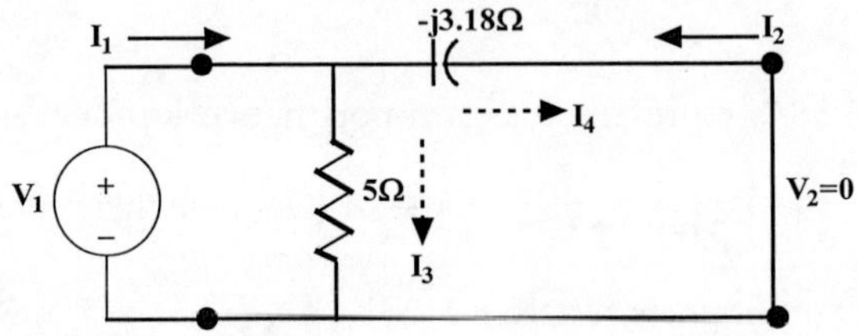

Fig. 5.57: Modified circuit of Example 5.16 when the output port is short circuited

Step2: Setting the voltage at the input port as zero or the input port is short circuited.

With this concept, the circuit is modified and presented in Fig.5.58. By observing the circuit, the input port short circuited which makes the 5Ω resistor in the circuit absurd or meaningless. Hence the circuit is modified as shown in Fig.5.59. In order to find the Y parameters, it is assumed that the current I_2 is divided into current I_5 and I_6. By applying current division technique, these currents are found as follows.

$$I_6 = I_2 * \frac{10}{10 - j3.18} = I_2 * \frac{10}{10.49\angle -17.64^0} = 0.953\angle 17.64^0\, I_2$$

$$I_5 = I_2 * \frac{-j3.18}{10 - j3.18} = I_2 * \frac{3.18\angle -90^0}{10.49\angle -17.64^0} = 0.303\angle -72.36^0\, I_2$$

By observing this circuit, it is understood that the current I_6 is opposing the current I_1. Hence I_1 = - I_6. Therefore,

$I_1 = -0.953\angle 17.64^0\, I_2 = 0.953\angle -162.36^0\, I_2$

Voltage across the 10Ω resistor is V_2, which is equal to I_5*10. Hence, $V_2 = 3.03\angle -72.36^0\, I_2$

$$\frac{I_2}{V_2} = \frac{1}{3.03\angle -72.36^0} = 0.33\angle 72.36^0$$

It is equal to $Y_{22} = \frac{I_2}{V_2} = 0.33\angle 72.36^0\, \upsilon$

$$V_2 = 3.03\angle -72.36^0 I_2 = 3.03\angle -72.36^0\left(\frac{1}{0.953\angle -162.36^0}\right) I_1 = 3.179\angle 90^0 I_1$$

$$Y_{12} = \frac{I_1}{V_2} = 0.314\angle -90^0 \,\upsilon$$

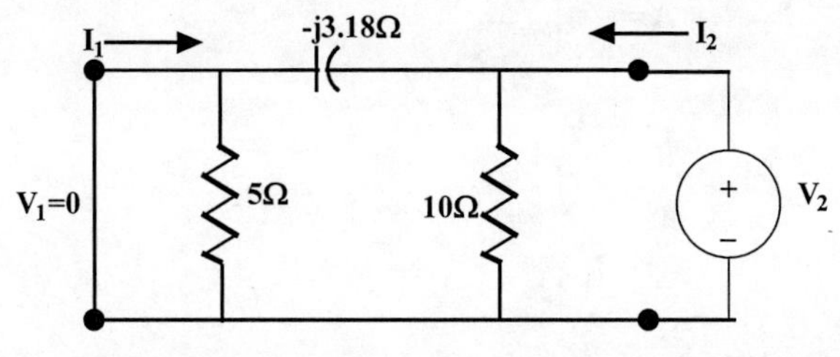

Fig. 5.58: Circuit of Example 5.16 when the output port is short circuited

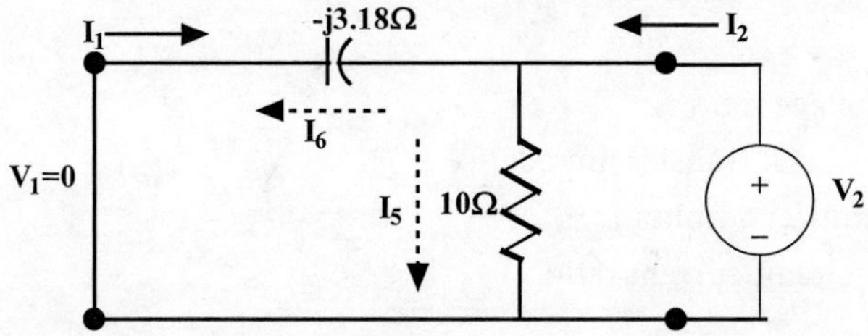

Fig. 5.59: Modified circuit of Example 5.16 when the output port is short circuited

Y parameters of the given circuit is $\begin{bmatrix} Y_{11} & Y_{12} \\ Y_{21} & Y_{22} \end{bmatrix} = \begin{bmatrix} 0.373\angle 57.55^0 & 0.314\angle -90^0 \\ 0.314\angle -90^0 & 0.33\angle 72.36^0 \end{bmatrix}$

By observing the Z and Y parameters, it is noticed that $Z_{12} = Z_{21}$ and $Y_{12} = Y_{21}$. Hence the given circuits are linear.

5.6.3 Transmission Parameters

Both the input port and output port have voltages and currents, any voltages and currents are considered as independent variables. Based on those, many parameters could be obtained. In transmission parameters, voltage and current at the input port are considered as dependent variables. These are expressed in terms of output port voltage and output port current linearly by the equations (5.79) and (5.80). These equations could be arranged in matrix form in equation (5.81). The equivalent two port networks of these parameters are presented as shown in Fig. 5.60. By looking at this figure, the output port current is going away from the circuit. In electric power transmission, the input side is considered as the power generation and the output side is considered as consumers. Hence the current at the output port supplies to the consumers. Hence it is represented as $(-I_2)$, since it is opposite of the earlier assumptions. These equations (5.79) and (5.80) provide information about how an electric network transmits the voltages and currents from source to the consumer end. Hence it is named as transmission parameter. These parameters are very useful in power transmission networks and communication networks. These parameters are also called ABCD parameters or T parameters.

$$V_1 = A_{V2} - B_{I2} \tag{5.79}$$

$$I_1 = C_{V2} - D_{I2} \tag{5.80}$$

$$\begin{bmatrix} V_1 \\ I_1 \end{bmatrix} = \begin{bmatrix} A & B \\ C & D \end{bmatrix} \begin{bmatrix} V_2 \\ -I_2 \end{bmatrix} \tag{5.81}$$

$$A = \frac{V_1}{V_2}\Big|_{I_2=0} \tag{5.82}$$

$$B = -\frac{V_1}{I_2}\Big|_{V_2=0} \tag{5.83}$$

$$C = \frac{I_1}{V_2}\Big|_{I_2=0} \tag{5.84}$$

$$D = -\frac{I_1}{I_2}\Big|_{V_2=0} \tag{5.85}$$

A = Open-circuit voltage ratio
B = Negative short-circuit transfer impedance
C = Open-circuit transfer admittance
D = Negative short-circuit current ratio

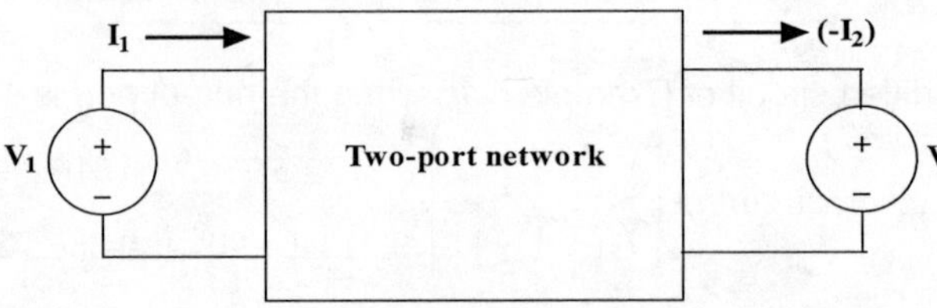

Fig. 5.60: A two-port network model for the transmission parameters

These parameters are calculated like Z and Y parameters based on the conditions described. For the reciprocal circuits, $Z_{12} = Z_{21}$ and $Y_{12} = Y_{21}$. The ABCD parameters of the reciprocal circuits are AD - BC = 1).

Example 5.17:

Determine the ABCD parameters of the circuit shown in Fig.5.28.

Solutions:

The circuit shown in Fig.5.28 is represented as two-port network model as shown in Fig.5.61. Solve this problem by short circuit and open circuit the output port sequentially.

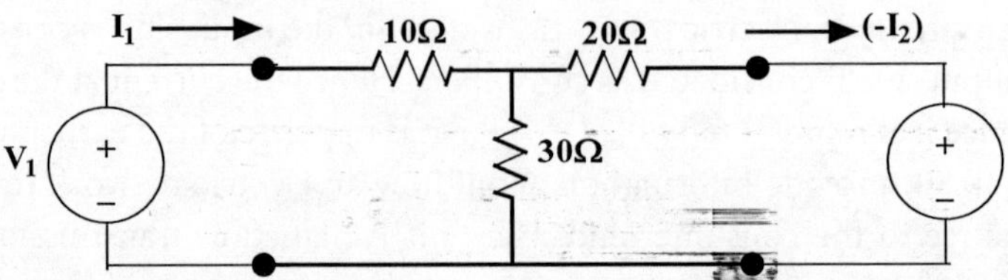

Fig. 5.61: Circuit of Example 5.17 rearranged as two-port network pattern

Step1: Setting the voltage at the output port as zero or the output port is short circuited.

With this concept, the circuit is modified and presented in Fig.5.62. In order to find the two-port parameters, the circuit is solved with node analysis with the node potential V_A, the concern circuit is drawn in Fig.5.63.

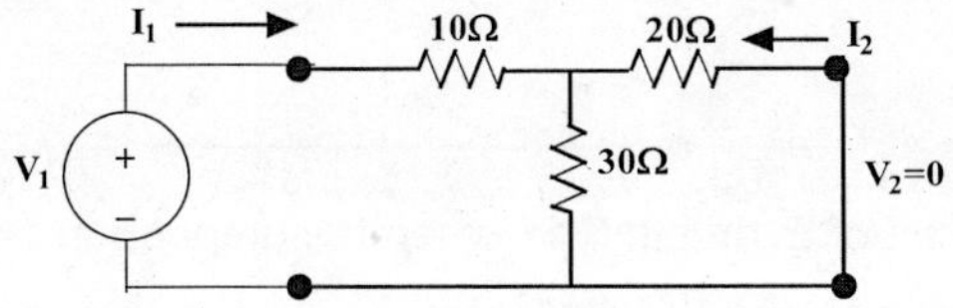

Fig. 5.62: Circuit of Example 5.17 when the output port is short circuited

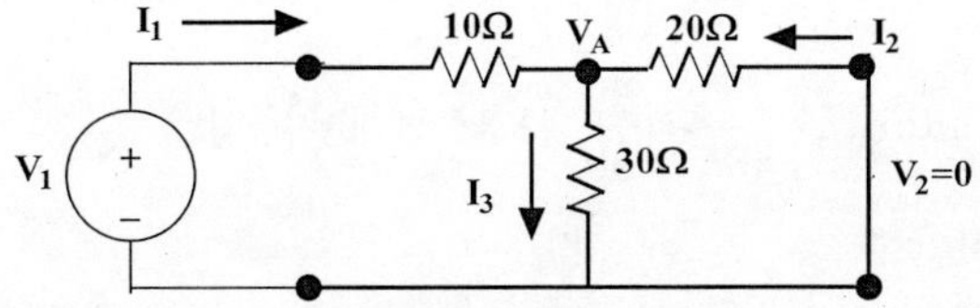

Fig. 5.63: Circuit of Example 5.17 when the output port is short circuited

Apply the node analysis concepts at V_A node, $I_1 + I_2 = I_3$ (5.86)

The equation (5.67) is written as, $\frac{V_1 - V_A}{10} + \frac{0 - V_A}{20} = \frac{V_A}{30}$

$$\frac{V_1}{10} = \frac{V_A}{30} + \frac{V_A}{20} + \frac{V_A}{10}$$

Hence, $V_A = 0.546\ V_1$.

Whereas the current flowing in 10Ω resistor is $I_1 = \frac{V_1 - V_A}{10} = \frac{V_1 - 0.546V_1}{10} = 0.0454V_1$

The current flowing in 20Ω resistor is $B = -\frac{V_1}{I_2} = 36.63\Omega$

$$I_1 = 0.0454V_1 = 0.0454\left(-\frac{1}{0.0273}\right)I_2 = -1.663I_2$$

With this, $B = -\frac{V_1}{I_2} = 36.63\Omega$

$$I_1 = 0.0454V_1 = 0.0454\left(-\frac{1}{0.0273}\right)I_2 = -1.663I_2$$

Hence $D = -\frac{I_1}{I_2} = 1.663$

Step2: Setting the current at the output port as zero or the output port is open circuited.

With this concept, the circuit is modified and presented in Fig.5.64. This section is similar to find the z parameters (Ref. section 5.5.1, Fig.5.30). By observing the circuit, it is understood that there is no current flows in the 20Ω resistor and the input current I_1 flows through 10Ω and 30Ω resistors only.

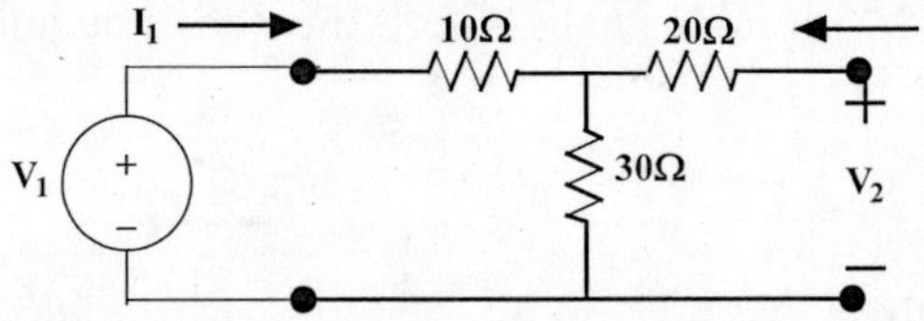

Fig. 5.64: Circuit of Example 5.17 when the output port is open circuited

Apply the loop analysis concepts to the first loop, $V_1 - 10I_1 - 30I_1 = 0;\ V_1 = 40I_1$

The voltage across the output terminals is V_2 which is equal to the voltage across 30Ω. Since the current flows in the 20Ω resistor is zero.

Hence, $V_2 = 30I_1$. From this, $A = \frac{V_1}{V_2} = \frac{40}{30} = 1.33$ is obtained.

$$C = \frac{I_1}{V_2} = \frac{1}{30} = 0.033$$

ABCD parameters of the given circuit is $\begin{bmatrix} A & B \\ C & D \end{bmatrix} = \begin{bmatrix} 1.333 & 36.63 \\ 0.033 & 1.663 \end{bmatrix}$

Verify that the circuit is reciprocal or not:

For reciprocal circuit, AD – BC = 1

AD – BC = (1.333)(1.663) – (36.63)(0.033) = 1

Hence the given circuit is reciprocal.

Example 5.18:

Determine the T parameters pf the circuit shown in Fig.5.32.

Solutions:

Solve this problem by short circuiting the output port and open circuiting the output port sequentially.

Step1: Setting the voltage at the output port as zero or the output port is short circuited.

With this concept, the circuit is modified and presented in Fig.5.52. In order to find the admittance parameters, the circuit is solved by the node analysis. For analysis, the branch current and node potentials are identified and marked in Fig.5.53. By applying KCL at nodes V_A and V_B, the linear equations are formed in (5.69) and (5.70) respectively. The values obtained by that calculation are given as follows.

$$I_2 = -0.018\ V_1$$

$$I_1 = 0.043\ V_1$$

$$B = -\frac{V_1}{I_2} = 55.55\Omega$$

$$D = -\frac{I_1}{I_2} = -\frac{0.043V_1}{-0.018V_1} = 2.388$$

Step2: Setting the voltage at the output port as zero or the output port is short circuited.

This concept is similar to the problem 5.14 (Ref. Section 5.5.1).

Based on equation (5.59), $i_2 = \frac{30}{65} i_1$ which is substituted in equation (5.58).

Hence, $V_1 = 26.15 i_1$.

$i_1 = I_1$ is as per circuit diagram shown in Fig. 5.63. $V_1 = 26.15 I_1$.

$$\frac{V_1}{I_1} = 26.15\Omega$$

The voltage across the output terminals is V_2 which is equal to the voltage across 15Ω. Since the current flows in the 5Ω resistor is zero.

Hence, $V_2 = 15 i_2$.

$$V_2 = 15 i_2 = 15 * \left(\frac{30}{65}\right) i_1 = \frac{90}{13} i_1 = \frac{90}{13} I_1$$

The T parameters or ABCD parameters,

$$A = \frac{V_1}{V_2} = \frac{26.15 I_1}{(90/13) I_1} = 3.777$$

$$C = \frac{I_1}{V_2} = \frac{13}{90} = 0.1444$$

T or ABCD parameters of the given circuit is $\begin{bmatrix} A & B \\ C & D \end{bmatrix} = \begin{bmatrix} 3.777 & 55.55 \\ 0.144 & 2.388 \end{bmatrix}$

Verify that the circuit is reciprocal or not:

For reciprocal circuit, AD – BC=1

$$AD - BC = (3.777)(2.388) - (55.55)(0.144) = 1$$

Hence the given circuit is reciprocal.

Example 5.19:

Determine the ABCD parameters pf the circuit shown in Fig.5.38. Let us assume that the source frequency is 50Hz.

Solutions:

The circuit shown in Fig.5.38 is represented as two-port network model as shown in Fig.5.39 with all impedances in Ω. $X_C = \left(\frac{1}{c\omega}\right) = \left(\frac{1}{0.001 * 2 * \pi * 50}\right) = 3.18\Omega$. Solve this problem by making the output ports as open and short circuited.

Step1: Setting the voltage at the output port as zero or the output port is short circuited.

With this concept, the circuit is modified and presented in Fig.5.57. By observing the circuit, the output port short circuited which makes the 10Ω resistor in the circuit absurd or meaningless. Hence the circuit is modified as shown in Fig.5.57. In order to find the ABCD parameters, it is assumed that the current I_1 is divided into current I_3 and I_4. By applying current division technique, these currents are found as follows.

$$I_4 = I_1 * \frac{5}{5 - j3.18} = I_1 * \frac{5}{5.93\angle -32.45^0} = 0.843\angle 32.45^0 \, I_1$$

$$I_3 = I_1 * \frac{-j3.18}{5 - j3.18} = I_1 * \frac{3.18\angle -90^0}{5.93\angle -32.45^0} = 0.536\angle -57.55^0 \, I_1$$

By observing this circuit, it is understood that the current I_4 is opposing the current I_2. Hence $I_2 = -I_4$. Therefore,

$$I_2 = -0.843\angle 32.45^0 \, I_1 = \angle 0.843 - 147.55^0 \, I_1$$

Voltage across the 5Ω resistor is V_1, which is equal to $I_3 * 5$. Hence, $V_1 = 2.68\angle -57.55^0 \, I_1$

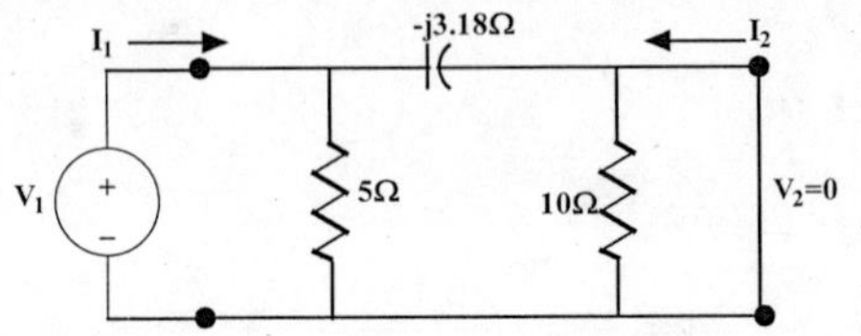

Fig. 5.56: Circuit of Example 5.16 when the output port is short circuited (Repeated)

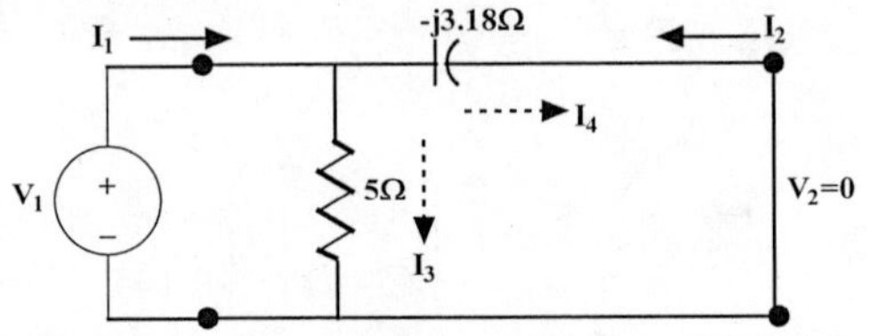

Fig. 5.57: Modified circuit of Example 5.16 when the output port is short circuited (Repeated)

Step2: Setting the current at the output port as zero or the output port is open circuited.

With this concept, the circuit is modified and presented in Fig.5.40. By observing the circuit, the input current I_1 divides to 5Ω and (–j3.18Ω)/10Ω. In order to realize this, the circuit is redrawn as shown in Fig.5.41. Now it is understood that the current flowing in (–j3.18) Ω and 10Ω at the output side remain the same. In order to find the current through these branches are found by the current division technique.

$$I_{3.18\Omega \ or 10\Omega} = I_1 * \left(\frac{5}{5 - j3.18 + 10}\right) = I_1 * \left(\frac{5}{15 - j3.18}\right) = \left(\frac{5}{15.33\angle -11.96^0}\right) I_1 = 0.326\angle 11.96^0 \, I_1$$

$$I_{5\Omega} = I_1 * \left(\frac{-j3.18 + 10}{5 - j3.18 + 10}\right) = I_1 * \left(\frac{10 - j3.18}{15 - j3.18}\right) = \left(\frac{10.49\angle -17.64^0}{15.33\angle -11.96^0}\right) I_1 = 0.684\angle -5.68^0 \, I_1$$

Voltage across the 10Ω resistor is V_2, which is equal to $I_{10\Omega} * 10$. Hence, $V_2 = 3.26\angle 11.96^0 \, I_1$

$$\frac{V_2}{I_1} = 3.26\angle 11.96^0 \, \Omega$$

$$C = \frac{I_1}{V_2} = \frac{1}{3.26\angle 11.96^0} = 0.306\angle -11.96^0\,\upsilon$$

Voltage across the 5Ω resistor is V_1, which is equal to $I_{5\Omega}$*5. Hence, $V_1 = 3.42\angle -5.68^0\, I_1$

$$A = \frac{V_1}{V_2} = \frac{3.42\angle -5.68^0\, I_1}{3.26\angle 11.96^0\, I_1} = 1.049\angle -17.64^0$$

ABCD or T parameters of the given circuit is $\begin{bmatrix} A & B \\ C & D \end{bmatrix} = \begin{bmatrix} 1.049\angle -17.64^0 & 3.179\angle 270^0 \\ 0.306\angle -11.96^0 & 1.186\angle 327.55^0 \end{bmatrix}$

Verify that the circuit is reciprocal or not:

For reciprocal circuit, AD – BC = 1

AD – BC = $(1.049\angle -17.64^0)(1.186\angle 327.55^0) - (3.179\angle 270^0)(0.306\angle -11.96^0)$

$= (1.244\angle 309.91^0) - (0.971\angle 258.04^0)$

$= 0.798 - j0.952 + 0.201 + j0.949 = 1$

Hence the given circuit is reciprocal.

5.6.4 Hybrid Parameters

Some circuits do not have either Z or Y parameters. In transmission parameters, voltage at the input port and current at the output port are considered as dependent variables. Hence, the input voltage and output port currents are linearly related with the input port currents and output port voltages sequentially by the equations (5.87) and (5.88). These equations could be arranged in matrix form in equation (5.89).

$$V_1 = h_{11}\, I_1 + h_{12}\, V_2 \tag{5.87}$$

$$I_2 = h_{21}\, I_1 + h_{22}\, V_2 \tag{5.88}$$

$$\begin{bmatrix} V_1 \\ I_2 \end{bmatrix} = \begin{bmatrix} h_{11} & h_{12} \\ h_{21} & h_{22} \end{bmatrix} \begin{bmatrix} I_1 \\ V_2 \end{bmatrix} \tag{5.89}$$

$$h_{11} = \frac{V_1}{I_1}\Big|_{V_2=0} \tag{5.90}$$

$$h_{12} = \frac{V_1}{V_2}\Big|_{I_1=0} \tag{5.91}$$

$$h_{21} = \frac{I_2}{I_1}\Big|_{V_2=0} \tag{5.92}$$

$$h_{22} = \frac{I_2}{V_2}\Big|_{I_1=0} \tag{5.93}$$

By looking the equations (5.90) to (5.93), h_{11}, h_{12}, h_{21} and h_{22} are the impedance, voltage gain, current gain and the admittance respectively. Hence these parameters are called hybrid parameters.

h_{11} = Short-circuit input impedance

h_{12} = Open-circuit reverse voltage gain (Since it is the ratio of input voltage to output voltage)

$h21$ = Short-circuit current gain

h_{22} = Open-circuit output admittance

These parameters are calculated as Z and Y parameters based on the conditions described. For the reciprocal circuits, $Z_{12} = Z_{21}$ and $Y_{12} = Y_{21}$. The h parameters of the reciprocal circuits are $h_{12} = -h_{21}$). The equivalent circuit of the hybrid parameter circuit is represented in Fig. 5.65.

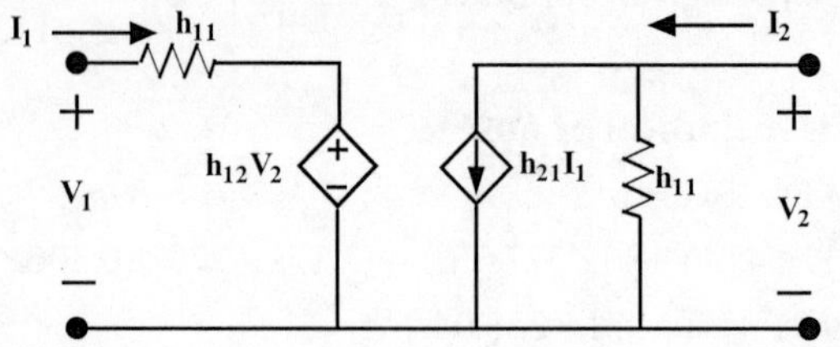

Fig. 5.65: Equivalent circuit of the h parameter of a two-port network

Example 5.20:

Determine the hybrid parameters of the circuit shown in Fig.5.28.

Solutions:

The circuit shown in Fig.5.28 is represented as a two-port network model as shown in Fig.5. 29 (Ref. Section 5.5.1, Example 5.11). Solve this problem by short circuit the output port and open circuit the input port sequentially to obtain the h parameters.

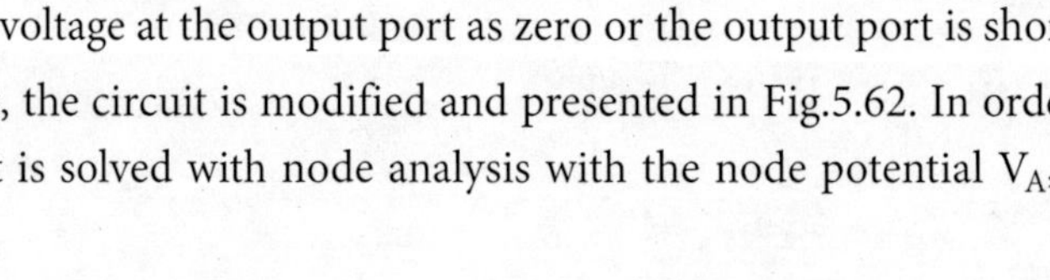

Fig. 5.29: Circuit of Example 5.11 rearranged as two-port network pattern (repeated)

Step1: Setting the voltage at the output port as zero or the output port is short circuited.

With this concept, the circuit is modified and presented in Fig.5.62. In order to find the two-port parameters, the circuit is solved with node analysis with the node potential V_A, the concern circuit is drawn in Fig.5.63.

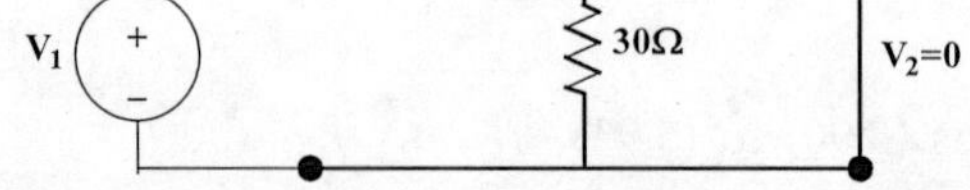

Fig. 5.62: Circuit of Example 5.17 when the output port is short circuited (repeated)

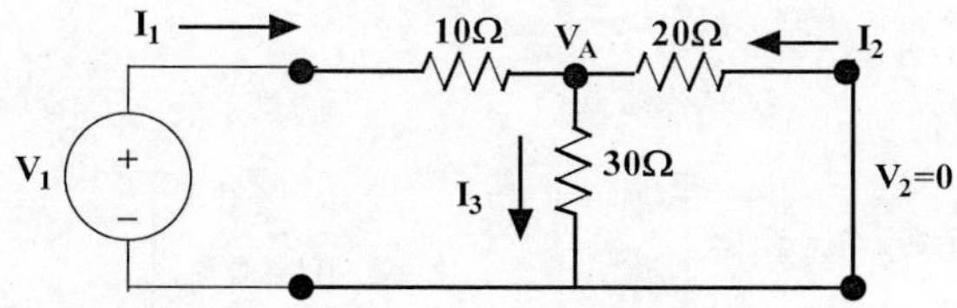

Fig. 5.63: Circuit of Example 5.17 when the output port is short circuited (repeated)

Apply the node analysis concepts at VA node, $I_1 + I_2 = I_3$ (5.94)

The equation (5.94) is written as, $\frac{V_1 - V_A}{10} + \frac{0 - V_A}{20} = \frac{V_A}{30}$

$$\frac{V_1}{10} = \frac{V_A}{30} + \frac{V_A}{20} + \frac{V_A}{10}$$

Hence, $V_A = 0.546\ V_1$.

Whereas the current flowing in 10Ω resistor is $I_1 = \frac{V_1 - V_A}{10} = \frac{V_1 - 0.546V_1}{10} = 0.0454V_1$

The current flowing in 20Ω resistor is $I_2 = \frac{0 - V_A}{20} = \frac{-0.546V_1}{20} = -0.0273V_1$

With this, $h_{11} = \frac{V_1}{I_1} = \frac{1}{0.0454} = 22.06\Omega$

$$I_1 = 0.0454V_1 = 0.0454\left(-\frac{1}{0.0273}\right)I_2 = -1.663I_2$$

Hence $h_{21} = \frac{I_2}{I_1} = -\left(\frac{1}{1.663}\right) = -0.601$

Step2: Setting the current at the input port as zero or the input port is open circuited.

With this concept, the circuit is modified and presented in Fig.5.31. By observing the circuit, it is understood that there is no current flows in the 10Ω resistor and the output current I_2 flows through 20Ω and 30Ω resistors only.

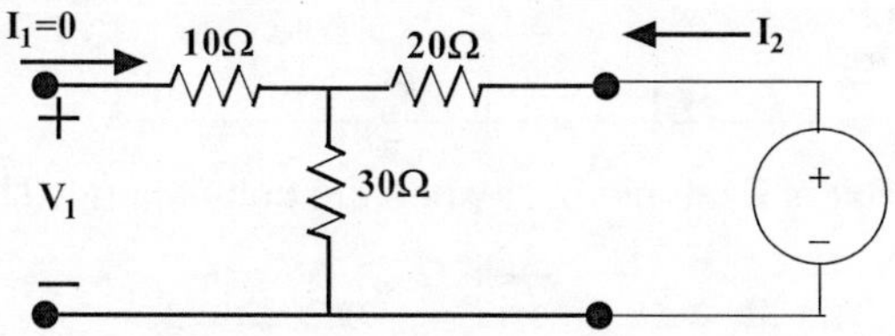

Fig. 5.31: Circuit of Example 5.11 when the input port is open circuited (repeated)

Apply the loop analysis concepts to the second loop, $V_2 - 20I_2 - 30I_2 = 0$; $V_2 = 50I_2$

The voltage across the input terminals is V_1 which is equal to the voltage across 30Ω. Since the current flows in the 10Ω resistor is zero. Hence, $V_1 = 30I_2$.

From this, $h_{12} = \frac{V_1}{V_2} = \frac{30I_2}{50I_2} = 0.6$ is obtained.

where, $h_{22} = \frac{I_2}{V_2} = \frac{1}{50} = 0.02\text{v}$

h parameters of the given circuit is $\begin{bmatrix} h_{11} & h_{12} \\ h_{21} & h_{22} \end{bmatrix} = \begin{bmatrix} 22.06 & 0.6 \\ -0.601 & 0.02 \end{bmatrix}$

Verify that the circuit is reciprocal or not:

For reciprocal circuit, $h_{12} = -h_{21}$

Hence the given circuit is reciprocal.

Example 5.21:

Determine the hybrid parameters of the circuit shown in Fig.5.66.

Solutions:

Solve this problem by short circuit the output port and open circuit the input port sequentially to obtain the *h* parameters.

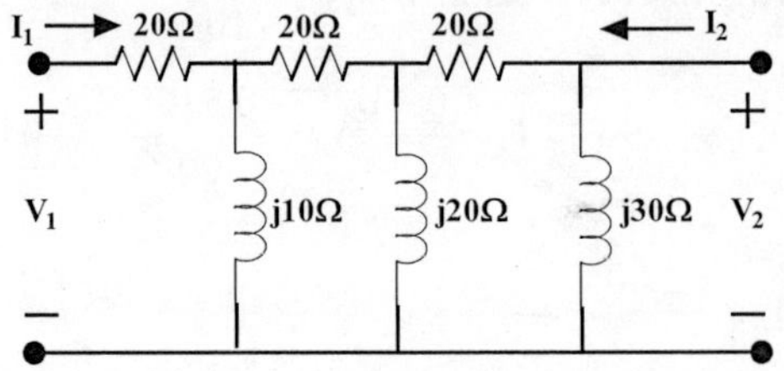

Fig. 5.66: Circuit of Example 5.21 to find the h parameters

Step1: Setting the voltage at the output port as zero or the output port is short circuited.

With this concept, the circuit is modified and presented in Fig.5.67. By looking this circuit, it is understood that the impedance (j30Ω) is negligible, since it is across the short-circuited terminals. The modified circuit is drawn in Fig.5.68.

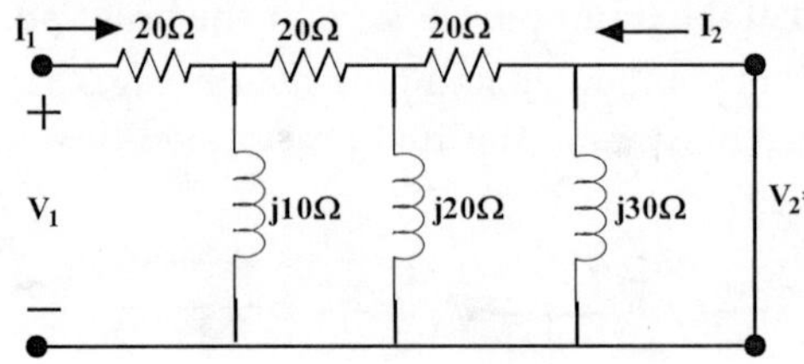

Fig. 5.67: Circuit of Example 5.21 when the output port is short circuited

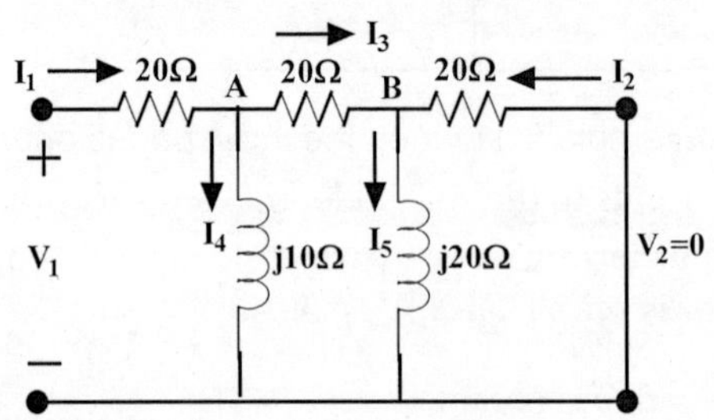

Fig. 5.68: Modified Circuit of Example 5.21 when the output port is short circuited

Apply the node analysis concepts at node A, $I_1 = I_4 + I_3$ (5.95)

The equation (5.95) is written as, $\frac{V_1 - V_A}{20} = \frac{V_A}{j10} + \frac{V_A - V_B}{20}$

$$\frac{V_1}{20} = \frac{V_A}{20} + \frac{V_A}{j10} + \frac{V_A}{20} - \frac{V_B}{20} \quad (5.96)$$

Apply the node analysis concepts at node B, $I_3 + I_2 = I_5$ (5.97)

The equation (5.97) is written as,

$$\frac{V_A - V_B}{20} + \frac{0 - V_B}{20} = \frac{V_B}{j20}$$

$$\frac{V_A}{20} = \frac{V_B}{j20} + \frac{V_B}{20} + \frac{V_B}{20} \quad (5.98)$$

Hence, $V_A = (2 - j1)V_B$ in (5.67a)

$$V_1 = V_B(1 - j6)$$

$$V_A = V_1 \frac{(2 - j1)}{(1 - j6)}$$

$$V_1 - V_A = \frac{-(1 + j5)}{(1 - j6)} V_1$$

Whereas the current flowing in 20Ω resistor is $I_1 = \frac{V_1 - V_A}{20} = \frac{-(1 + j5)}{20 * (1 - j6)} V_1$

With this, $h_{11} = \frac{V_1}{I_1} = (22.28 + j8.465)\Omega$

The current flowing in (output side) 20Ω resistor is $I_2 = \frac{0 - V_B}{20} = \frac{-V_1}{20 * (1 - j6)} = \frac{-I_1(22.28 + j8.465)}{20 * (1 - j6)}$

Hence $h_{21} = \frac{I_2}{I_1} = 0.038 - j0.192$

Step2: Setting the current at the input port as zero or the input port is open circuited.

With this concept, the circuit is modified and presented in Fig.5.69. By observing the circuit, it is understood that there is no current flows in the 20Ω resistor and hence, the circuit is modified as shown in Fig.5.70.

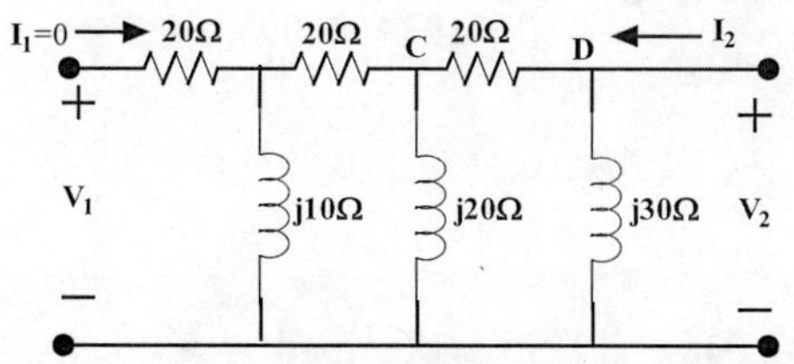

Fig. 5.69: Circuit of Example 5.21 when the input port is open circuited

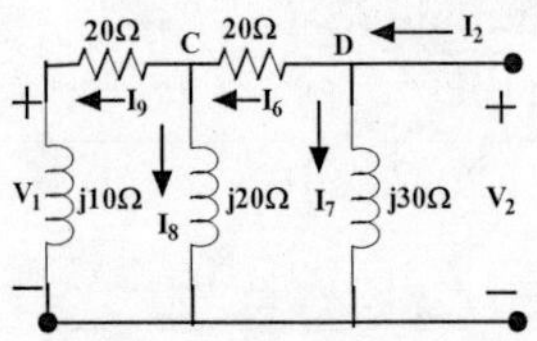

Fig. 5.70: Modified Circuit of Example 5.21 when the input port is open circuited

Apply the node analysis concepts at node C, $I_6 = I_9 + I_8$ (5.99)

The equation (5.99) is written as, $\frac{V_D - V_C}{20} = \frac{V_C}{j20} + \frac{V_C}{20 + j10}$

$$\frac{V_D}{20} = \frac{V_C}{20} + \frac{V_C}{j20} + \frac{V_C}{20 + j10} \quad (5.100)$$

$$V_D = (1.8 - j1.4)V_C$$

$$V_C = (0.3462 + j0.2692)V_D$$

Apply the node analysis concepts at node D, $I_2 = I_6 + I_7$ (5.101)

The equation (5.101) is written as, $I_2 = \frac{V_D - V_C}{20} + \frac{V_D}{j30}$

$$I_2 = \frac{-V_C}{20} + \frac{V_D}{j30} + \frac{V_D}{20} \quad (5.102)$$

$$I_2 = -(0.05)V_C + V_D(0.05 - j0.033)$$

$$I_2 = -(0.3462 + j0.2692)(0.05)V_D + V_D(0.05 - j0.033)$$

$$I_2 = (0.0327 - j0.0465)V_D$$

From Fig.5.98, it is understood that $V_D = V_2$

$$I_2 = (0.0327 - j0.0465)V_2$$

where, $h_{22} = \frac{I_2}{V_2} = (0.0327 - j0.0465)\text{v}$

The voltage across the input terminals is V_1 which is equal to the voltage across j10Ω. Hence, $V_1 = j10I_9$.

Whereas the current flowing in $(20 + j10)\Omega$ resistor is

$$I_9 = \frac{V_C}{20 + j10} = \frac{(0.3462 + j0.2692)V_2}{20 + j10} = (0.0192 + j0.0038)V_2$$

$$V_1 = j10(0.0192 + j0.0038)V_2$$

$$V_1 = (-0.038 + j0.192)V_2$$

From this, $h_{12} = \frac{V_1}{V_2} == (-0.038 + j0.192)$ is obtained.

h parameters of the given circuit is $\begin{bmatrix} h_{11} & h_{12} \\ h_{21} & h_{22} \end{bmatrix} = \begin{bmatrix} (22.28 + j8.465) & (-0.038 + j0.192) \\ (0.038 - j0.192) & (0.0327 - j0.0465) \end{bmatrix}$

Verify that the circuit is reciprocal or not:

For reciprocal circuit, $h_{12} = -h_{21}$

Hence the given circuit is reciprocal.

5.6.5 Inter-Relationship between two Port Network Parameters

All parameters are written related with same input and output terminals of a two-port network. Hence there is a relationship between these parameters.

A. Relation between Z and Y Parameters:

The impedance parameters and admittance parameters are written with respect to their input and output variables in equations (5.103) and (5.104) respectively.

$$\begin{bmatrix} V_1 \\ V_2 \end{bmatrix} = \begin{bmatrix} Z_{11} & Z_{12} \\ Z_{21} & Z_{22} \end{bmatrix} \begin{bmatrix} I_1 \\ I_2 \end{bmatrix} \tag{5.103}$$

$$\begin{bmatrix} I_1 \\ I_2 \end{bmatrix} = \begin{bmatrix} Y_{11} & Y_{12} \\ Y_{21} & Y_{22} \end{bmatrix} \begin{bmatrix} V_1 \\ V_2 \end{bmatrix} \tag{5.104}$$

The equation (5.103) could be modified as the equation (5.105). With the inverse of the matrix, the equation (5.105) is written as equation (5.106).

$$\begin{bmatrix} I_1 \\ I_2 \end{bmatrix} = \begin{bmatrix} Z_{11} & Z_{12} \\ Z_{21} & Z_{22} \end{bmatrix}^{-1} \begin{bmatrix} V_1 \\ V_2 \end{bmatrix} \tag{5.105}$$

$$\begin{bmatrix} Z_{11} & Z_{12} \\ Z_{21} & Z_{22} \end{bmatrix}^{-1} = \frac{1}{Z_{11}Z_{22} - Z_{12}Z_{21}} \begin{bmatrix} Z_{22} & -Z_{12} \\ -Z_{21} & Z_{11} \end{bmatrix}$$

$$\begin{bmatrix} I_1 \\ I_2 \end{bmatrix} = \frac{1}{Z_{11}Z_{22} - Z_{12}Z_{21}} \begin{bmatrix} Z_{22} & -Z_{12} \\ -Z_{21} & Z_{11} \end{bmatrix} \begin{bmatrix} V_1 \\ V_2 \end{bmatrix} \tag{5.106}$$

By comparing the equations (5.104) and (5.106), the admittance parameters could be related with z parameters as presented in equation (5.107).

$$\left. \begin{aligned} Y_{11} &= \frac{Z_{22}}{Z_{11}Z_{22} - Z_{12}Z_{21}} \\ Y_{12} &= \frac{-Z_{12}}{Z_{11}Z_{22} - Z_{12}Z_{21}} \\ Y_{21} &= \frac{-Z_{21}}{Z_{11}Z_{22} - Z_{12}Z_{21}} \\ Y_{22} &= \frac{Z_{11}}{Z_{11}Z_{22} - Z_{12}Z_{21}} \end{aligned} \right\} \tag{5.107}$$

In vice versa, the open circuit parameters or the Z parameters are needed to represent in the terms of Y parameters as presented in equation (5.108).

$$\left.\begin{aligned} Z_{11} &= \frac{Y_{22}}{Y_{11}Y_{22} - Y_{12}Y_{21}} \\ Z_{12} &= \frac{-Y_{12}}{Y_{11}Y_{22} - Y_{12}Y_{21}} \\ Z_{21} &= \frac{-Y_{21}}{Y_{11}Y_{22} - Y_{12}Y_{21}} \\ Z_{22} &= \frac{Y_{11}}{Y_{11}Y_{22} - Y_{12}Y_{21}} \end{aligned}\right\} \tag{5.108}$$

B. Relation between ABCD or T and Hybrid or h Parameters:

The transmission parameters and hybrid parameters are written with respect to their input and output variables in equations (5.109) and (5.110) respectively. These matrices are represented as the equation format in equations (5.111) to (5.114).

$$\begin{bmatrix} V_1 \\ I_1 \end{bmatrix} = \begin{bmatrix} A\ B \\ C\ D \end{bmatrix} \begin{bmatrix} V_2 \\ -I_2 \end{bmatrix} \tag{5.109}$$

$$\begin{bmatrix} V_1 \\ I_2 \end{bmatrix} = \begin{bmatrix} h_{11} & h_{12} \\ h_{21} & h_{22} \end{bmatrix} \begin{bmatrix} I_1 \\ V_2 \end{bmatrix} \tag{5.110}$$

$$V_1 = AV_2 - BI_2 \tag{5.111}$$

$$I_1 = CV_2 - DI_2 \tag{5.112}$$

$$V_1 = h_{11}\, I_1 + h_{12}\, V_2 \tag{5.113}$$

$$I_2 = \text{h}_{21}\, I_1 + \text{h}_{22}\, V_2 \tag{5.114}$$

Using equation (5.113), find the current that is mentioned in equation (5.115) which is substituted in equation (5.111).

$$I_1 = \frac{I_2 - h_{22}V_2}{h_{21}} \tag{5.115}$$

$$V_1 = h_{11}\frac{I_2 - h_{22}V_2}{h_{21}} + h_{12}V_2$$

$$V_1 = \left(h_{12} - \frac{h_{11}h_{22}}{h_{21}}\right)V_2 - \left(-\frac{h_{11}}{h_{21}}\right)I_2 \tag{5.116}$$

By comparing the equations (5.111) and (5.116), the relations between the ABCD parameters in terms of h parameters as below.

$$A = \left(h_{12} - \frac{h_{11}h_{22}}{h_{21}}\right) = \frac{-\Delta_h}{h_{21}};\ B = \left(-\frac{h_{11}}{h_{21}}\right)$$

Where, $\Delta h = h_{11} h_{22} - h_{21} h_{12}$

The equation (5.114) is modified as shown in equation (5.117).

$$I_1 = \left(-\frac{h_{22}}{h_{21}}\right) V_2 - \left(-\frac{1}{h_{21}}\right) I_2 \tag{5.117}$$

By comparing the equations (5.112) and (5.117), the relations between the ABCD parameters in terms of h parameters as below.

$$C = \left(-\frac{h_{22}}{h_{21}}\right); D = \left(-\frac{1}{h_{21}}\right)$$

Hence, ABCD parameters in terms of h parameters are given in equation (5.118).

$$\left.\begin{aligned} A &= \left(-\frac{\Delta_h}{h_{21}}\right) \\ B &= \left(-\frac{h_{11}}{h_{21}}\right) \\ C &= \left(-\frac{h_{22}}{h_{21}}\right) \\ D &= \left(-\frac{1}{h_{21}}\right) \end{aligned}\right\} \tag{5.118}$$

In vice versa, the hybrid parameters or the h parameters are represented in the terms of ABCD or T parameters. By modifying the equation (5.112), the equation (5.119) is obtained which is compared with (5.114).

$$I_2 = \left(-\frac{1}{D}\right) I_1 + \left(\frac{C}{D}\right) V_2 \tag{5.119}$$

$$h_{21} = \left(-\frac{1}{D}\right); h_{22} = \left(\frac{C}{D}\right)$$

The equation (5.119) is substituted in equation (5.111) and compared with (5.113). The relation between the h and ABCD parameters are given in equation (5.120).

$$V_1 = AV_2 - B\left[\left(-\frac{1}{D}\right) I_1 + \left(\frac{C}{D}\right) V_2\right]$$

$$V_1 = \left(\frac{B}{D}\right) I_1 + \left(\frac{AD - BC}{D}\right) V_2]$$

$$h_{11} = \left(\frac{B}{D}\right); h_{12} = \left(\frac{\Delta_T}{D}\right)$$

Where, $\Delta T = AD - BC$

$$\left.\begin{aligned} h_{11} &= \left(\frac{B}{D}\right) \\ h_{12} &= \left(\frac{\Delta_T}{D}\right) \\ h_{21} &= \left(-\frac{1}{D}\right) \\ h_{22} &= \left(\frac{C}{D}\right) \end{aligned}\right\} \tag{5.120}$$

C. Relation between ABCD or T and Z Parameters:

The Z parameters and T parameters' two-port input-output relating equations are written as follows.

$$V_1 = AV_2 - BI_2$$

$$I_1 = CV_2 - DI_2$$

$$V_1 = Z_{11}\, I_1 + Z_{12}\, I_2$$

$$V_2 = Z_{21}\, I_1 + Z_{22}\, I_2$$

V_2 of the Z parameter's equation could be modified as below.

$$I_1 = \frac{V_2 - Z_{22}I_2}{Z_{21}}$$

This equation is compared with I1 of ABCD parameters equation. Hence, $C = \frac{1}{Z_{21}}; D = \frac{Z_{22}}{Z_{21}}$. It is substituted in V_1 of the Z parameter's equation.

$$V_1 = Z_{11}\left(\frac{V_2 - Z_{22}I_2}{Z_{21}}\right) + Z_{12}I_2 = \frac{Z_{11}}{Z_{21}}V_2 - \frac{Z_{11}Z_{22} - Z_{12}Z_{21}}{Z_{21}}I_2$$

The above equation is compared with V_1 of ABCD parameters equation. Hence, $A = \frac{Z_{11}}{Z_{21}}; B = \frac{\Delta_Z}{Z_{21}}$.

Where $\Delta Z = Z_{11}\, Z_{22} - Z_{12}\, Z_{21}$

$$\left.\begin{aligned} A &= \frac{Z_{11}}{Z_{21}} \\ B &= \frac{\Delta_Z}{Z_{21}} \\ C &= \frac{1}{Z_{21}} \\ D &= \frac{Z_{22}}{Z_{21}} \end{aligned}\right\} \tag{5.121}$$

I_1 of the T parameter's equation could be modified as below.

$$V_2 = \frac{I_1 + DI_2}{C}$$

This equation is compared with V_2 of the Z parameters equation. Hence, $Z_{21} = \frac{1}{C}; Z_{22} = \frac{D}{C}$. It is substituted in V_1 of the T parameter's equation.

$$V_1 = A\left(\frac{I_1 + DI_2}{C}\right) - BI_2 = \frac{A}{C}I_1 + \frac{AD\_BC}{C}I_2$$

The above equation is compared with V1 of the Z parameters equation. Hence, $Z_{11} = \frac{A}{C}; Z_{12} = \frac{\Delta_T}{C}$.

Where $\Delta_T = AD - BC$

$$\left.\begin{aligned} A &= \frac{Z_{11}}{Z_{21}} \\ B &= \frac{\Delta_Z}{Z_{21}} \\ C &= \frac{1}{Z_{21}} \\ D &= \frac{Z_{22}}{Z_{21}} \end{aligned}\right\} \quad (5.122)$$

D. Relation between ABCD or T and Y Parameters:

$$V_1 = AV_2 - BI_2$$

$$I_1 = CV_2 - DI_2$$

$$I_1 = Y_{11}\,V_1 + Y_{12}\,V_2$$

$$I_2 = Y_{21}\,V_1 + Y_{22}\,V_2$$

I_2 of the Y parameter's equation could be modified as below.

$$V_1 = \frac{I_2 - Y_{22}V_2}{Y_{21}}$$

This equation is compared with V_1 of ABCD parameters equation. Hence, $A = -\frac{Y_{22}}{Y_{21}}; B = -\frac{1}{Y_{21}}$. It is substituted in I_1 of the Y parameter's equation.

$$I_1 = Y_{11}\left(\frac{I_2 - Y_{22}V_2}{Y_{21}}\right) + Y_{12}V_2 = -\frac{Y_{11}Y_{22} - Y_{12}Y_{21}}{Y_{21}}V_2 - (-\frac{Z_{11}}{Z_{21}})I_2$$

The above equation is compared with the I_1 of the ABCD parameters equation. Hence, $C = -\frac{\Delta_Y}{Y_{21}}; D = -\frac{Y_{11}}{Y_{21}}$. Where $\Delta_Y = Y_{11}\,Y_{22} - Y_{12}\,Y_{21}$

$$\left.\begin{aligned} A &= -\frac{Y_{22}}{Y_{21}} \\ B &= -\frac{1}{Y_{21}} \\ C &= -\frac{\Delta_Y}{Y_{21}} \\ D &= -\frac{Y_{11}}{Y_{21}} \end{aligned}\right\} \tag{5.123}$$

I_2 of the T parameter's equation could be modified as below.

$$I_2 = \frac{AV_2 - V_1}{B}$$

This equation is compared with I_2 of the Y parameters equation. Hence, $Y_{21} = -\frac{1}{B}; Y_{22} = \frac{A}{B}$. It is substituted in I_1 of the T parameter's equation.

$$I_1 = AV_2 - B\left(\frac{AV_2 - V_1}{B}\right) = \frac{D}{B}V_1 + \left(-\frac{AD - BC}{B}\right)V_2$$

The above equation is compared with I_1 of the Y parameters equation. Hence, $Y_{21} = -\frac{1}{B}; Y_{22} = \frac{A}{B}$.

Where $\Delta_T = AD - BC$

$$\left.\begin{aligned} Y_{11} &= \frac{D}{B} \\ Y_{12} &= -\frac{\Delta_T}{B} \\ Y_{21} &= -\frac{1}{B} \\ Y_{22} &= \frac{A}{B} \end{aligned}\right\} \tag{5.124}$$

E. Relation between hybrid and Z Parameters:

The hybrid parameters and Z parameters' two-port input-output relating equations are written as follows.

$$V_1 = h_{11}\,I_1 + h_{12}\,V_2$$

$$I_2 = h_{21} I_1 + Y_{22} V_2$$

$$V_1 = Z_{11} I_1 + Z_{12} I_2$$

$$V_2 = Z_{21} I_1 + Z_{22} I_2$$

By rearranging the equations of Z parameters, the following equations are obtained. Those are compared with h parameter equations and the relations are obtained.

$$I_2 = \left(-\frac{Z_{21}}{Z_{22}}\right) I_1 + \left(\frac{1}{Z_{22}}\right) V_2$$

$$V_1 = \left(\frac{\Delta_Z}{Z_{22}}\right) I_1 + \left(\frac{Z_{12}}{Z_{22}}\right) V_2$$

$$\left.\begin{aligned} h_{11} &= \frac{\Delta_Z}{Z_{22}} \\ h_{12} &= \frac{Z_{12}}{Z_{22}} \\ h_{21} &= \frac{-Z_{21}}{Z_{22}} \\ h_{22} &= \frac{1}{Z_{22}} \end{aligned}\right\} \quad (5.125)$$

By rearranging the equations of h parameters, the following equations are obtained. Those are compared with Z parameter equations and the relations are obtained.

$$V_1 = \left(\frac{h_{11}h_{22} - h_{12}h_{21}}{h_{22}}\right) I_1 + \left(\frac{h_{12}}{h_{22}}\right) I_2$$

$$V_2 = \left(\frac{-h_{21}}{h_{22}}\right) I_1 + \left(\frac{1}{h_{22}}\right) I_2$$

$$\left.\begin{aligned} Z_{11} &= \frac{\Delta_h}{h_{22}} \\ Z_{12} &= \frac{h_{12}}{h_{22}} \\ Z_{21} &= \frac{-h_{21}}{h_{22}} \\ Z_{22} &= \frac{1}{h_{22}} \end{aligned}\right\} \quad (5.126)$$

F. Relation between hybrid and Y Parameters:

The hybrid parameters and Y parameters' two-port input-output relating equations are written as follows.

$$V_1 = h_{11} I_1 + h_{12} V_2$$

$$I_2 = h_{21} I_1 + Y_{22} V_2$$

$$I_1 = Y_{11} V_1 + Y_{12} V_2$$

$$I_2 = Y_{21} V_1 + Y_{22} V_2$$

By rearranging the equations of Y parameters, the following equations are obtained. Those are compared with h parameter equations and the relations are obtained.

$$I_2 = \left(\frac{Y_{21}}{Y_{11}}\right) V_1 + \left(\frac{\Delta_Y}{Y_{11}}\right) V_2$$

$$V_1 = \left(\frac{1}{Y_{11}}\right) I_1 + \left(-\frac{Y_{12}}{Y_{11}}\right) V_2$$

$$\left.\begin{aligned} h_{11} &= \frac{1}{Y_{11}} \\ h_{12} &= \frac{-Y_{12}}{Y_{11}} \\ h_{21} &= \frac{Y_{21}}{Y_{11}} \\ h_{22} &= \frac{\Delta_Y}{Y_{11}} \end{aligned}\right\} \tag{5.127}$$

By rearranging the equations of h parameters, the following equations are obtained. Those are compared with Y parameter equations and the relations are obtained.

$$V_1 = \left(\frac{h_{11}}{h_{21}}\right) I_2 + \left(-\frac{h_{11}h_{22} - h_{12}h_{21}}{h_{21}}\right) V_2$$

OR

$$I_2 = \left(\frac{\Delta_h}{h_{11}}\right) V_2 + \left(\frac{h_{21}}{h_{11}}\right) V_1$$

$$I_1 = \left(\frac{1}{h_{11}}\right) V_1 + \left(-\frac{h_{12}}{h_{11}}\right) V_2$$

$$\left.\begin{aligned} Y_{11} &= \frac{1}{h_{11}} \\ Y_{12} &= -\frac{h_{12}}{h_{11}} \\ Y_{21} &= \frac{h_{21}}{h_{11}} \\ Y_{22} &= \frac{\Delta_h}{h_{11}} \end{aligned}\right\} \quad (5.128)$$

The relations among all parameters needed to be obtained as the same procedure discussed above, these are presented as in Table. 5.1.

Table 5.1: Relations among the two-port network parameters

Parameters	Z	Y	ABCD	h
Z	$\begin{bmatrix} Z_{11} & Z_{12} \\ Z_{21} & Z_{22} \end{bmatrix}$	$\begin{bmatrix} \frac{Y_{22}}{\Delta_Y} & \frac{-Y_{12}}{\Delta_Y} \\ \frac{-Y_{21}}{\Delta_Y} & \frac{Y_{11}}{\Delta_Y} \end{bmatrix}$	$\begin{bmatrix} \frac{A}{C} & \frac{\Delta_T}{C} \\ \frac{1}{C} & \frac{D}{C} \end{bmatrix}$	$\begin{bmatrix} \frac{\Delta_h}{h_{22}} & \frac{h_{12}}{h_{22}} \\ \frac{-h_{21}}{h_{22}} & \frac{1}{h_{22}} \end{bmatrix}$
Y	$\begin{bmatrix} \frac{Z_{22}}{\Delta_Z} & \frac{-Z_{12}}{\Delta_Z} \\ \frac{-Z_{21}}{\Delta_Z} & \frac{Z_{11}}{\Delta_Z} \end{bmatrix}$	$\begin{bmatrix} Y_{11} & Y_{12} \\ Y_{21} & Y_{22} \end{bmatrix}$	$\begin{bmatrix} \frac{D}{B} & -\frac{\Delta_T}{B} \\ -\frac{1}{B} & \frac{A}{B} \end{bmatrix}$	$\begin{bmatrix} \frac{1}{h_{11}} & -\frac{h_{12}}{h_{11}} \\ \frac{h_{21}}{h_{11}} & \frac{\Delta_h}{h_{11}} \end{bmatrix}$
ABCD	$\begin{bmatrix} \frac{Z_{11}}{Z_{21}} & \frac{\Delta_Z}{Z_{21}} \\ \frac{1}{Z_{21}} & \frac{Z_{22}}{Z_{21}} \end{bmatrix}$	$\begin{bmatrix} -\frac{Y_{22}}{Y_{21}} & -\frac{1}{Y_{21}} \\ -\frac{\Delta_Y}{Y_{21}} & -\frac{Y_{11}}{Y_{21}} \end{bmatrix}$	$\begin{bmatrix} A & B \\ C & D \end{bmatrix}$	$\begin{bmatrix} \left(-\frac{\Delta_h}{h_{21}}\right) & \left(-\frac{h_{11}}{h_{21}}\right) \\ \left(-\frac{h_{22}}{h_{21}}\right) & \left(-\frac{1}{h_{21}}\right) \end{bmatrix}$
h	$\begin{bmatrix} \frac{\Delta_Z}{Z_{22}} & \frac{Z_{12}}{Z_{22}} \\ \frac{-Z_{21}}{Z_{22}} & \frac{1}{Z_{22}} \end{bmatrix}$	$\begin{bmatrix} \frac{1}{Y_{11}} & \frac{-Y_{12}}{Y_{11}} \\ \frac{Y_{21}}{Y_{11}} & \frac{\Delta_Y}{Y_{11}} \end{bmatrix}$	$\begin{bmatrix} \left(\frac{B}{D}\right) & \left(\frac{\Delta_T}{D}\right) \\ \left(-\frac{1}{D}\right) & \left(\frac{C}{D}\right) \end{bmatrix}$	$\begin{bmatrix} h_{11} & h_{12} \\ h_{21} & h_{22} \end{bmatrix}$

$\Delta_Z = Z_{11}Z_{22} - Z_{12}Z_{21}$; $\Delta_Y = Y_{11}Y_{22} - Y_{12}Y_{21}$; $\Delta_h = h_{11}h_{22} - h_{12}h_{21}$; $\Delta_T = AD - BC$

5.6.6 Inter Connection of Two Port Networks

A large and complex network will be divided into multiple sub-networks. These sub-networks are joined like series, parallel and cascade connection in order to form the original large network. In order to find the two port parameters of the large network, there is a necessity of finding a method to evaluate if these networks are in series or parallel or cascade connection. Figure 5.71, Fig. 5.72 and Fig. 5.73 are drawn for the two sub-networks connected in series, parallel and cascade connection respectively.

A. Series connection of sub-networks:

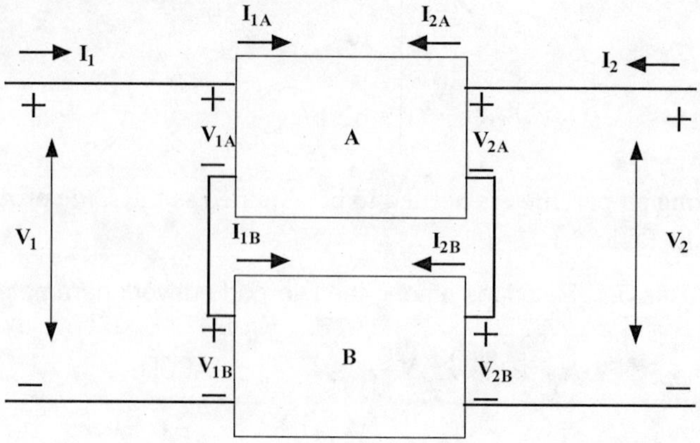

Fig. 5.71: Series interconnection of two two-port networks (A and B)

Fig.5.71 represents two series connected sub-networks. The Z parameter equations for these networks are presented in equations (5.129) to (5.132).

$$V_{1A} = Z_{11A} I_{1A} + Z_{12A} I_{2A} \quad (5.129)$$

$$V_{2A} = Z_{21A} I_{1A} + Z_{22A} I_{2A} \quad (5.130)$$

$$V_{1B} = Z_{11B} I_{1B} + Z_{12}B I_{2B} \quad (5.131)$$

$$V_{2B} = Z_{21B} I_{1B} + Z_{22B} I_{2B} \quad (5.132)$$

Since these sub-networks are in series. the current in these networks are the same. Hence the input currents I_{1A} and I_{1B} are equal and that is equal to I_1. Similarly it is applicable to the output current also. Also, the input voltage (V_{1A}, V_{1B}) and output voltages (V_{2A}, V_{2B}) are added in order to obtain the original network input and output voltages (V_1, V_2) respectively and that are presented in equations (5.133) and (5.134). The original network's two-port Z parameters are written in equations (5.135) and (5.136). By comparing the input and output voltage parameters of the original network with the series connected networks, the Z parameter of the original network is represented in terms of the Z parameters of these sub-networks as in equation (5.137).

$$V_1 = V_{1A} + V_{1B} = (Z_{11A} + Z_{11B})I_1 + (Z_{12A} + Z_{12B})I_2 \quad (5.133)$$

$$V_2 = V_{2A} + V_{2B} = (Z_{21A} + Z_{21B})I_1 + (Z_{22A} + Z_{22B})I_2 \quad (5.134)$$

$$V_1 = Z_{11} I_1 + Z_{12} I_2 \quad (5.135)$$

$$V_2 = Z_{21} I_1 + Z_{22} I_2 \quad (5.136)$$

$$\begin{bmatrix} Z_{11} & Z_{12} \\ Z_{21} & Z_{22} \end{bmatrix} = \begin{bmatrix} Z_{11A} + Z_{11B} & Z_{12A} + Z_{12B} \\ Z_{21A} + Z_{21B} & Z_{22A} + Z_{22B} \end{bmatrix} = \begin{bmatrix} Z_{11A} & Z_{12A} \\ Z_{21A} & Z_{22A} \end{bmatrix} + \begin{bmatrix} Z_{11B} & Z_{12B} \\ Z_{21B} & Z_{22B} \end{bmatrix}$$

$$[Z] = [Z_A] + [Z_B] \quad (5.137)$$

The equation (5.137) presents that the z parameters for the original network are the sum of the z parameters for the individual sub-networks. This equation or the concepts could be extended to the n series connected sub-networks. In order to find other parameters of the two-port network, the Table 5.1 is used.

B. Parallel connection of sub-networks:

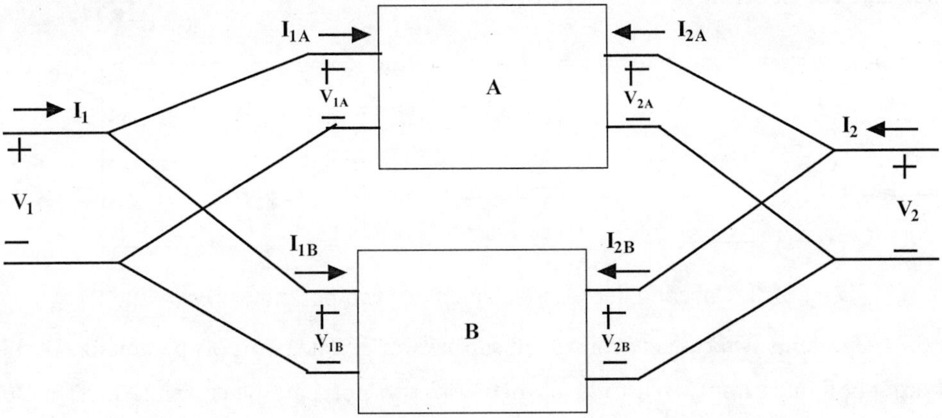

Fig. 5.72: Parallel interconnection of two two-port networks (A and B)

Fig. 5.72 represents two parallel connected sub-networks with two port parameters. The Y parameter equations for these networks are presented in equations (5.138) to (5.141).

$$I_{1A} = Y_{11A}\, V_{1A} + Y_{12A}\, V_{2A} \tag{5.138}$$

$$I_{2A} = Y_{21A}\, V_{1A} + Y_{22A}\, V_{2A} \tag{5.139}$$

$$I_{1B} = Y_{11B}\, V_{1B} + Y_{12B}\, V_{2B} \tag{5.140}$$

$$I_{2B} = Y_{21B}\, V_{1B} + Y_{22B}\, V_{2B} \tag{5.141}$$

Since these sub-networks are in parallel, the voltage in these networks is equal or remains the same. Hence the input voltages V_{1A} and V_{1B} are equal and that is equal to V_1. Similarly it is applicable to the output voltages also. The input currents (I_{1A}, I_{1B}) and output currents (I_{2A}, I_{2B}) are added in order to obtain the original network input and output currents (I_1, I_2) respectively and that are presented in equations (5.142) and (5.143). The original network's two-port Y parameters are written in equations (5.109) and (5.145). By comparing the input and output voltage parameters of the original network with the series connected networks, the Z parameter of the original network is represented in terms of the Z parameters of these sub-networks as in equation (5.146).

$$I_1 = I_{1A} + I_{1B} = (Y_{11A} + Y_{11B})V_1 + (Y_{12A} + Y_{12B})V_2 \tag{5.142}$$

$$I_2 = I_{2A} + I_{2B} = (Y_{21A} + Y_{21B})V_1 + (Y_{22A} + Y_{22B})V_2 \tag{5.143}$$

$$I_1 = Y_{11}\, V_1 + Y_{12}\, V_2 \tag{5.144}$$

$$I_2 = Y_{21}\, V_1 + Y_{22}\, V_2 \tag{5.145}$$

$$\begin{bmatrix} Y_{11} & Y_{12} \\ Y_{21} & Y_{22} \end{bmatrix} = \begin{bmatrix} Y_{11A} + Y_{11B} & Y_{12A} + Y_{12B} \\ Y_{21A} + Y_{21B} & Y_{22A} + Y_{22B} \end{bmatrix} = \begin{bmatrix} Y_{11A} & Y_{12A} \\ Y_{21A} & Y_{22A} \end{bmatrix} + \begin{bmatrix} Y_{11B} & Y_{12B} \\ Y_{21B} & Y_{22B} \end{bmatrix}$$

$$[Y] = [Y_A] + [Y_B] \tag{5.145a}$$

The equation (5.145a) presents that the Y parameters for the original network are the sum of the Y parameters for the individual sub-networks. This equation or the concepts could be extended to the n series connected sub-networks. In order to find other parameters of the two-port network, Table 5.1 is used.

C. Cascaded connection of sub-networks:

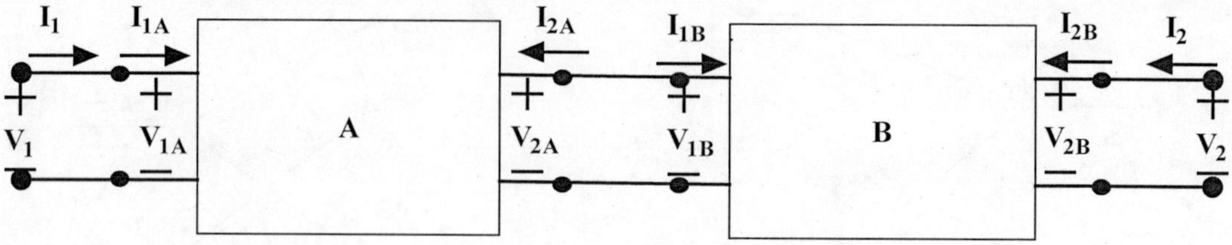

Fig. 5.73: Cascade interconnection of two two-port networks (A and B)

Fig.5.73 represents two cascade connected sub-networks. The two-port parameters of the original network are to be found using two cascade sub-network's two-port parameters. The ABCD or T parameter equations for these networks are presented in equations (5.145) to (5.148). The original network's two-port ABCD parameters are written in equations (5.149) - (5.150).

$$V_{1A} = A_A\, V_{2A} - B_A\, I_{2A} \tag{5.145}$$

$$I_{1A} = C_A\, V_{2A} - D_A\, I_{2A} \tag{5.146}$$

$$V_{1B} = A_B\, V_{2B} - B_B\, I_{2B} \tag{5.147}$$

$$I_{1B} = C_B\, V_{2B} - D_B\, I_{2B} \tag{5.148}$$

$$\left.\begin{aligned} V_1 &= AV_2 - BI_2 \\ I_1 &= CV_2 - DI_2 \end{aligned}\right\} \tag{5.149}$$

$$\left.\begin{bmatrix} V_1 \\ I_1 \end{bmatrix} = \begin{bmatrix} A & B \\ C & D \end{bmatrix}\begin{bmatrix} V_2 \\ -I_2 \end{bmatrix}\right\} \tag{5.150}$$

By observing these sub-networks, the following points are noted down.

$$V_1 = V_{1A};\ V_{2A} = V_{1B};\ V_{2B} = V_2\, I_1 = I_1 A;\ I_{2A} = -I_{1B};\ I_{2B} = I_2\} \tag{5.151}$$

Based on the equation (5.151), the equations (5.145) – (5.148) are modified and presented in equations (5.152) and (5.153). Those equations are presented in matrix form in (5.154) and (5.155) respectively.

$$\left.\begin{aligned} V_1 &= A_A V_{1B} + B_A I_{1B} \\ I_1 &= C_A V_{1B} + D_A I_{1B} \end{aligned}\right\} \tag{5.152}$$

$$\left.\begin{aligned} V_{1B} &= A_B V_2 - B_B I_2 \\ I_{1B} &= C_B V_2 - D_B I_2 \end{aligned}\right\} \tag{5.153}$$

$$\left.\begin{bmatrix} V_1 \\ I_1 \end{bmatrix} = \begin{bmatrix} A_A & B_A \\ C_A & D_A \end{bmatrix} \begin{bmatrix} V_{1B} \\ I_{1B} \end{bmatrix}\right\} \tag{5.154}$$

$$\left.\begin{bmatrix} V_{1B} \\ I_{1B} \end{bmatrix} = \begin{bmatrix} A_B & B_B \\ C_B & D_B \end{bmatrix} \begin{bmatrix} V_2 \\ -I_2 \end{bmatrix}\right\} \tag{5.155}$$

$$\left.\begin{bmatrix} V_1 \\ I_1 \end{bmatrix} = \begin{bmatrix} A_A & B_A \\ C_A & D_A \end{bmatrix} \begin{bmatrix} A_B & B_B \\ C_B & D_B \end{bmatrix} \begin{bmatrix} V_2 \\ -I_2 \end{bmatrix}\right\} \tag{5.156}$$

By comparing the input and output voltage parameters of the original network with the cascade connected networks, the ABCD or T parameter of the original network is presented in equation (5.150) and the simplified ABCD parameters of these sub-networks as in equation (5.157).

$$[T] = [T_A] \cdot [T_B] \tag{5.157}$$

The equation (5.157) presents that the ABCD or T parameters for the original network are the product of the ABCD or T parameters for the individual sub-networks. This equation or the concepts could be extended to the n cascade connected sub-networks. In order to find other parameters of the two-port network, Table 5.1 is used.

Example 5.22:

Find $\dfrac{V_2}{V_S}$ of the circuit shown in Fig.5.74.

Solutions:

The circuit shown in Fig.5.74 is a two-port network which consists of two two-port networks in series. First, both circuits' Z parameters are to be calculated. The circuit A consists of (-j45Ω) and (j10Ω) and 20Ω, that forms the T network and is shown in Fig. 5.75. The circuit B consists of (-j20Ω) and (j40Ω) and 50Ω, this also forms the T network which is shown in Fig. 5.76.

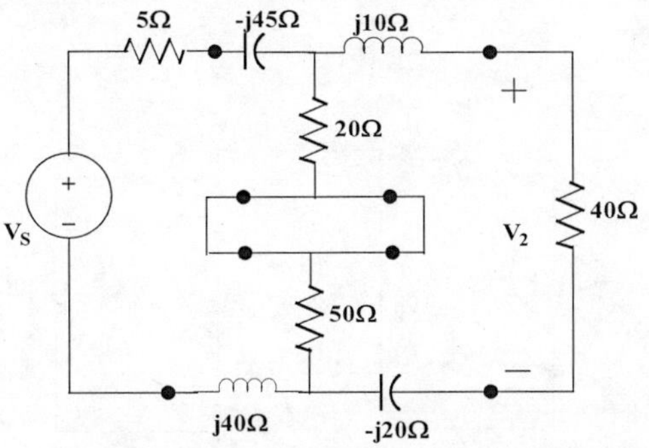

Fig. 5.74: Circuit of Example 5.22

Step1: Finding the Z parameters of Circuit A

The circuit A is shown in Fig.5.75. By applying the concepts of Z parameters, the equations are to be written as follows and the corresponding Z parameters are found.

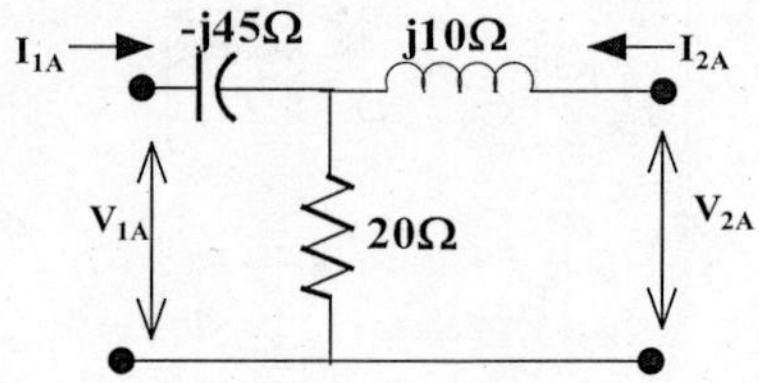

Fig. 5.75: Circuit A of Example 5.22

$$V_{1A} = Z_{11A}\, I_{1A} + Z_{12A}\, I_{2A}$$
$$V_{2A} = Z_{21A}\, I_{1A} + Z_{22A}\, I_{2A}$$

$$Z_A = \begin{bmatrix} Z_{11A} & Z_{12A} \\ Z_{21A} & Z_{22A} \end{bmatrix} = \begin{bmatrix} 20 - j45 & 20 \\ 20 & 20 + j10 \end{bmatrix}$$

Step2: Finding the Z parameters of Circuit B

The circuit B is shown in Fig.5.76. By applying the concepts of Z parameters, the equations are to be written as follows and the corresponding Z parameters are found.

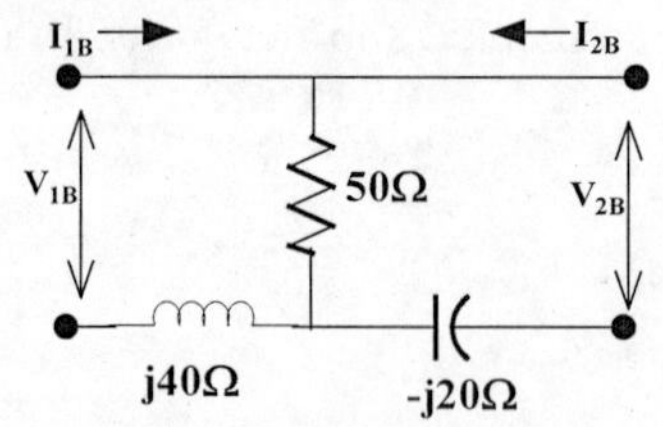

Fig. 5.76: Circuit B of Example 5.22

Step3: Finding the source voltage in terms of the output voltage

The circuit shown in Fig.574 is represented as a two-port network model as shown in Fig.5. 77 with their equivalent Z parameters. Both circuits A and B are in series, Z parameters are added as per the interconnection of networks.

$$Z_A + Z_B = \begin{bmatrix} 70 - j5 & 70 \\ 70 & 70 - j10 \end{bmatrix}$$

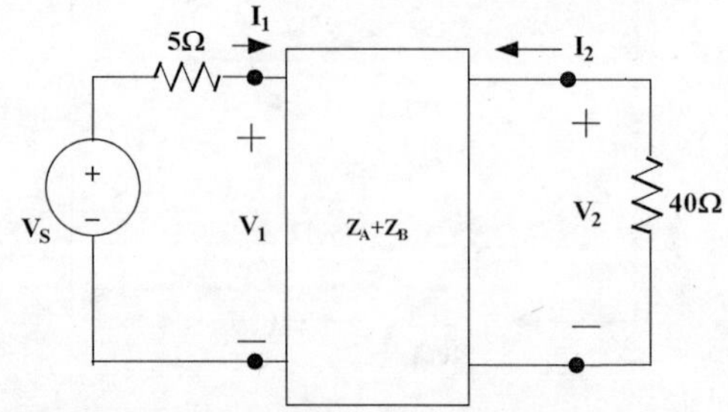

Fig. 5.77: Circuit of Example 5.22 with the equivalent circuit parameters

$$V_1 = Z_{11} I_1 + Z_{12} I_2$$
$$V_2 = Z_{21} I_1 + Z_{22} I_2$$

As per the circuit shown in Fig. 5.77, the input source voltage and the output voltage could be written as

$$V_1 = V_S - 5I_1; \; V_2 = -40I_2$$

Hence, $I_2 = -\frac{V_2}{40}$. Both I_2 and Z_{11}, Z_{12} are substituted in the Z parameter equation of V_1 and the equation (5.158) is obtained.

$$V_1 = (70 - j5)\, I1 + 70I_2$$
$$V_S - 5I_1 = (70 - j5)I_1 + 70(-\frac{V_2}{40}) \quad (5.158)$$

With Z_{21}, Z_{22} are substituted in the Z parameter equation of V_2 and the equation (5.159) is obtained.

$$V_2 = 70I_1 + (70 - j10)(-\frac{V_2}{40}) \quad (5.159)$$

From this, $I_1 = \frac{90.55\angle -6.34^0}{1400} V_2$ obtained, which is substituted in equation (5.158).

$$VS = (1.28 - j0.856)\, V_2$$
$$\frac{V_S}{V_2} = (1.28 - j0.856) = 1.539\angle -33.77^0$$

Example 5.23:

Find the admittance parameters of the circuit shown in Fig.5.78.

Solutions:

The circuit shown in Fig.5.78 is a two-port network which consists of two two-port networks in parallel. First, both circuits' Y parameters are to be calculated. The circuit A consists of (-j5S), (j5S) and 1S, that forms the T network and is shown in Fig. 5.79. The circuit B consists of (-j10S), (2S) and 2S, this forms the ⊠ network which is shown in Fig. 5.82.

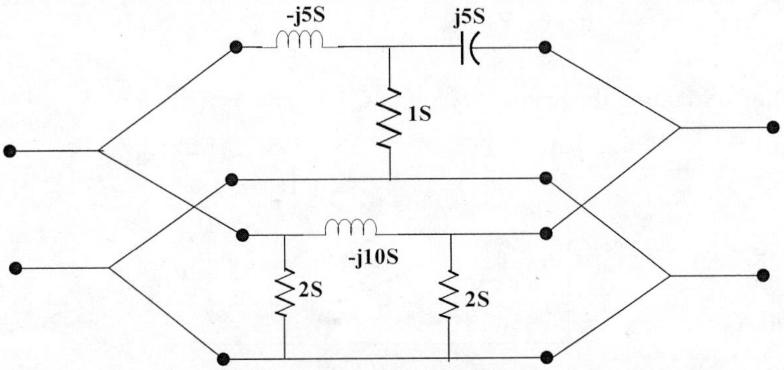

Fig. 5.78: Circuit of Example 5.23

Step1: Finding the Y parameters of Circuit A

The circuit A is shown in Fig.5.79. By applying the concepts of Z parameters, the equations are to be written as follows and the corresponding Z parameters are found.

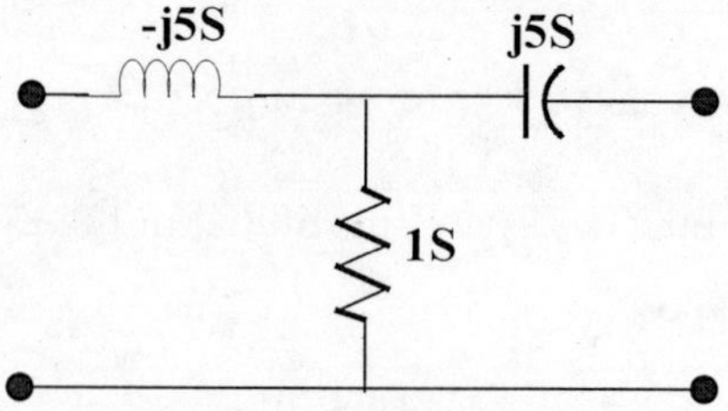

Fig. 5.79: Circuit A of Example 5.23

$$I_{1A} = Y_{11A}\, V_{1A} + Y_{12A}\, V_{2A}$$

$$I_{2A} = Y_{21A}\, V_{1A} + Y_{22A}\, V_{2A}$$

By short-circuiting the secondary terminals, the circuit 5.108 looks like 5.109. From which Y_{11A} and Y_{21A} are found.

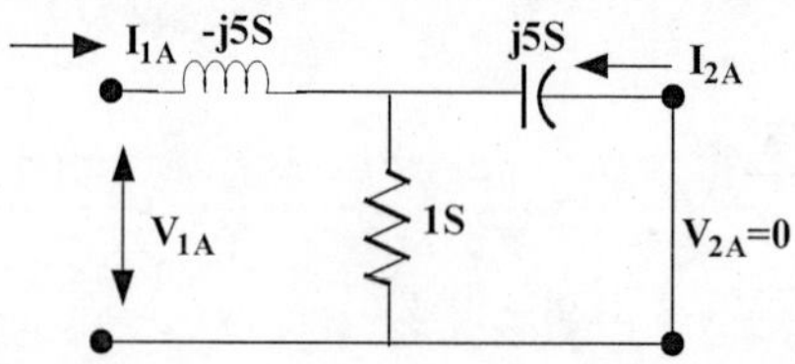

Fig. 5.80: Circuit A of Example 5.23 with input-output parameters and the output port short-circuited

The equivalent parameters of the circuit are to be calculated. (1S) and (j5S) are in parallel. Hence (1+*j*5)S is obtained. (–*j*5S) and (1+*j*5)S are in parallel. The result value is (25-j5)S.

Hence $Y_{11A} = (25 - j5)$S.

$$I_{2A} = \frac{-I_{1A} * \left(\frac{1}{1}\right)}{\left(\frac{1}{1}\right) + \left(\frac{1}{j5}\right)}$$

$$I_{1A} = -(1 - j0.2) I_{2A}$$

As $Y_{11A} = \frac{I_{1A}}{V_{1A}}; Y_{21A} = \frac{I_{2A}}{V_{1A}}$, hence $Y_{21A} = -25\text{S}$

Similarly, short-circuiting the primary terminals, Y_{12A} and Y_{22A} are found.

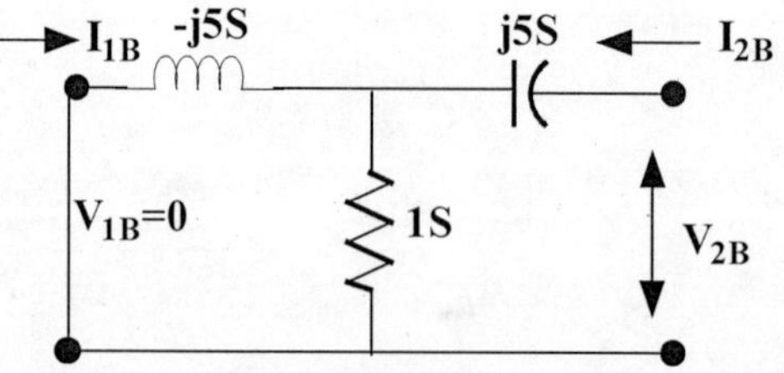

Fig. 5.81: Circuit A of Example 5.23 with input-output parameters and the input port short-circuited

By performing the calculation similar to the output short-circuited, the Y_{12A} =–25S and Y_22A = (25–j5)S are calculated.

$$Y_A = \begin{bmatrix} Y_{11A} & Y_{12A} \\ Y_{21A} & Y_{22A} \end{bmatrix} = \begin{bmatrix} 25-j5 & -25 \\ -25 & 25-j5 \end{bmatrix}$$

Step2: Finding the Y parameters of Circuit B

The circuit B is shown in Fig.5.82. By applying the concepts of Y parameters, the equations are to be written as follows and the corresponding Y parameters are found.

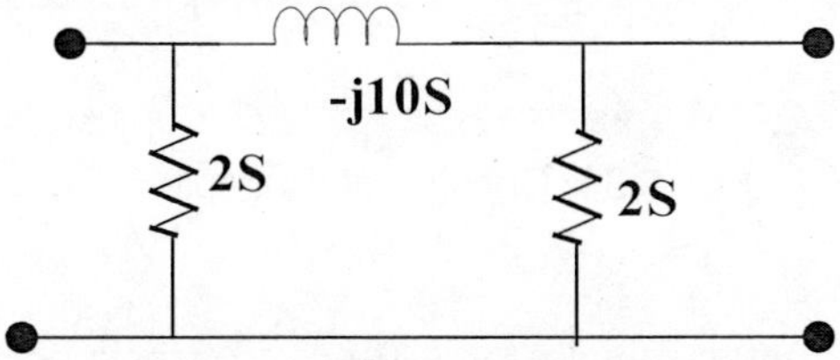

Fig. 5.82: Circuit B of Example 5.22

$$I_{1B} = Y_{11B}\,V_{1B} + Y_{12B}\,V_{2B}$$
$$I_{2B} = Y_{21B}\,V_{1B} + Y_{22B}\,V_{2B}$$

$$Y_A = \begin{bmatrix} Y_{11A} & Y_{12A} \\ Y_{21A} & Y_{22A} \end{bmatrix} = \begin{bmatrix} 25-j5 & -25 \\ -25 & 25-j5 \end{bmatrix}$$

Step3: Finding the Y parameters of the original network

As per the interconnection concept, circuit A and circuit B are connected in parallel. The Y parameters of the individual circuit are found. Hence the original circuit's Y parameter is the addition of the individual circuit's Y parameters.

$$Y = Y_A + Y_B = \begin{bmatrix} 27-j15 & -25+j10 \\ -25+j10 & 27-j15 \end{bmatrix}$$

Example 5.24:

Find the two port parameters of the circuit shown in Fig.5.83.

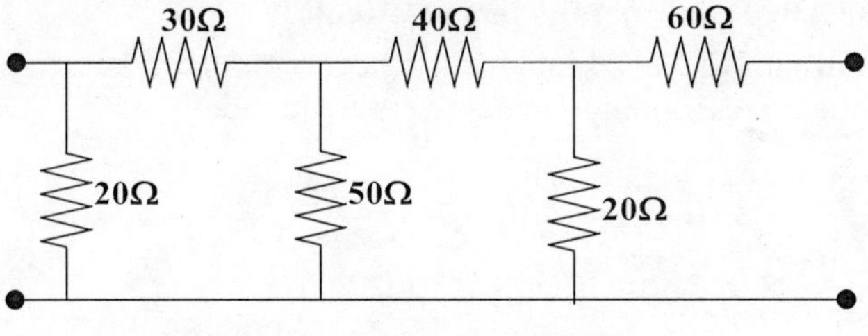

Fig. 5.83: Circuit of Example 5.24

Solutions:

The circuit shown in Fig.5.83 is a two-port network which consists of two two-port networks in cascade connection. Hence, both circuits ABCD or T parameters are to be calculated. The circuit A consists of (30Ω), (50Ω) and (20Ω), that forms the π network and is shown in Fig. 5.84. The circuit B consists of (40Ω), (60Ω) and (20Ω), this forms the T network which is shown in Fig. 5.85.

$$V_1 = AV_2 - BI_2 \quad I_1 = CV_2 - DI_2$$

Step1: Finding the ABCD or T parameters of Circuit A

The circuit A is shown in Fig.5.84. By applying the concepts of T parameters, the equations are to be written as follows and the corresponding T parameters are found.

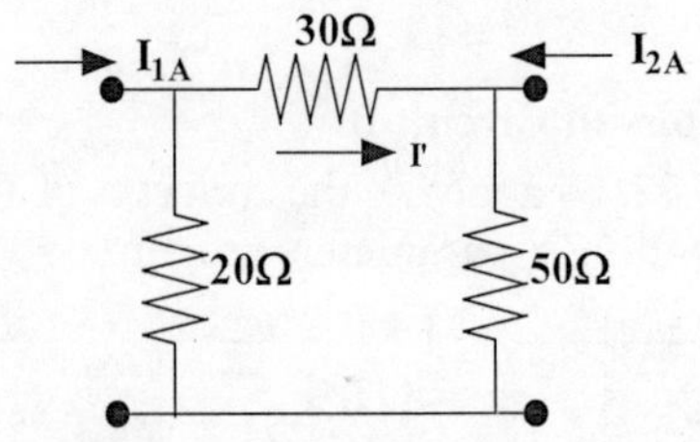

Fig. 5.84: Circuit A of Example 5.24

$$V_{1A} = A_A\, V_{2A} - B_A\, I_{2A}\, I_{1A} = C_A\, V_{2A} - D_A\, I_{2A}$$

By open circuiting the output terminals, $V_{1A} = 16 I_{1A}; I^{'} = \frac{I_{1A} * 20}{100}; V_{2A} = 50 I^{'}$

From this, $A_A = 1.6$; $C_A = 0.1$

By short circuiting the output terminals, $V_{2A} = 0; V_{1A} = 12 I_{1A}; I^{''} = \frac{-I_{1A} * 20}{50}; V_{1A} = -30 I^{''}$

From this, $B_A = 30$; $D_A = 2.5$

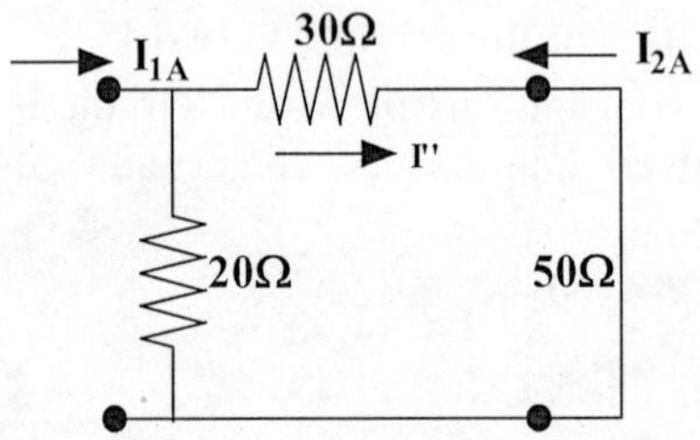

Fig. 5.85: Circuit A of Example 5.24 with short circuiting the output terminals

$T_A = [A_A\, B_A\, C_A\, D_A] = [1.6300.12.5]$

Step2: Finding the ABCD or T parameters of Circuit B

The circuit A is shown in Fig.5.86. By applying the concepts of T parameters, the equations are to be written as follows and the corresponding T parameters are found.

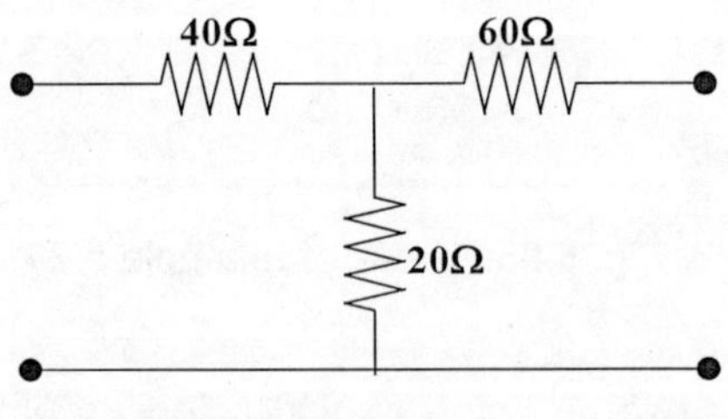

Fig. 5.86: Circuit B of Example 5.24

$$V_{1B} = A_B\, V_{2B} - B_B\, I_{2B}\, I_{1B} = C_B\, V_{2B} - D_B\, I_{2B}$$

By open circuiting the output terminals, $V_{1B} = 60 I_{1B}$; $V_{2B} = 60 I_{1B}$

From this, $A_B = 3$; $C_B = 0.05$

By short circuiting the output terminals, $V_{2B} = 0; V_{1B} = 55I_{1B}; I''' = \frac{-I_{1A} * 20}{80}; V_{1A} = -220I_{2B}$

From this, $B_A = 220$; $D_A = 4$

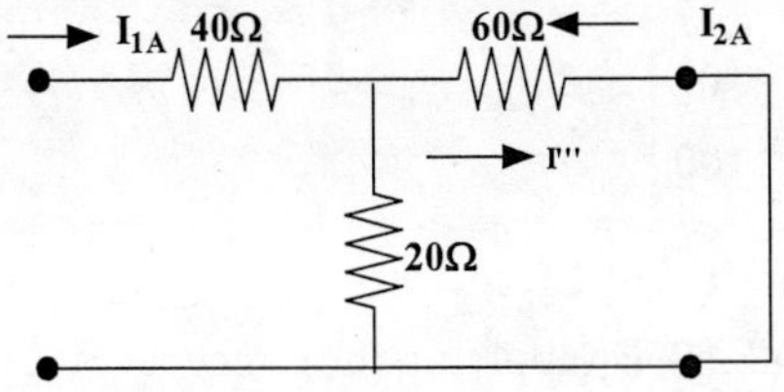

Fig. 5.87: Circuit B of Example 5.24 with short circuiting the output terminals

$$T_B = \begin{bmatrix} A_B & B_B \\ C_B & D_B \end{bmatrix} = \begin{bmatrix} 3 & 220 \\ 0.05 & 4 \end{bmatrix}$$

Step3: Finding the T parameters of the original network

$$T = T_A * T_B = \begin{bmatrix} 6.3 & 472 \\ 0.425 & 32 \end{bmatrix}$$

As per the interconnection concept, the circuit A and the circuit B are connected in cascade connection. The T parameters of the individual circuit are found. Hence the original circuit's T parameter is the multiplication of the individual circuit's T parameters.

Example 5.25:

Find $\frac{V_0}{V_S}$ for the given two-port network shown in Fig.5.115. Let $Y_{12} = Y_{21} = 0$; $Y_{11} = 2mS$ and $Y_{22} =$ 10mS.

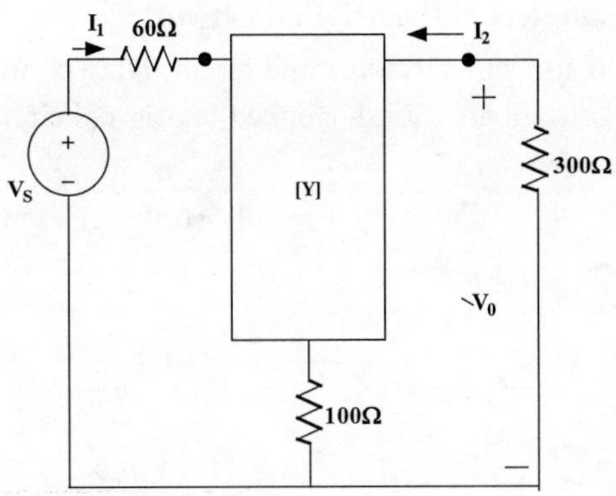

Fig. 5.88: Circuit of Example 5.25

Solutions:

The circuit shown in Fig.5.88 is a two-port network which consists of two two-port networks in series. First, both circuits' Z parameters are to be calculated. The circuit A's Y parameters are given. From that, Z parameters are to be found. The circuit B consists of (100Ω) only which is shown in Fig. 5.89.

Step1: Finding the Z parameters of Circuit A

The circuit A of the Figure shown in Fig.5.88 has [Y] parameters that is converted into Z parameters.

$$Y_A = \begin{bmatrix} Y_{11A} & Y_{12A} \\ Y_{21A} & Y_{22A} \end{bmatrix} = \begin{bmatrix} 0.002 & 0 \\ 0 & 0.001 \end{bmatrix}$$

$$Z_A = \begin{bmatrix} Z_{11A} & Z_{12A} \\ Z_{21A} & Z_{22A} \end{bmatrix} = \begin{bmatrix} 500 & 0 \\ 0 & 100 \end{bmatrix}$$

Step2: Finding the Z parameters of Circuit B

The circuit B is shown in Fig.5.89. By applying the concepts of Z parameters, the equations are to be written as follows and the corresponding T parameters are found.

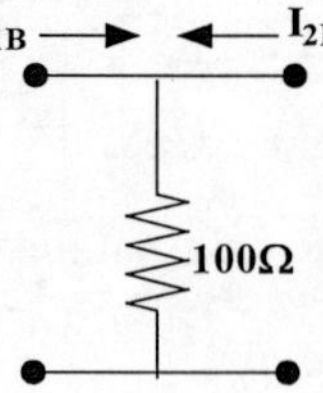

Fig. 5.89: Circuit B of Example 5.25

By open circuiting the output terminals, V1B = $100I_{1B}$; $V_{2B} = 100I_{1B}$

From this, $Z_{11B} = 100\Omega$; $Z_{21B} = 100\Omega$

By open circuiting the input terminals, $V_{1B} = 100I_{2B}$; $V_{2B} = 100I_{2B}$

From this, $Z_{22B} = 100\Omega$; $Z_{12B} = 100\Omega$

$$Z_B = \begin{bmatrix} Z_{11B} & Z_{12B} \\ Z_{21B} & Z_{22B} \end{bmatrix} = \begin{bmatrix} 100 & 100 \\ 100 & 100 \end{bmatrix}$$

Step3: Finding the Z parameters of the original network

As per the interconnection concept, circuit A and circuit B are connected in series connection. The Z parameters of the individual circuit are found. Hence the original circuit's Z parameter is the addition of the individual circuit's Z parameters.

$$Z = Z_A + Z_B = \begin{bmatrix} 600 & 100 \\ 100 & 200 \end{bmatrix}$$

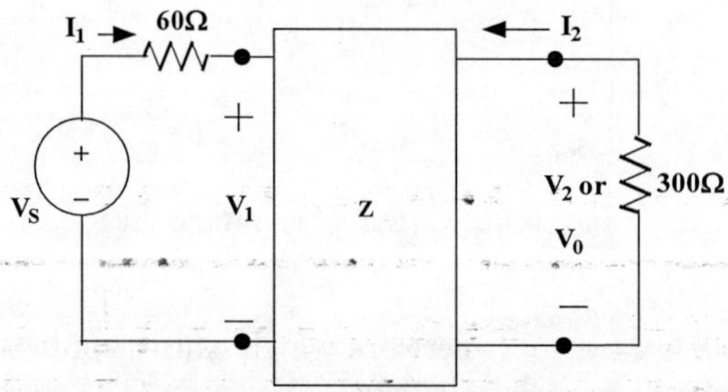

Fig. 5.90: Circuit of Example 5.25 with the equivalent circuit parameters

$$V_1 = Z_{11}\, I_1 + Z_{12}\, I_2$$

$$V_2 = Z_{21}\, I_1 + Z_{22}\, I_2$$

As per the circuit shown in Fig. 5.90, the input source voltage and the output voltage could be written as

$$V_1 = V_S - 60I_1;\ V_0 = -300I_2$$

Hence, $I_2 = -\frac{V_0}{300}$. Both I_2 and Z_{11}, Z_{12} are substituted in the Z parameter equation of V_1 and the equation (5.160) is obtained.

$$V1 = (600)\, I_1 + 100I_2$$

$$V_S - 60I_1 = 600I_1 + 100(-\frac{V_0}{300})$$

With Z_{21}, Z_{22} are substituted in the Z parameter equation of V_2 and the equation (5.161) is obtained.

$$V_2 = 100I_1 + 200(-\frac{V_0}{300}) \qquad (5.161)$$

$$V_0 = 100I_1 + 200(-\frac{V_0}{300}) \qquad (5.162)$$

By solving both equations (5.160) and (5.162),

$$\frac{V_0}{V_S} = 0.09395$$

UNIT SUMMARY

1. The basic definitions of the network topology are described.
2. The formation of the tie-set matrix and cut-set matrix is explained.
3. The network solution is found using the tie-set analysis and cut-set analysis.
4. The concept of the two-port networks and the different two-port network parameters are discussed. Many examples are solved to find the two-port parameters of the sample network.
5. The interconnection between the two-port network parameters is obtained and the same is presented in table format.
6. The networks are interconnected in series, parallel or cascade mode. The method of finding the two-port networks of such interconnected networks are described and the example problems are solved.

EXERCISES

Multiple Choice Questions

1. The graph of a network has 8 nodes and 5 independent loops. The number of branches of the graph is ....

   (a) 10 (b) 11

   (c) 12 (d) 13

2. The graph associated with an electric network has 7 branches and 5 nodes. The number of independent KCL equations and the number of KVL equations are ....

   (a) 2 and 5 (b) 5 and 2

   (c) 3 and 4 (d) 4 and 3

3. The number of the trees in a network shown in Fig. 5.91.

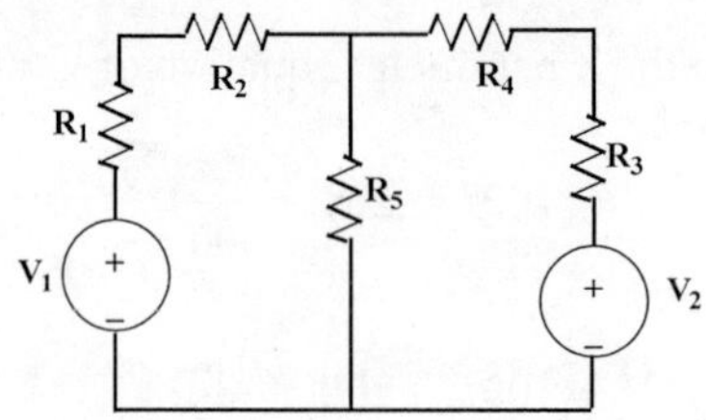

Fig. 5.91: Circuit for question 3

   (a) 2 (b) 5

   (c) 3 (d) 1

4. The graph of an electrical network has N nodes and B branches. The number of links (L) with respect to the choice of a tree is given by.......

   (a) B–N+1 (b) B+N

   (c) N–B+1 (d) N–2B+1

5. The admittance parameters of the passive resistive two-port network shown in Fig. 5.92 is $y_{11} = 5S$; $y_{22} = 1S$; $y_{21} = y_{12} = -2.5S$. The power delivered to the load resistor RL in Watts is ....

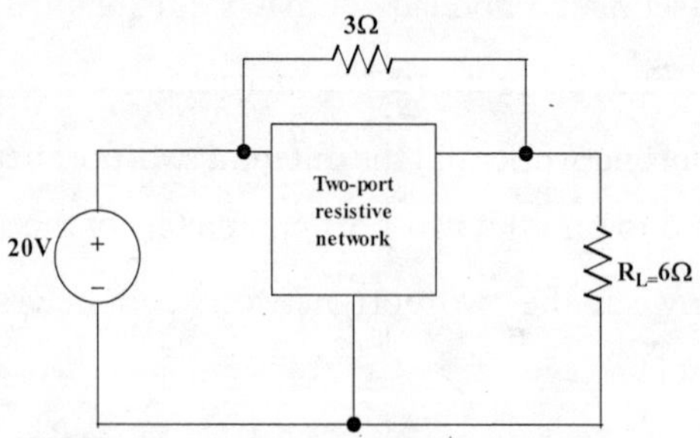

Fig. 5.92: Circuit for question 5

   (a) 452.5 (b) 238

   (c) 232.12 (d) 145.25

6. In the two-port network shown in Fig.5.93, the h_{11} parameters in Ω is ....

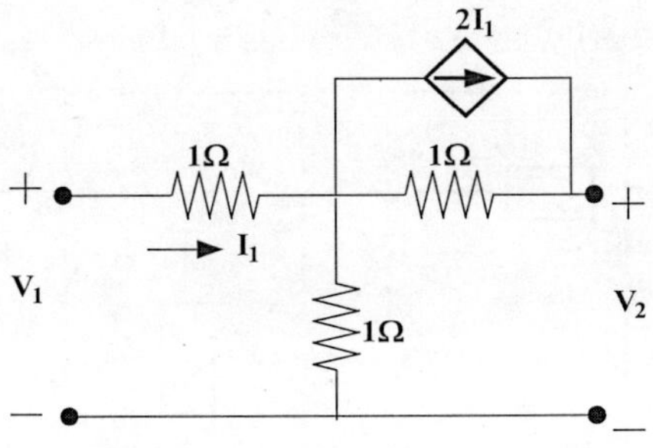

Fig. 5.93: Circuit for question 6

(a) 0.25 (b) 0.5

(c) 0.75 (d) 0.85

7. The Z parameters of the network shown in Fig.5.94 are Z_{11} = 40Ω; Z_{12} = 60Ω; Z_{21} = 80Ω; Z_{22} = 100Ω. The average power delivered to the load resistance (20Ω) is in watts.............

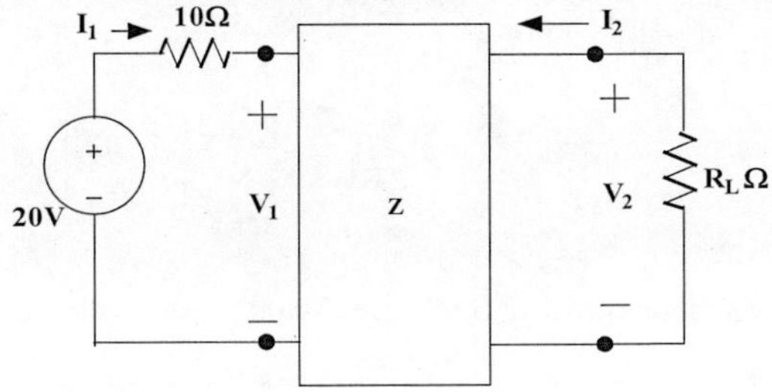

Fig. 5.94: Circuit for question 7

(a) 35.55 (b) 28.45

(c) 24.35 (d) 18.25

8. The two-port network P shown in Fig.5.95 has input port 11' and output port 22'. It has an impedance matrix Z with parameters denoted by *Zij*. A 1Ω resistor is connected in series with the network at port 1 as shown in the figure. The impedance of the modified two-port network is .....

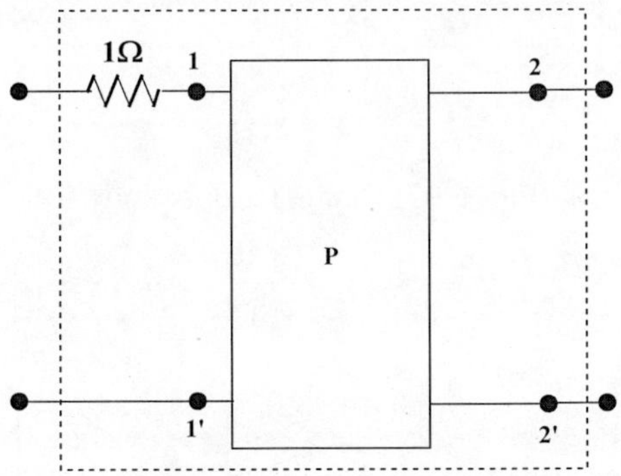

Fig. 5.95: Circuit for question 8

9. Two networks are connected in cascade as shown in Fig. 5.96. With usual notations the equivalent A, B, C and D constants are obtained. Given that, C = $0.025\angle 45^0$, the value of Z is ....

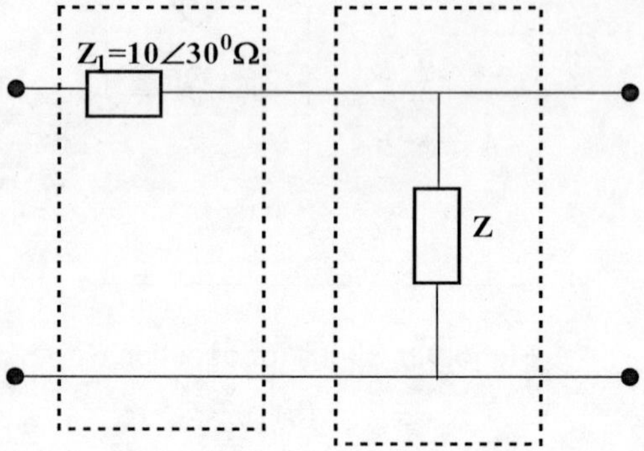

Fig. 5.96: Circuit for question 9

(a) $\begin{bmatrix} Z_{11}+1 & Z_{12}+1 \\ Z_{21} & Z_{22}+1 \end{bmatrix}$ (b) $\begin{bmatrix} Z_{11}+1 & Z_{12} \\ Z_{21} & Z_{22}+1 \end{bmatrix}$

(c) $\begin{bmatrix} Z_{11}+1 & Z_{12} \\ Z_{21} & Z_{22} \end{bmatrix}$ (d) $\begin{bmatrix} Z_{11}+1 & Z_{12} \\ Z_{21}+1 & Z_{22} \end{bmatrix}$

10. For the two-port network shown in Fig.5.97, the Z matrix is given by ....

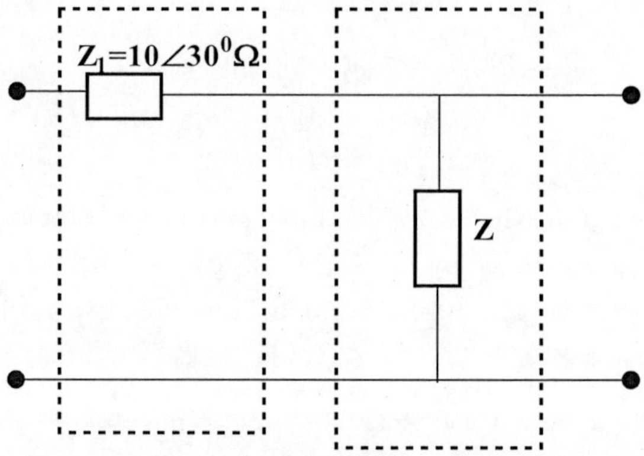

Fig. 5.97: Circuit for question 10

(a) $\begin{bmatrix} Z_1 & Z_1+Z_2 \\ Z_1+Z_2 & Z_2 \end{bmatrix}$ (b) $\begin{bmatrix} Z_1 & Z_1 \\ Z_1+Z_2 & Z_2 \end{bmatrix}$

(c) $\begin{bmatrix} Z_1 & Z_2 \\ Z_2 & Z_1+Z_2 \end{bmatrix}$ (d) $\begin{bmatrix} Z_1 & Z_1 \\ Z_1 & Z_1+Z_2 \end{bmatrix}$

ANSWERS															
1.	(c)	2.	(d)	3.	(d)	4.	(a)	5.	(b)	6.	(a)	7.	(a)	8.	(c)
9.	(b)	10.	(d)												

Short Answer Questions:

1. Define: Network, Oriented graph, Tree and Link.
2. Describe the procedure to obtain the tie-set matrix.
3. Describe the procedure to obtain the cut-set matrix.
4. Define co-tree.
5. What is the fundamental cut-set? Explain with simple examples.
7. List the two-port network parameters.
8. Write the relation between Z parameters to Y parameters.
9. Write the relation between ABCD parameters to Z parameters.
10. If the two networks are connected in series/ parallel/ cascade connection, provide the resultant network's Z parameters.
11. Find the Y parameters for the circuit shown in Fig. 5.98.

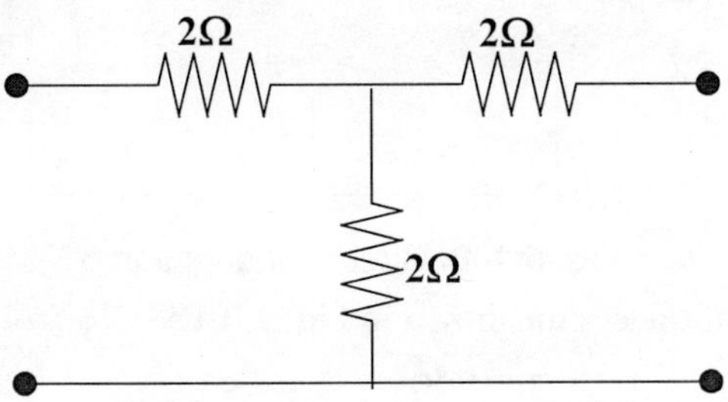

Fig. 5.98: Circuit for question 10

Long Answer Questions:

1. Explain the procedure for obtaining the tie-set matrix.
2. Explain the procedure for obtaining the cut-set matrix.
3. Derive the method to find the possible trees of the network.
4. Explain the two-port networks and its parameter descriptions.
5. Discuss the method for obtaining the Z/Y/T/H parameters for any two-port network.
6. Discuss the network interconnections and explain the procedure for obtaining the two-port parameters of those interconnected networks.
7. Derive the Y parameters from the given Z parameters.
8. Derive the ABCD or T parameters from the H parameters.
9. Obtain the interconnections between the Z parameters and the H parameters.

Numerical Problems

1. Find all possible trees of the network shown in Fig.5.99.

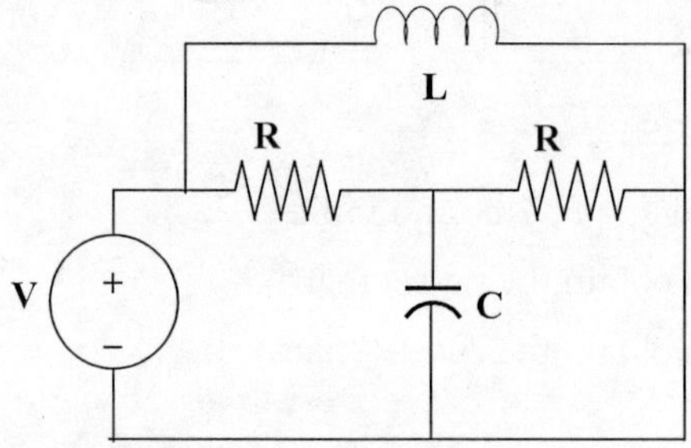

Fig. 5.99: Circuit for question 1

2. Find the cut-set and tie-set matrix of the circuit shown in Fig.5.99, if the R=10Ω, Inductive reactance is (j10Ω), the capacitive reactance is (-j20Ω) and the source voltage is $10\angle -30^0$ V.
3. Find the branch currents of the circuit shown in Fig.5.100 using (*a*) tie-set matrix (*b*) cut-set matrix.

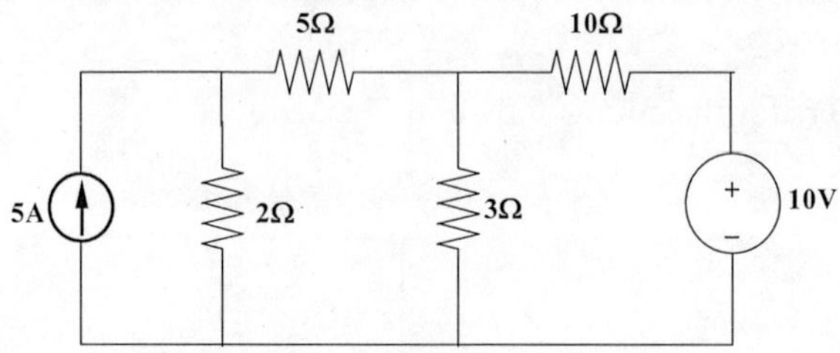

Fig. 5.100: Circuit for question 3

4. Find the Z parameters of the circuit shown in Fig. 5.101.

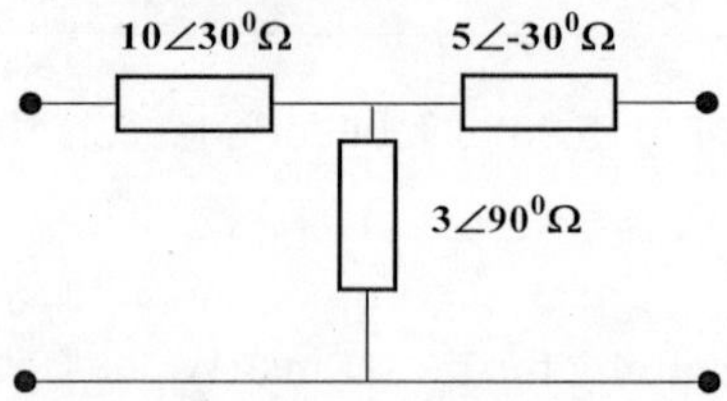

Fig. 5.101: Circuit for question 4

5. Find the ABCD parameters of the network shown in Fig. 5.102.

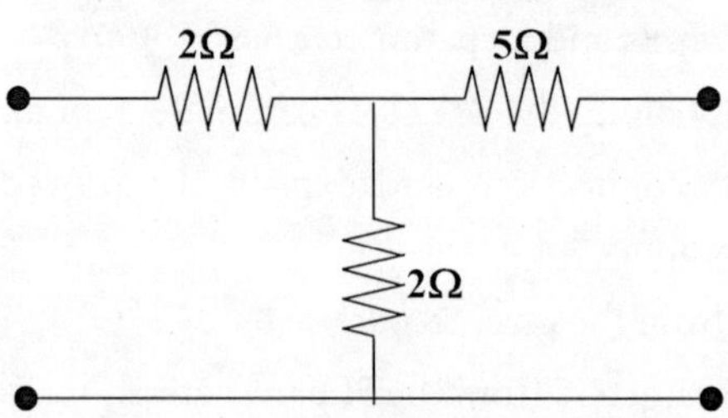

Fig. 5.102: Circuit for question 5

6. The Z parameter of the two-port network shown in Fig.5.103 is given below. The impedances are represented in Ω. Find the value of R_b in Ω, if the $Z_b = (R_b + 1j\omega)\Omega$.

$$\begin{bmatrix} 2j\omega & j\omega \\ j\omega & 3+2j\omega \end{bmatrix}$$

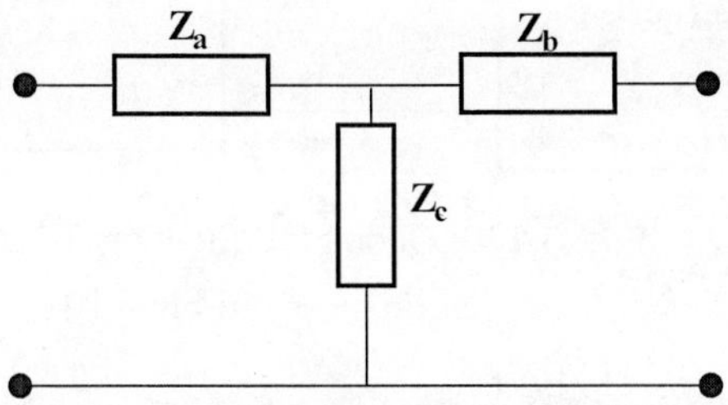

Fig. 5.103: Circuit for question 6

7. Find the Z parameter of the two-port network shown in Fig.5.104. Also calculate other two-port parameters of this network.

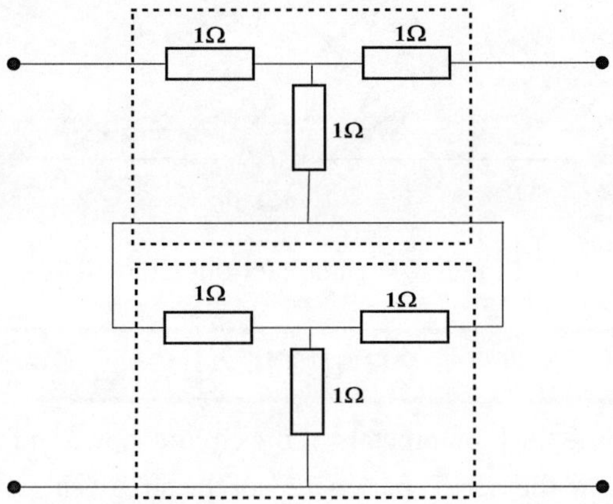

Fig. 5.104: Circuit for question 7

8. Find all two-port parameters of the network shown in Fig.5.105.

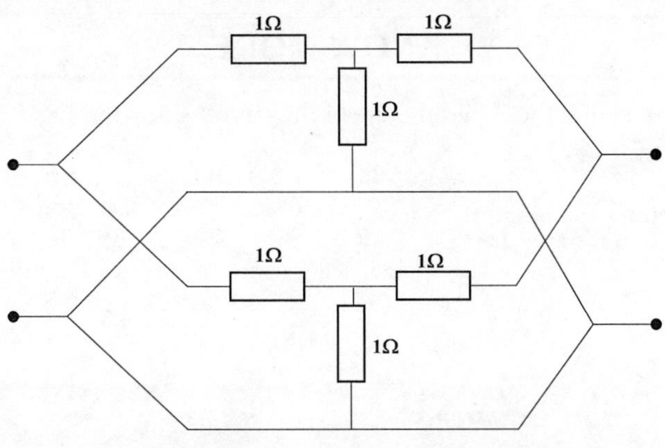

Fig. 5.105: Circuit for question 8

9. For the cascade network shown in Fig.5.106, find all two-port network parameters. Where the network 1 and network 2 remain same and the network is shown in Fig. 5.106.

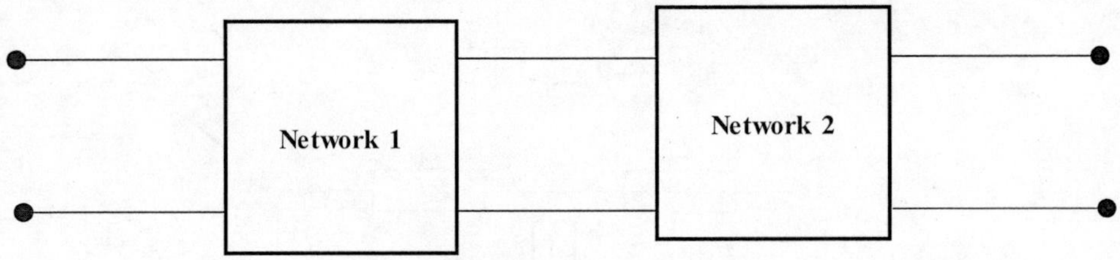

Fig. 5.106: Circuit for question 9

10. Find the Z and Y parameters of the network shown in Fig.5.107.

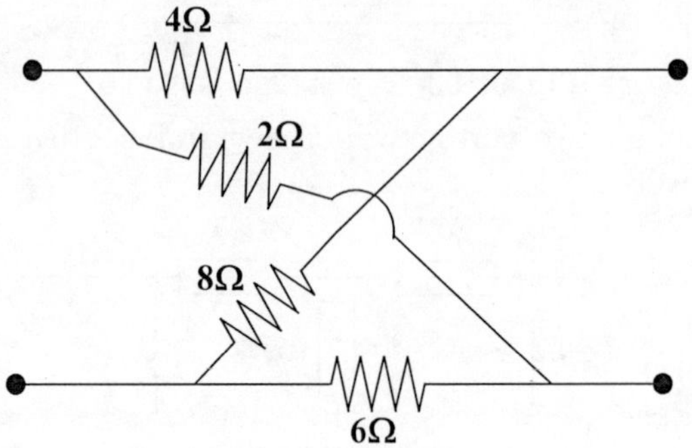

Fig. 5.107: Circuit for question 10

PRACTICAL

1. Use LTSpice to determine the h parameters of the circuit shown in Fig. 5.32 of Example 5.12.
2. Use LTSpice to determine the ABCD parameters of the circuit shown in Fig. 5.32 of Example 5.12.
3. Use LTSpice to determine all two-port parameters of the circuit shown in Fig. 5.32 of Example 5.12.

KNOW MORE

1. Use LTSpice to determine the Z parameters of the circuit shown in Fig. 5.32 of Example 5.12.

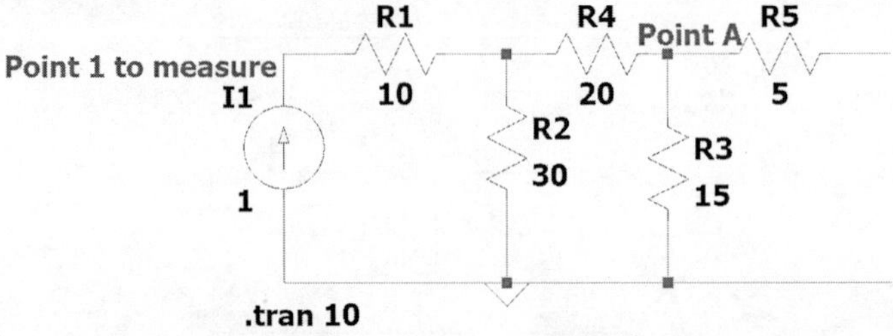

Fig. 5.108: Circuit for question 1 (When the output port is open circuited)

- The $Z_{11} = \frac{V_1}{I_1}$ and $Z_{21} = \frac{V_2}{I_1}$ are found when the output port is open circuited.
- V_1 and V_2 are measured at point 1 and point A in the circuit shown in Fig. 5.108. The plot corresponding to that is shown in Fig.5.109 and Fig. 5.110. The input current I_1 is equal to 1A. Hence obtained values of the Z parameters Z_{11} and Z_{21} are 26.15Ω and 6.92Ω respectively.

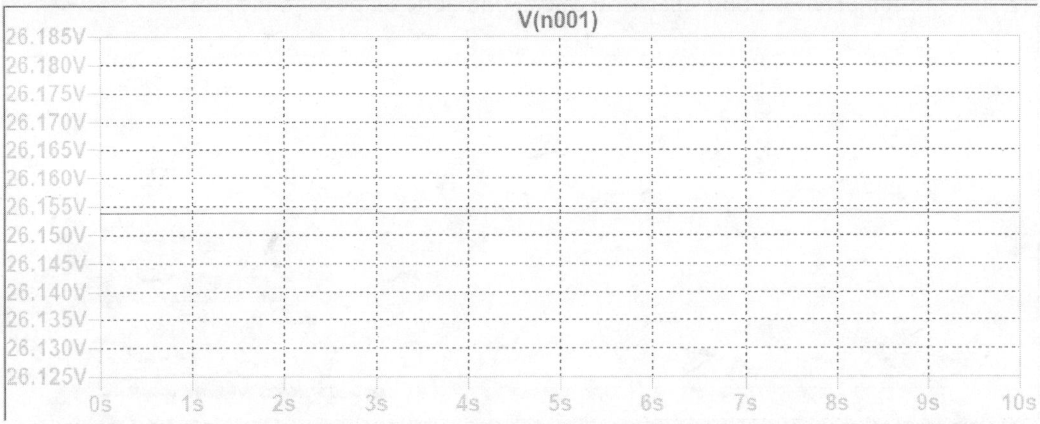

Fig. 5.109: Circuit for question 1 (Voltage at Point 1 of the circuit shown in Fig. 5.108)

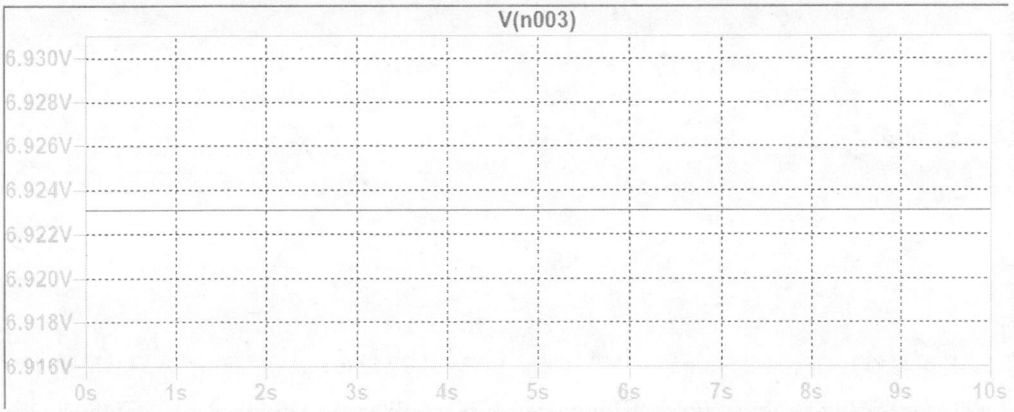

Fig. 5.110: Circuit for question 1 (Voltage at Point A of the circuit shown in Fig. 5.108)

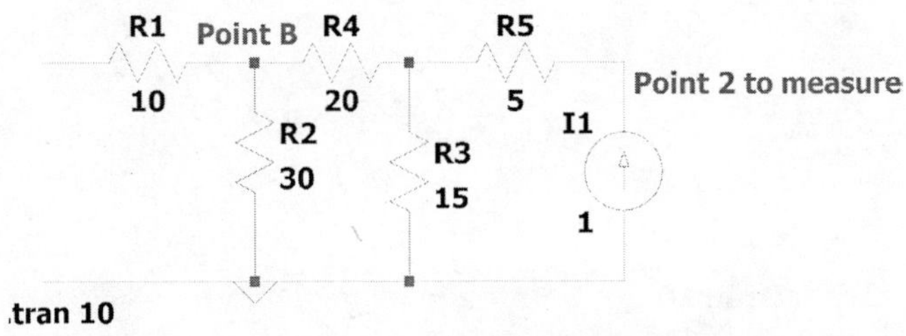

Fig. 5.111: Circuit for question I (Fig. 5.108)

- The $Z_{12} = \frac{V_1}{I_2}$ and $Z_{22} = \frac{V_2}{I_2}$ are found when the input port is open circuited.
- V_1 and V_2 are measured at point 2 and point B in the circuit shown in Fig. 5.111. The plot corresponding to that is shown in Fig.5.112 and Fig. 5.113. The input current I_2 is equal to 1A. Hence obtained values of the Z parameters Z_{12} and Z_{22} are 6.92Ω and 16.54Ω respectively. The results are compared with the analytical calculation done in the example 5.12.

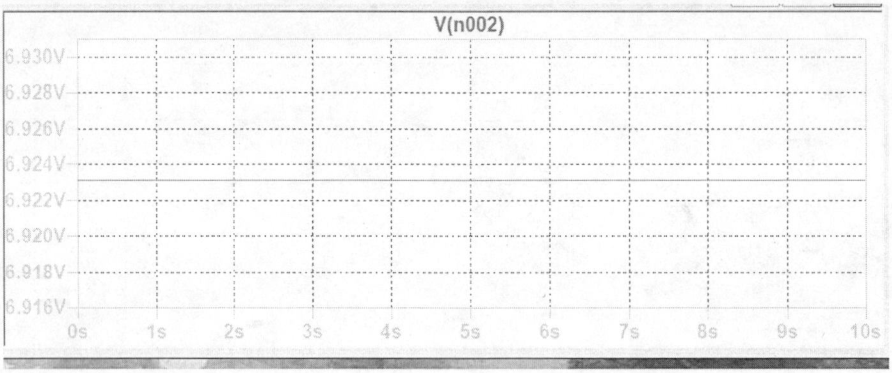

Fig. 5.112: Circuit for question 1 (Voltage at Point B of the circuit shown in Fig. 5.111)

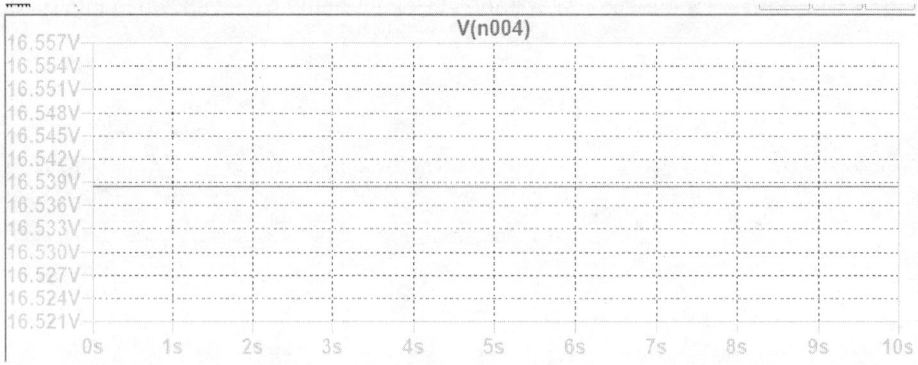

Fig. 5.113: Circuit for question 1 (Voltage at Point 2 of the circuit shown in Fig. 5.111)

2. Use LTSpice to determine the Y parameters of the circuit shown in Fig. 5.32 of Example 5.15.

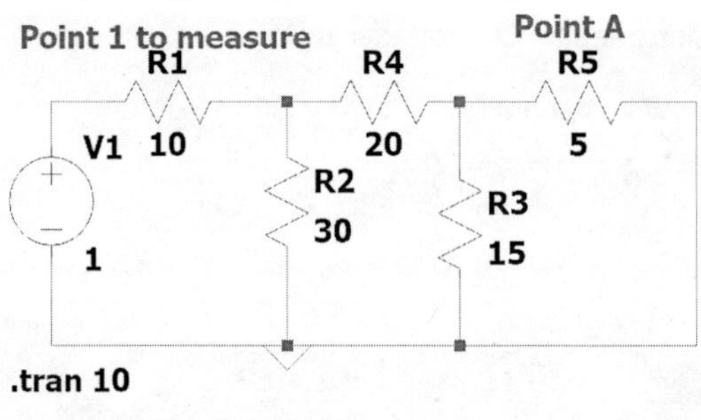

Fig. 5.114: Circuit for question 2 (When the output port is short circuited)

- The $Y_{11} = \frac{I_1}{V_1}$ and $Y_{21} = \frac{I_2}{V_1}$ are found when the output port is short circuited.
- I_1 and I_2 are measured at point 1 (current through R_1) and point A (current through R_5) in the circuit shown in Fig. 5.114. The plot corresponding to that is shown in Fig.5.115 and Fig. 5.116. The input voltage V_1 is equal to 1V. Hence obtained values of the Y parameters Y_{11} and Y_{21} are 0.043v and 0.018v respectively.

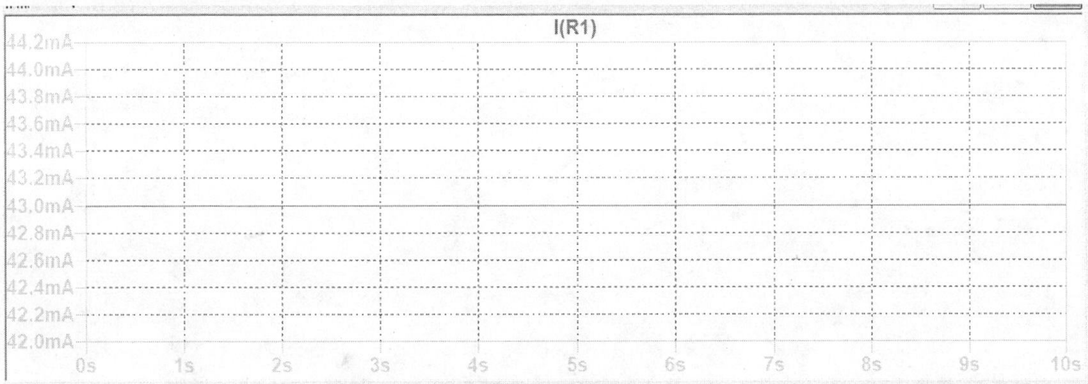

Fig. 5.115: Circuit for question 2 (Current through the resistance R1 of the circuit shown in Fig. 5.114)

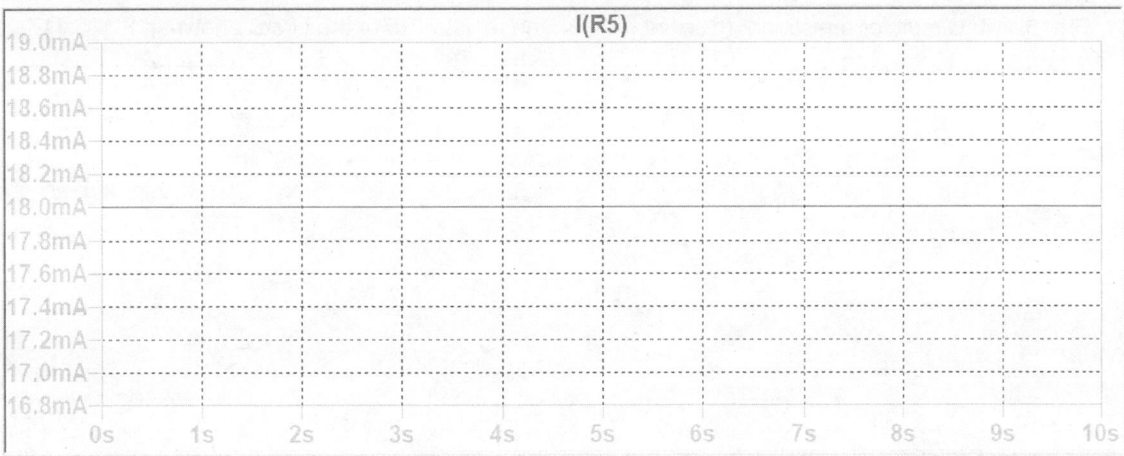

Fig. 5.116: Circuit for question 2 (Current through the resistance R5 of the circuit shown in Fig. 5.114)

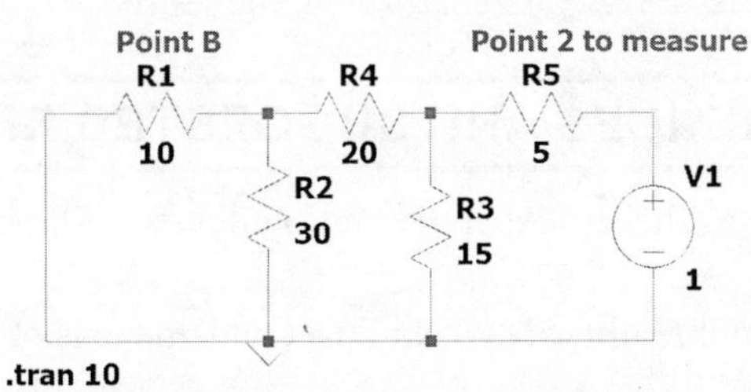

Fig. 5.117: Circuit for question 2

- The $Y_{12} = \frac{I_1}{V_2}$ and $Y_{22} = \frac{I_2}{V_2}$ are found when the input port is short circuited.
- I_2 and I_1 are measured at point 2 (current through R5) and point B (current through R1) in the circuit shown in Fig. 5.117. The plot corresponding to that is shown in Fig.5.118 and Fig. 5.119. The output voltage V_2 is equal to 1V. Hence obtained values of the Y parameters Y_{12} and Y_{22} are 0.018v and 0.068v respectively.

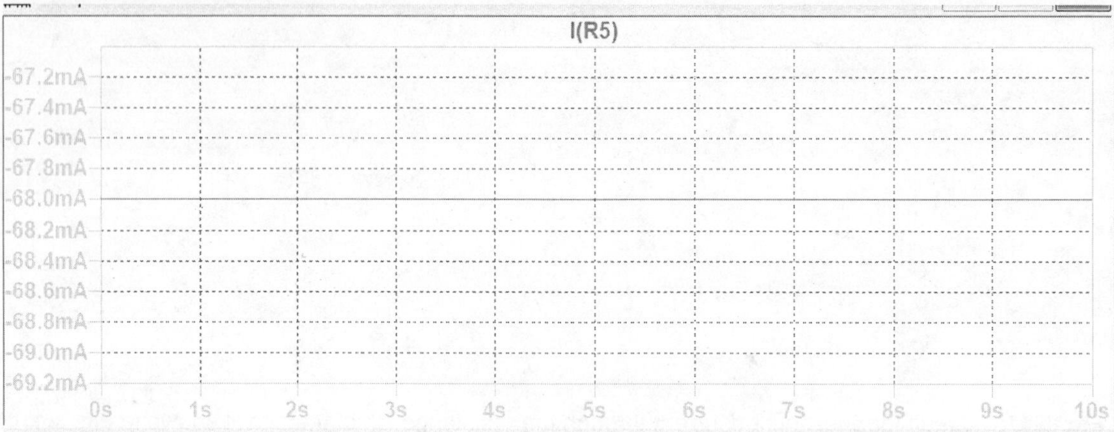

Fig. 5.118: Circuit for question 2 (Current through the resistor R5 of the circuit shown in Fig. 5.117)

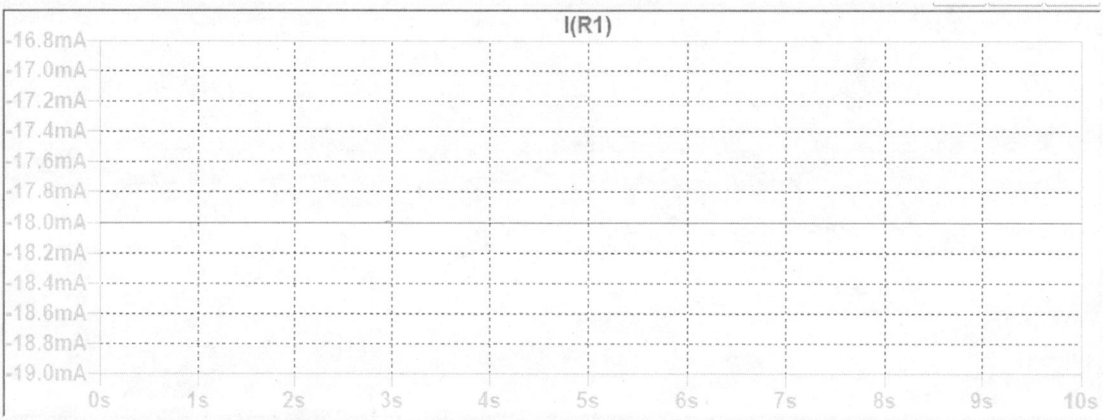

Fig. 5.119: Circuit for question 2 (Current through the resistor R1 of the circuit shown in Fig. 5.117)

REFERENCES AND SUGGESTED READINGS

1. A Chakrabarti, "Circuit Theory (Analysis and Synthesis)", Dhanpat Rai & Co, 7th Edition, 6th Edition, August 2010.
2. Charles K Alexander and Mathew N O Sadiku, "Fundamentals of Electric Circuits", McGraw Hill Publishers, 5th Edition, 2012.

CO AND PO ATTAINMENT TABLE

Course outcomes (COs) for this course can be mapped with the programme outcomes (POs) after the completion of the course and a correlation can be made for the attainment of POs to analyze the gap. After proper analysis of the gap in the attainment of POs necessary measures can be taken to overcome the gaps.

Table for CO and PO attainment

Course Outcomes	Attainment of Programme Outcomes (1- Weak Correlation; 2- Medium Correlation; 3- Strong Correlation)						
	PO-1	PO-2	PO-3	PO-4	PO-5	PO-6	PO-7
CO-1	-	-	-	-	-	-	-
CO-2	-	-	-	-	-	-	-
CO-3	-	-	-	-	-	-	-
CO-4	-	-	-	-	-	-	-
CO-5	-	-	-	-	-	-	-

The data filled in the above table can be used for gap analysis.

INDEX

P

Q

R

S

T

U

V

Y

Z

❑❑